CURRENT RESEARCH IN
EARTHQUAKE PREDICTION I

DEVELOPMENTS IN EARTH AND PLANETARY SCIENCES

CURRENT RESEARCH IN EARTHQUAKE PREDICTION I

Edited by Tsuneji Rikitake

Developments in Earth and Planetary Sciences

02

Center for Academic Publications Japan / Tokyo

D. Reidel Publishing Company / Dordrecht · Boston · London

Library of Congress Cataloging in Publication Data

DATA-APPEARS ON SEPARATE CARD

ISBN 90-277-1133-X

Published by Center for Academic Publications Japan, Tokyo, in co-publication with D. Reidel Publishing Company, P. O. Box 17, 3300 AA Dordrecht.

Sold and distributed in Japan, China, Korea, Taiwan, Indonesia, Cambodia, Laos, Malaysia, Philippines, Thailand, Vietnam, Burma, Pakistan, India, Bangla Desh, Sri Lanka by Center for Academic Publications Japan, 4-16, Yayoi 2-chome, Bunkyo-ku, Tokyo 113, Japan.

Sold and distributed in the U.S.A. and Canada by Kluwer Boston Inc., 190 Old Derby Street, Hingham, MA 02043, U.S.A.

Sold and distributed in all other countries by Kluwer Academic Publishers Group, P. O. Box 322, 3300 AH Dordrecht Holland.

Printed in Japan

PREFACE

Studies related to the earth and planets along with their surroundings are of great concern for modern scientists. Global geodynamics as represented by the plate tectonics has now become one of the most powerful tools by which we can study the causes of earthquakes, volcanic eruptions, mountain formation, and the like. Various missions sent out to the space, manned or unmanned, brought out geoscientific features of the moon, Mars, Venus, and other planets. Earthquake prediction that was the business of astrologers and fortune-tellers some twenty years ago, has now grown up to be an important science. A number of destructive earthquakes were successfully forecast in the People's Republic of China.

In the light of the above-mentioned and other accomplishments in geosciences, we feel that it is a good thing to publish a series of monographs which review selected topics of earth and planetary sciences. We are of course well aware of the fact that similar monographs have been and will be published from overseas publishers. The series, which we plan to publish, will therefore put stress on Japanese work. But we hope that the series will also include review articles by distinguished overseas authors.

The series, which is named the "Developments in Earth and Planetary Sciences (DEPS)" will be published by the Center for Academic Publications Japan and the D. Reidel Publishing Company. It is my great pleasure to work as the Editor of the series. I should like to have comments on what subjects we shall choose in the future publications. I shall be greatly obliged if anyone would suggest suitable subjects and potential authors for the series to me.

Tsuneji Rikitake
Editor

PREFACE TO VOLUME 2

Earthquake prediction is one of the most important subjects of present-day earth science. As the Editor of DEPS, I, therefore, believe that it is appropriate for me to publish reviews of current advances in the various aspects of earthquake prediction research. Although I have already published a book entitled "Earthquake Prediction" in 1976, the development of earthquake prediction research has since been so rapid that there are many things to be added to the text to-day.

First of all, I thought that it would be a good thing to invite experts in various disciplines of earthquake prediction to write up-to-date reviews on the respective disciplines. I was fortunate in obtaining a number of manuscripts from eminent researchers working on prediction-oriented geodesy, seismology, electric and magnetic study, and so on. These are the articles by Drs. Dambara, Hamada, and Honkura.

In addition to such detailed reviews on each discipline, I also thought that an overall view of the present status of earthquake prediction work should be presented at least as an introductory chapter. In view of this I undertook to write an article for such a review although I am further planning to publish a much more extensive review in the succeeding issues of DEPS.

DEPS is essentially an international publication, so that I tried to invite a number of overseas contributions. In response to my invitation, I was fortunate enough to have two such manuscripts in this volume: One is Dr. Wyss's review on the U.S. earthquake prediction work, and the other is an article on the sociological aspect of earthquake prediction by Drs. Hutton, Sorensen, and Mileti. The latter contribution is highly timely because much attention is now being paid to conversion of earthquake predicton information into earthquake warning as I shall briefly mention in the introductory article.

I took the liberty of including two short articles of mine in this volume. They are concerned with the 1978 Near Izu-Oshima Island earthquake (Izu-Oshima Kinkai earthquake). I think that the characteristics of earthquake precursors revealed in relation to the earthquake is of current interest to everyone.

In the course of the publishing work, Mr. K. Oshida of the Center for Academic Publications Japan has worked very hard. I am very grateful to him for his services. A part of the publishing cost is defrayed from the funds given to me by the Ministry of Education, Science, and Culture of Japan. I am grateful to the Ministry.

T. Rikitake
Editor, DEPS

CONTENTS

ELECTRIC AND MAGNETIC APPROACH TO EARTHQUAKE
PREDICTION   Y. HONKURA 301

PRACTICAL APPROACH TO EARTHQUAKE PREDICTION AND WARNING

Tsuneji RIKITAKE

Department of Applied Physics, Faculty of Science, Tokyo Institute of Technology, Ookayama, Meguro-ku, Tokyo, Japan

Abstract Earthquake prediction programs are at the moment conducted very intensively in Japan, the U.S.A., the U.S.S.R. and the People's Republic of China. As a result of such intensive programs, it has become clear that many earthquakes are accompaneid by some sort of pre-earthquake symptoms or precursors. First of all, the regionalization of a large earthquake may be possible on the basis of analyses of historical earthquakes and of monitoring of crustal strain as disclosed by repetition of geodetic surveys at least for inter-plate earthquakes.

It is also made clear by the programs that we sometimes observe many a precursory effect such as anomalous land-uplift, tilting of the ground, changes in seismic wave velocities, changes in geomagnetic field and earth resistivity and so on. An analysis of these precursors lead us to their practical classification.

It turns out that the precursor time of a certain class of precursors depends on the magnitude of the main shock, this class of precursor being called the precursor of the first kind. Why we have this kind of precursor may be accounted for by the physical procedure of a dilatancy model. The larger the spatial extent of this kind of precursor is, the larger is the earthquake magnitude of the main shock. The logarithmic precursor time of this class of precursor is linearly correlated to the magnitude of the main shock.

Meanwhile, it appears that there are precursors of the second and third kinds.   The former, which involves sudden sea retreat, resistivity change and the like, has a precursor time scattered around 2–3 hours, while the latter, typically observed for foreshocks and ground tiltings, has a rather broad time spectrum centered at several days.

A practicable algorithm for the prediction of a damaging earthquake can be formed by taking the possible regionalization and the characteristics of actual precursors into account.   It is really wonderful to see the current development of earthquake prediction algorithm because no concrete idea about any practicable approach to a prediction existed when we started earthquake prediction programs some 15 years ago. Although the algorithm that an imminent earthquake prediction can in general be made after following regionalization, long- and medium-term predictions sounds attractive, the point that precursors are apt to be contaminated by noise must seriously be taken into account in an actual prediction.

Much of the difficulties in issuing an earthquake warning has been discussed elsewhere even when an earthquake prediction has been achieved.   Only a very brief account on this point is presented in this article. But the author believes that essentials for the problem is covered here.

Many of the important points related to earthquake prediction and warning is discussed in this article mostly in relation to Japanese problems because the author is most concerned with them.   The most urgent problem for Japan is to forecast as accurately as possible the occurrence of the Tokai earthquake, which is expected to occur sooner or later, and to have a magnitude 8 or thereabout.   The case history of intended prediction of that earthquake may provide an instructive aspect of earthquake prediction and warning in general.

1. Introduction

A detailed review on earthquake prediction was presented by RIKITAKE (1976 a) in a book entitled " Earthquake Prediction " and published only a few years ago.   In the preface of the book, the author wrote "A number of friends of his said that it would not be the right time to write a book on the subject because earthquake prediction study is rapidly developing and relevant data are accumulating with an immense speed at the moment."

It now turns out, truly enough, that earthquake prediction study has been developing so rapidly in recent years that many more data pertinent to the subject have to be added to what were written in the book although the present author believes that the basic idea about earthquake prediction has not substantially changed. It is really surprising that research in earthquake prediction has reached a stage which enables us to presume a possible way for approaching the problem of actual earthquake prediction. Such achievements have been accomplished chiefly through intensive programs on earthquake prediction in Japan, the U.S.A., the U.S.S.R., and China.

It is the aim of this review report to summarize the development of earthquake prediction study in recent years, and to bring out a possible and practical approach to actual prediction of an earthquake as a logical conclusion of recent developments. This article also provides a supplement to the author's recent review on earthquake prediction and warning (RIKITAKE, 1978 a) which was written for nonspecialist readers.

Although a review of national programs in the four major countries will be briefly presented in Section 2, emphasis will be put on Japanese achievements throughout the article. The mode of earthquake occurrence and its interpretation on the basis of present-day theory of plate tectonics will be presented in Sections 3 and 4 with special reference to the Japan Islands, a typical island arc and its surroundings. It will be made clear why great earthquakes of magnitude 8 or over so frequently hit island arcs such as Japan in those chapters.

Japan has a long history covering 2,000 years or so. Many historical documents that describe great earthquakes in the past are available in Japan, so that the space-time pattern of occurrence of great earthquakes throughout Japan's history can be brought to light fairly clearly. As will be mentioned in Section 5, the probability of a great earthquake occurring in certain parts of Japan can be estimated on the basis of a statistical analysis of historical seismicity in some cases. Similar estimates of probability will be extended to a number of subduction zones along the Pacific Ocean margins.

Japan is famous for accurate geodetic surveys covering it. Accumulation of crustal strain can be monitored by making differences in horizontal position of triangulation station between two triangulation or trilateration surveys conducted at two epochs separated a suitable interval of time. Similar information can be obtained from repetition of levelling surveys. On the other hand, the ultimate strain to rupture of the earth's crust can be inferred from geodetic surveys around seismic faults before and after an earthquake.   Comparison between accumulated strain and statistical distribution of ultimate strain enables us to estimate the probability of the crust to break.   Such a probability provides a long-term prediction of occurrence of a great earthquake as will be shown in Section 6.

In order to gain access to an actual prediction of an earthquake, it is required to detect a symptom, that positively indicates the occurrence of an earthquake coming sooner or later, in addition to a long-term prediction as mentioned above.   A fairly detailed review of new data relevant to such symptoms, usually called earthquake precursors, will be attempted in Section 7.   It will be pointed out that there are varieties of precursors.   Most of such precursors are based on geophysical and geochemical effects of various disciplines differing from one another, but anomalous animal behavior may sometimes provide a kind of precursor.

Precursor data accumulated so far will be classified into three classes in Section 8.   First of all, it will be clearly shown that there are precursors of which the precursor time depends on the magnitude of the main shock.   The larger the magnitude is, the longer is the precursor time.   It will next be shown that we sometimes observe imminent precursors having a precursor time amounting to a few hours or thereabout. No magnitude-dependence of precursor time is known for this class of precursor.   In addition to these two classes, it appears that there is a kind of precursor having a considerably broad spectrum of precursor time, the peak of the histogram of precursor time taking place around a time range of several days.   The characteristics of animal behavior precursor, should it be a precursor at all, will also be described.

Sections 9 and 10 will be reserved for a review on the possible algorithm of earthquake prediction and the present-day policy concerning actual issuance of earthquake prediction to the public.   Although only a few legislations have so far been made, every effort toward establishing ways and means for earthquake warning of practical use is now under consideration by national and local governments at least in Japan and the State of California in the U.S.A.   A number of successful as well as unsuccessful earthquake warnings have been reported from the People's Republic of China where issuance of earthquake warning may not be as difficult as that in capitalistic countries because of the powerful control of the public by the government.

The Prediction Council for the anticipated Tokai earthquake (Tokai literally means the east sea, and denotes the area in Central Japan facing the Pacific Ocean), of which the magnitude is believed to exceed 8 from various reasons, has now been formed in the Coordinating Committee for Earthquake Prediction (CCEP), that acts as the headquarters of earthquake prediction in Japan. Being composed of several university professors working in Tokyo, the task of the council is to judge whether or not an earthquake will occur in the Tokai area on the basis of geophysical and geochemical data telemetercd to the Japan Meteorological Agency (JMA) in Tokyo. Such activities related to earthquake prediction will also be stated in Section 10.

2. National and International Activities Related to Earthquake Prediction

2.1 Japanese program

When we launched a national program on earthquake prediction in Japan in 1965, most researchers put stress on acquiring basic data for earthquake prediction by observing whatever geophysical elements that should be thought to have some relation to premonitory effect of an earthquake. Frankly speaking, they were not quite certain where all this would lead to when the program had been launched.

It is fortunate that the guideline of the Japanese national program proposed by the so-called " blue-print " of earthquake prediction (TSUBOI et al., 1962) was in general directed in the right direction. The blue-print had been prepared after much debate among scores of Japanese seismologists. A good many data relevant to crustal strain accumulation and earthquake precursor have been obtained through the Japanese national program over the 13 years' period since its beginning.

The highlights of the Japanese achievements under the program are as follows;

1) A precise network composed of about 6,000 first- and second-order triangulation stations will soon been established by the Geographical Survey Institute (GSI). In order to monitor crustal strain accumulation, trilateration surveys by means of a geodimeter, an electro-optical instrument, will be repeated over the network every 5 years.

2) Funds for performing levelling surveys along the 1st-order levelling routes, 20,000 km in total length, every 5 years have been allocated to GSI. Surveys over important areas will be repeated with much shorter intervals of time.

3) The original plan to establish tide gauge stations along the coast

of the Japan Islands at intervals of approximately 100 km has been accomplished by constructing some 20 tide gauge stations in addition to those already existing.

4) Thirteen additional observatories for continuous observation of crustal movement have been constructed under the national program, so that 19 observatovies are now in operation. These observatories are equipped with tiltmeters and strainmeters. JMA has developed a bore-hole type strainmeter which can measure changes in the volume of the earth's crust. Twelve strainmeters of this type have been set up along the Pacific coast of Central Japan.

5) Dramatic modernization of JMA's routine network of seismic observation has recently been completed. It will become possible to monitor almost all earthquake of magnitude (M) 3 or over occurring in and around Japan in the immediate future. National universities have constructed 19 micro-earthquake observatories, each having its own satellite stations connected to a regional center by telemetering channels. Real-time processing of seismic data has now become popular. The National Research Center for Disaster Prevention set up a bore-hole seismograph at a depth of 3,510 m at a point 28 km north of Tokyo. As the sensitivity of seismic observation can be made higher than that on the ground surface by a factor of 1,000, the instrument provides a powerful means by which we can monitor very small earthquakes occurring in the vicinity of Tokyo. Two more deep-well seismographs will shortly be added at different sites around Tokyo.

A submarine seismic observation system for on-line real-time observation has been developed by JMA, and set up off Omaezaki Pt., on the Pacific coast of Central Japan where a possible occurrence of a large-scale earthquake of magnitude 8 or so has been feared in recent years.

6) A series of experiments for observing changes in seismic wave velocities, if any, have been conducted by observing seismic waves generated by an explosion on an island about 100 km south of Tokyo. Such explosions have been made with an interval of approximately 1 year. Although it is aimed at detecting changes in seismic wave velocity being propagated through the Sagami Bay, which was the seat of the 1923 Kanto earthquake $(M=7.9)$, no marked indication of velocity change has been found over the 10-year period of observation.

7) Intensive search for active faults in the field has been made by geologists and geographers, and a rating of faults in regard to their

potential for possible production of a great earthquake was attempted. Even the Median Line, a large-scale tectonic line running through the Kyusu-Shikoku-Kii Peninsula area has proved to be an active fault.

8) Accurate observation of changes in the geomagnetic field has been one of the important disciplines of the Japanese national programs. 14 observatories equipped with a proton precession magnetometer were constructed under the program. An unusually sensitive resistivity variometer has been set up at an observatory some 60 km south of Tokyo. The variometer has often observed coseismic as well as preseismic crustal strains in terms of resistivity change.

9) For the purpose of providing basic ideas of earthquake prediction, much attention has been paid on behaviors of rock samples under stresses which finally result in a rupture of the sample. High-pressure instruments for tri-axial compression and similar ones were installed at a number of national universities.

10) Important bearing of gravity, underground water, radon content and the like on earthquake prediction has become recognized in the course of development of the national program. Promotion of earthquake prediction study in these disciplines has now been emphasized.

11) Several U.S.-Japan seminars on earthquake prediction were organized under the framework of the U.S.-Japan Scientific Cooperation. Frequent exchanges of researchers and instruments were also made between the two countries. A few seismology missions visited China which in turn sent missions to Japan in 1975 and 1978. Exchanges of seismologists also took place between the U.S.S.R. and Japan.

2.2 American program

Prediction-oriented study on earthquake had been almost non-existing in the U.S.A. until 1964 when the 1st seminar on earthquake prediction between the U.S.A. and Japan was held in Tokyo. The importance of earthquake prediction was recognized by the public when a great earthquake of magnitude 8.4 struck Alaska immediately after the seminar in March, 1964.

In 1965, a 10-year program on earthquake prediction research was proposed by PRESS et al. (1965). Unfortunately, the program was not put into practice because of various difficulties. Seismologists and geophysicists started, however, to conduct earthquake prediction study very intensively with funds from other sources. Much stress has naturally been put on observation of various disciplines along the San Andreas fault and its surroundings where most of Californian earthquakes are occurring. A fairly dense network of microearthquake observation telemetered to the National Center for Earthquake Research (NCER), an institute belonging to the U.S. Geological Survey (USGS), in Menlo Park near San Francisco was established by 1970.

A nation-wide earthquake prediction program was officially set forth in 1973 as a part of the Earthquake Hazards Reduction Program resulting in a more intensive study of the subject. The program was developed so extensively that an array of 70 tiltmeters, 15 magnetometers and 11 strainmeters had been set up along the San Andreas fault by 1974 in addition to the seismic observation network already mentioned. Observations of fault creep, underground water and the like were also intensified.

An extensive ground uplift, amounting to 25 cm at maximum during 1960–1975, was found along the San Andreas fault centering on Palmdale about 70 km north of Los Angeles, California (CASTLE et al., 1976). Being called the Palmdale Bulge, the anomaly has become a target of intensive studies of various disciplines aiming at possible prediction, should it be connected to an earthquake at all. By the middle of 1977, the uplift seems to have extended to the California-Arizona border although no further uplift has developed around Palmdable.

Bilateral cooperation in the field of earthquake prediction study has been taking place between the U.S.A. and the U.S.S.R. in recent years as a part of scientific cooperation in general between the two nations. Short- and long-term exchanges of seismologists were made.

2. 3   Soviet program

It is understood that an earthquake prediction program has been underway in the Kazakh, Kirghiz, Uzbek, Tajik, and Turkmen Republics of the U.S.S.R. Middle Asia as well as in Kamchatka. Outstanding premonitory effects such as changes in seismic wave velocities, focal mechanism, underground electric resistivity, geomagnetic field, water level and radon content in underground water and so on have been found by our Soviet colleagues especially in the highly-seismic Garm area in the Tajik Republic where well-organized and concentrated effort toward earthquake prediction has long been made.

2. 4   Chinese program

A destructive earthquake of magnitude 7.3, that occurred near Haicheng, Liaoning Province, on Feb. 4, 1975, was successfully foretold for the first time in history, and many lives were saved by a warning to the public issued several hours prior to the main shock. Such an imminent prediction was made possible after a series of long-, medium-, and short-term predictions.

The success was a result of the intensive program on earthquake prediction since 1966 when strong earthquakes of magnitude 6.8 and 7.2 hit Hsingtai, Hopei Province, about 300 km southwest of Peking. The late Premier Chou En-lai visited the devastated area immediately after the quakes, and soon after his return to Peking, it is said that a large-scale national project for earthquake prediction was set out. It is also said that several hundred scientists and several thousand technicians are nowadays working on earthquake prediction at about 250 seismological observatories and 5,000 observation points. Participation of countless amateurs to the work seems to be strongly encouraged in China. Those amateurs observe not only geophysical and geochemical elements such as earth-currents, underground water, radon content, and the like with home-made apparatus but also anomalous animal behavior.

It is really surprising that China has risen to such an ambitious program from, frankly speaking, a not-highly-developed level of geophysics. Yet, it is more surprising that China has, to date, achieved a number of successful forecasts of destructive earthquakes. Among 5 destructive earthquakes of magnitude 7 or around in 1975–1976, 4 were predicted. They are the Lungling ($M=7.5$, 1976, Yunnan Province), Sangpan-Pingwu ($M=7.2$, 1976, Szechwan Province) and Yenyuan-Minglang ($M=6.9$, 1976, Yunnan-Szechwan border) earthquakes in addition to the afore-mentioned Haicheng earthquake. It is regrettable, however, that no imminent prediction of the Tangshan earthquake ($M=7.8$, 1976, Hopei Province), that occurred in an industrial area and killed hundred thousands of people, was put forward for some reasons including the intervention of the so-called gang of four.

In China, observation of every element which seems likely to have something to do with an earthquake precursor is encouraged. This approach is quite in contrast to that of our U.S. colleagues who usually put more stress on a possible model and tend to observe what is expected from the model.

2.5 *International activities on earthquake prediction*

An international Commission on Earthquake Prediction was formed as one of the activities of the International Association of Seismology and the Physics of the Earth's Interior (IASPEI) at the General Assembly of the International Union of Geodesy and Geophysics (IUGG) held in Zurich in 1967. The commission has organized a number of international symposia on earthquake prediction on occasions of IASPEI assemblies. Such symposia were especially useful for researchers in western countries because earthquake prediction work in the U.S.S.R. and other east European countries were brought to light by the papers presented at the symposia.

Although the commission desired to encourage more strengthened international cooperations of earthquake prediction work, no outstanding progress has been made through such a channel of scientific unions and associations which have no power for budgeting actual work. It appears that a more strongly funded organization is needed for promoting actual earthquake prediction work, especially in the developing countries.

An Intergovernmental Conference on the Assessment and Mitigation of Earthquake Risk was called by the United Nations Educational, Scientific and Cultural Organization (UNESCO) in February, 1976. According to the resolutions adopted at the conference, it was decided that UNESCO would hold an International Symposium on Earthquake Prediction in April, 1979. A preparatory committee for the conference met at the UNESCO Headquarters in July, 1977 and September, 1978. Various aspects of earthquake prediction were discussed from both the sides of natural and social sciences at the symposium with 240 participants from 40 countries. A panel meeting after the symposium raised a number of recommendations to UNESCO. It is interesting to note that a proposal that international test sites for earthquake prediction should be provided was adopted. Any team of scientists could go to a site with permission of the host country and conduct earthquake prediction work. As extremely large earthquakes are not frequent enough for practicing and improving earthquake prediction techniques in any one country, the proposed expedition would accelerate the development of earthquake prediction technology.

3. Japan Islands—An Example of Earthquake Country

The present state of the art of earthquake prediction will be reviewed

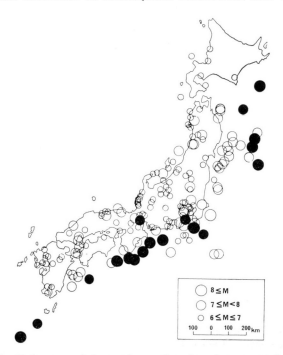

FIG. 1.   Epicenters of destructive earthquakes that occurred in and around
Japan during 1600–1977.   Only earthquakes of which the magnitude is equal
to or larger than 6 are shown.   The solid circles indicate great earthquakes
having a magnitude equal to or larger than 7.9.

in this report mainly in relation to Japan.   In this context, the author
should like to present an outline of the mode of earthquake occurrence
in and around the Japan Islands in this section.

Figure 1 shows the epicenters of destructive earthquakes that
occurred in and around Japan during a period from 1600 to 1977.
Earthquakes having a magnitude smaller than 6 are not shown in the
figure.   If one looks into Japan's whole history covering almost 2,000
years, many more earthquakes can be added to the distribution in Fig. 1
although the magnitudes and epicenters for earlier quakes may not be
as accurate as those for present-day quakes.

The solid circles in Fig. 1 indicate earthquakes of which the magnitude is equal to or larger than 7.9.   Such an earthquake will be called
a great earthquake in this article.   Looking at Fig. 1, it is noticeable
that almost all great earthquakes occurred off the Pacific coast of the
Japan Islands, the only exception being the Nobi earthquake ($M=7.9$,

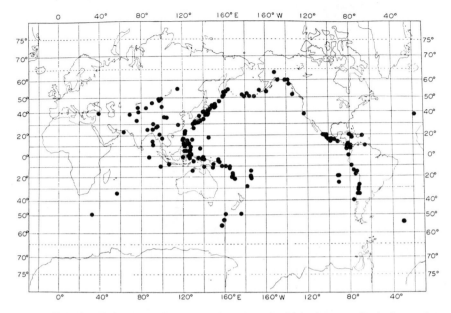

FIG. 2.   Epicenters of great earthquakes of which the magnitude is equal to or larger than 7.9 since 1896.

1891) that occurred in the middle of Honshu, the main island.

It is interesting to compare the mode of occurrence of great earthquakes in and around Japan to that for the whole earth.   World's great earthquakes since 1896 are shown on the map as shown in Fig. 2.   Judging from the fact that the great earthquakes tend to occur along a belt encircling the Pacific margin, it is no surprise that the Japan Islands are repeatedly struck by great earthquakes.   It appears that most of these great earthquakes are related to subduction of ocean plates as will be discussed in Section 4.

Japan's seismicity is manifested a little better by plotting epicenters of earthquakes of which the magnitude is smaller than 6.   Tsuboi (1958) plotted epicenters of 3147 conspicuous and moderately conspicuous earthquakes during the period 1900–1950 as reproduced in Fig. 3. According to JMA, an earthquake is said to be conspicuous when it is felt over an area having an epicentral distance of 300 km or larger. Similarly, a moderately conspicuous earthquake is defined for those having an epicentral distance of 200 km or larger.   In Fig. 3, we observe an enormous concentration of earthquake foci along the Hokkaido—

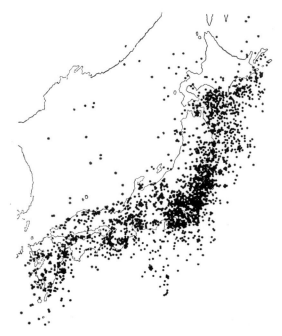

FIG. 3.   Epicenters of 3147 conspicuous and moderately conspicuous earth-quakes during 1900–1950 (TSUBOI, 1958).

Sanriku (an area in the northeastern Honshu facing the sea)—Kanto (an area surrounding Tokyo) coastal belt.   The only area which is free from earthquakes is the northeastern portion of Hokkaido.   Most earthquakes in the Japan Sea are deep focus ones having a focal depth of several hundred kilometers.

The observation network for microearthquakes, that was established under the earthquake prediction program, has now been in operation. As one of the achievements by the network, Figure 4 shows the disribution of epicenters of earthquakes having a focal depth smaller than 30 km as determined by the Tohoku University network for the period from May 1 to July 31, 1977 as presented at a CCEP meeting.

It is demonstrated that the occurrence pattern of small earthquakes is approximately the same as that of fairly large earthquakes by comparing Fig. 4 to Fig. 3.   The microearthquake observation has also brought out a fact that earthquakes beneath the northeastern Japan are occurring along two separate planes dipping continent-ward with an angle of 40° or so.   The distance between the two planes amounts to

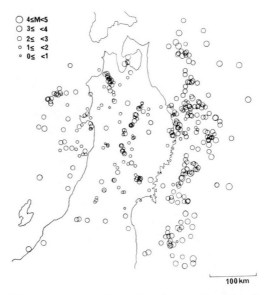

○ 4≤M<5
○ 3≤ <4
○ 2≤ <3
○ 1≤ <2
○ 0≤ <1

100 km

FIG. 4. Epicenters of small earthquakes in the Tohoku district, Japan. Only earthquakes having focal depth smaller than 30 km as determined by the Tohoku University network for the period from May 1 to July 31, 1977 are shown. This is presented at a meeting of CCEP.

30 km at a depth of 80–150 km (UMINO and HASEGAWA, 1975).   It is also made clear that most land earthquakes in the crust occur in a layer for which the P wave velocity is 6.0 km/s (TAKAGI and HASEGAWA, 1977).

4. Global Tectonics with Special Reference to Japan Islands

The present-day theory of plate tectonics claims that the earth is covered by a number of lithospheric plates, large and small, which are produced mostly at oceanic ridges. Hot substance carried up by convection currents in the mantle is cooled down underneath the axis of an oceanic ridge forming a new ocean floor which gradually spreads from there. Outstanding geomagnetic anomalies observed on the sea across the ridges can be accounted for by a hypothesis that assumes geomagnetic reversals and mantle convection at the same time. According to the hypothesis, it is quite natural to arrive at a conclusion that an ocean floor, which has alternations of magnetization in normal and reverse directions, is produced. The ocean floor magnetized in such a way can produce strong magnetic anomalies parallel to the ridge axis.

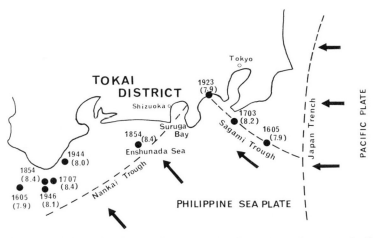

FIG. 5.   Schematic plate configuration, trench and trough system in Central Japan.   The historical great earthquakes since 1600 that occurred in association with the Nankai and Sagami troughs are shown.

As the occurrence time of reversals is known from palaeomagnetic study, it is possible to estimate the spreading rate which usually amounts to several centimeters a year.

Close examination of sea floor spreading brought forth an idea that the motion of the sea floor can be approximated by that of a rotating rigid plate slipping over the earth's surface about a certain axis.   Outstanding geophysical and geological events such as earthquakes, volcanic eruptions, mountain building, and the like occur mostly at interfaces between these plates.

In connection with the northwest Pacific margin, it is presumed that the Pacific plate spreading from the East Pacific Rise is descending beneath the Eurasia plate at the Aleutian-Kurile-Japan-Mariana trench system.   Such a phenomenon is called subduction.   Figure 5 shows the plate configuration and trench system in Central Japan.   There is evidence that the Philippine Sea plate, a subsidiary plate bounded by two deep sea-canyons, the Nankai and Sagami troughs, and the Izu-Mariana island arc, is moving northwestward.   In association with the subduction of the sea plate at the Nankai trough, the land plate is compressed and dragged downwards.   When the displacement of the land plate exceeds some ultimate value, rupture takes place at the interface of the two plates giving rise to a rebound of the land plate.   At the time of the rebound, the land, which has been compressed and dragged, suddenly expands seawards and upwards usually by several meters or so.

A huge mass of water then appears on the surface in association with the sudden uplift of sea bottom, and a tsunami is propagated when the mass consequently collapses. The strain energy stored in the compressed crust is released by the rupture being converted into the energy of seismic waves.

Great earthquakes, which have been repeatedly occurring in relation to the Nankai trough, are believed to be excited by the above-mentioned mechanism. Meanwhile, the earthquakes along the Sagami trough where the land and sea plates pass each other, are believed to be caused by a similar mechanism of rupture and rebound although the horizontal dragging of the land plate by the motion of sea plate is more predominating than the subduction in this case.

Great earthquakes off the Pacific coast of Hokkaido and Kurile Islands are also believed to be caused primarily by the subduction-rupture-rebound process at the Kurile-Japan trench.

The detailed mechanism of inland earthquakes in Japan is not quite clear although the primary cause is supposed to be based on the plate motions as stated above. The actual stress system in the earth's crust seems extremely complicated because of crustal heterogeneity. It is said, however, that the activity of such intra-plate earthquakes becomes high as the time of a great inter-plate earthquake approaches (e.g. UTSU, 1974).

5. History of Great Earthquakes

In the light of the mechanism of great earthquakes at subduction zones based on the global tectonics as discussed in the last two sections, it is not unreasonable to think that we are in a position to say something about the recurrence of great earthquakes in the future by analysing the occurrence pattern in the past because such occurrence pattern of great earthquakes is not entirely random in certain favorable cases.

First of all, let us examine the historical great earthquakes in Japan related to the Nankai trough (see Fig. 5). These earthquakes are listed in Table 1 respectively for the Nankai (South Sea) and Tokai (East Sea) zones. Although the magnitude of the 1360 earthquake is estimated as 7.0, UTSU (1974) mentioned that the actual magnitude must have been larger than 7.0. The space-time pattern of great earthquakes in the Nankai and Tokai zones is shown in Fig. 6. The rupture zones

TABLE 1.   Great earthquakes along the Nankai Trough.

Nankai zone				Tokai zone			
Year	Interval (year)	Epicenter	Magnitude	Year	Interval (year)	Epicenter	Magnitude
684		32.5°N 134.0°E	8.4	887		?	?
	203				209		
887		33.0 135.3	8.6	1096		34.2°N 137.3°E	8.4
	212				264		
1099		33.0 135.5	8.0	1360		33.4 136.2	7.0
	262				138		
1361		33.0 135.0	8.4	1498		34.1 138.2	8.6
	244				107		
1605		33.0 134.9	7.9	1605		34.3 140.4	(7.9)
	102				102		
1707		33.2 135.9	8.4	1707		?	(8.4)
	147				147		
1854		33.2 135.6	8.4	1854		34.1 137.8	8.4
	92				90		
1946		33.0 135.6	8.1	1944		33.7 136.2	8.0

are estimated on the basis of the spatial extent of aftershock area for recent earthquakes as are indicated with solid lines.   Meanwhile the rupture zones for historical earthquakes are guessed from historical documents, so that their extents as indicated with dashed lines are not very reliable.

When the data after 1600 are taken into acount, we immediately see from the figure that the mean return period of great earthquakes takes on a value around 100 years.   RIKITAKE (1976 b) applied a Weibull distribution analysis to the above set of data for the return period.

It is assumed that the probability of a return period to lie between t and $t+\varDelta t$ is given by $\lambda(t)\varDelta t$ on the condition that no great earthquake occurs prior to t, and that

$$\lambda(t) = Kt^m \qquad (1)$$

where $K>0$ and $m>-1$.

Let us denote a cumulative probability for the recurrence of a great earthquake during the period between 0 and t by $F(t)$, the last earthquake being assumed to have occurred at $t=0$.

Putting

$$R(t) = 1-F(t) \qquad (2)$$

a Weibull distribution function is assumed as

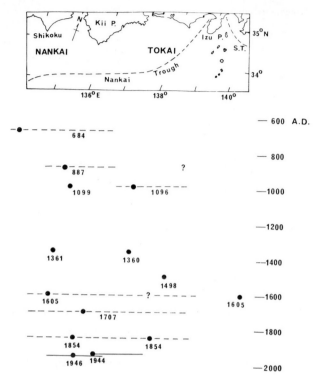

FIG. 6. Space-time distribution of great earthquakes in the Nankai-Tokai zone. The rupture zones are indicated with solidline segments. The segments shown with a dashed line are uncertain because they are estimated from historical documents (RIKITAKE, 1976b).

$$R(t) = \exp\left[-Kt^{m+1}/(m+1)\right]. \qquad (3)$$

Parameters K and m, that govern the distribution, can be determined by analysing actual data. When the parameters are known, the mean return period and its standard deviation are readily calculated. The mean return period and its standard deviation for the Nankai-Tokai zone are thus estimated as 117 and 35 years, respectively.

One of the most important points demonstrated in Fig. 6 is the fact that no great earthquake has occurred in the eastern portion of the Tokai zone since the last shock of 1854 despite the fact that we had two great earthquakes in the 1940's, i.e. the Tonankai ($M=8.0$, 1944) and Nankai ($M=8.1$, 1946) earthquakes in the western Tokai and the Nankai zone, respectively.

As the period exceeding the mean return period, 124 years say, has already past since the 1854 Tokai earthquake of magnitude 8.4, there is every reason to fear a possible recurrence of a great earthquake there. The cumulative probability of a great earthquake recurring there reaches 65% at present, according to an estimate based on the above-mentioned Weibull distribution.

Recurrence tendency of great earthquakes is also predominating along subduction zones such as Hokkaido-Kurile, Kamchatka, Aleutian-Alaska, Central and South Americas encircling the Pacific Ocean. It is possible to estimate cumulative probabilities of a great earthquake in these zones in a similar fashion as that for the Nankai-Tokai zone (RI-KITAKE, 1976 b). It is especially interesting to estimate the probability for Central America. The cumulative probability increases from 0 to 1 almost within several years' time, so that prediction is almost deterministic provided things are discussed with an allowance of several years. This is caused by the fact that the mean return period is estimated as 34.5 years with a standard deviation as small as 3.6 years.

Turning to the Sanriku (off the Pacific coast of the northeastern Honshu, Japan) zone, where extremely great earthquakes sometimes occur near the Japan trench, the recurrence tendency is not clear. This is also the case for great earthquakes along the Sagami trough and the San Andreas fault, California, where strike-slip movement predominates rather than subduction. It is not possible, therefore, to estimate the probability for recurrence in these zones.

6. Monitoring of Crustal Strain

Granting that most great earthquakes at a subduction zone is caused by the subduction-rupture-rebound process as described in the previous sections, one can readily think of a practicable way of gaining access to earthquake prediction. It is obvious that we can say something about earthquake occurrence if the extent of strain accumulation is known, and if the ultimate strain to rupture of the earth's crust is also known.

6. 1 Strain accumulation

In order to monitor accumulation of crustal strain, the most powerful means is the repetition of geodetic surveys. On making the difference in the result between the triangulation surveys before and after the Kanto earthquake ($M=7.9$, 1923), it is made clear that the triangula-

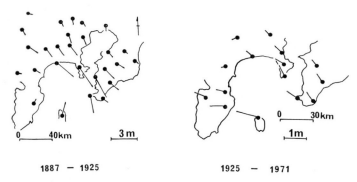

1887 — 1925   1925 — 1971

FIG. 7.   Left; Horizontal displacement of the first-order triangulation stations in the South Kanto area associated with the 1923 Kanto earthquake ($M=7.3$). Right; Horizontal displacement of the stations during the post-earthquake period from 1925 to 1971 (KAKIMI *et al.*, 1977).

tion stations in an area to the south and west of Tokyo (South Kanto) shifted 5–6 m in an southeast direction as can be seen in the left figure of Fig. 7 (KAKIMI *et al.*, 1977).   Many more examples of this kind are reproduced in the book by RIKITAKE (1976 a).

Triangulation survey, which is highly laborious and expensive, is now replaced by trilateration survey, in many cases by making use of a geodimeter, an electro-optical instrument for measuring a distance of 50 km with an accuracy of 1 cm or thereabout.

A recent trilateration survey made it clear that considerable extension and contraction of the distance between triangulation stations in the South Kanto area have taken place as can be seen in Fig. 8 (GEO-GRAPHICAL SURVEY INSTITUTE, 1972).   It is predominantly noticed that the distance between a station on Oshima Island and that on Izu Peninsula has shortened by 1 m during 1925–1971.   Meanwhile, the distances between the island and Boso and Miura Peninsulas elongated as much as several tens of centimeters.

The right figure of Fig. 7 shows the horizontal displacement of triangulation stations as calculated from the surveyed results.   It may be said that approximately one-third of the displacement associated with the Kanto earthquake has recovered during the said 46 years' period. Judging from the pattern of displacement, the leading portion of the crustal movement must have been caused by the northwest-ward motion of the Philippine Sea plate as already mentioned in Sections 3 and 4.

On the basis of the surveyed results as shown in Fig. 8, it is pos-

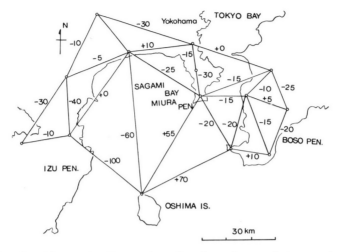

Fig. 8.   Changes in distance in units of centimeters between triangulation stations in the South Kantó area during 1925–1971 (Geographical Survey Institute, 1972).

sible to estimate strains for each triangle formed by any combination of three stations.   If we calculate the mean value of maximum shearing strain for triangles in the South Kanto, and if we divide that value by the time span, 46 years say, during which the strain has been stored, the mean rate of strain accumulation is estimated as 0.057×10^{-5}/year. Similarly, the mean rate of maximum shearing strain over the Suruga Bay (see Fig. 5) during 1931–1974 is estimated as 0.043×10^{-5}/year (Rikitake, 1977).

If it is assumed that crustal strain so far accumulated is released at the time of the last great earthquake in a certain area, and that crustal strain tends to be stored again since then, the accumulated amount of strain at any epoch in that area can be obtained from the strain rate as estimated in the above multiplied by the time span in years reckoned from the time origin which is taken at the time of the last earthquake.

6. 2   Ultimate strain

The ultimate strain to rupture of the earth's crust can be obtained by comparing triangulation (trilateration) or levelling surveys over the epicentral area before and after an earthquake.   It is particularly important in this context to obtain the diminishing rate of horizontal or vertical displacement as the distance from an earthquake fault increases.

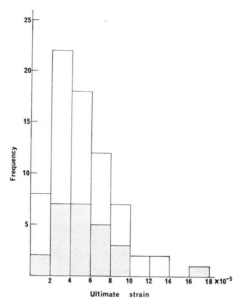

Fig. 9. Histogram of ultimate strain for the inland (shaded columns) and sub-duction zone (blank columns) earthquakes (Rikitake, 1976b).

Rikitake (1975 a) presented 26 examples of ultimate strain estimated for inland earthquakes. Similar estimate of ultimate strain can also be effected for inter-plate earthquakes based on some presumptions which may be physically acceptable (Rikitake, 1976 b).

In Fig. 9, the histogram of ultimate strain as estimated for inland earthquakes and inter-plate ones at subduction zones is shown. According to a Weibull distribution analysis of the distribution, the mean ulti-mate strain and its standard deviation are estimated as 4.4×10^{-5} and 1.7×10^{-5}, respectively. Although Tsuboi (1933) remarked that the ultimate strain of the earth's crust is of the order of 10^{-4}, it appears that many examples added later lead us to a value cited above. As the parameters of Weibull distribution fitting in the histogram in Fig. 9 are obtained by analysis, the probability for the earth's crust to break can at once be estimated provided the amount of strain accumulated there is given.

6. 3 Probability of earthquake occurrence

What have been stated in the last two Subsections enable us to esti-

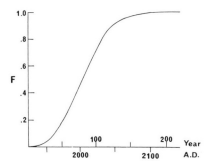

Fig. 10.   Culmulative probability for a great earthquake to recur in the South Kanto area (Rikitake, 1975a).

mate the probabilities of occurrence of a great earthquake in an area such as the South Kanto and Tokai areas where the strain accumulation can be estimated year by year.

As for the possible recurrence of the Kanto earthquake, the increase in the cumulative probability for recurrence is estimated as given by the curve in Fig. 10.   The probability at around 1980 is still as low as 20%, so that we feel that the recurrence of a great earthquake is a matter for serious discussion in the following century.

On the contrary, similar estimate of cumulative probability gives a value amounting to 89% at present in the Tokai area (Rikitake, 1977). Even if the errors involved in the geodetic surveys are processed so as to minimize the strain over the Suruga Bay area, the probability seems to exceed 60%.   It is also possible to estimate the probability of having a great earthquake within 10 years' time from a certain epoch on the condition that no earthquake has occurred during the period from the time origin to the epoch concerned.   Such a probability is estimated as 52% for 1976 provided the usual partition of errors of geodetic surveys is assumed (Rikitake, 1977).

It should therefore be emphasized that the probability of earthquake occurrence is several times as high as for the Tokai area compared to the South Kanto area.   This is the reason why it is feared that we would have a great earthquake in the Tokai area sooner or later, and probably sooner.

This kind of probability estimate can be performed for an area where well-organized geodetic surveys have been and will be conducted. Only areas along the San Andreas fault are possible candidates qualified for this kind of estimate besides the Japanese areas discussed above. RIKITAKE (1975 a, 1976 b) pointed out a high probability area where the 1857 Fort Tejon earthquake of magnitude 8 or so occurred. It is interesting to note that an outstanding land uplift has been found over the said area in recent years (CASTLE *et al.*, 1976).

It has been mentioned in the last and present chapters that the history of great earthquakes and monitoring of crustal strain sometimes lead us to an estimate of occurrence probability of an earthquake. This kind of information can hardly be said to be an earthquake prediction. This kind of thing is comparable to a long-term forecasting of weather. But such a long-term information is useful for nominating an area to an earthquake-threatened area where every effort toward predicting an earthquake should actually be made.

7. Earthquake Precursor

Statistics of great earthquakes in countries having a long history such as China, Japan, Turkey and the like and monitoring of crustal strain in areas over which geodetic surveys have often been repeated as stated respectively in Sections 5 and 6 may indicate potential areas where a large earthquake would occur sooner or later. As has already been pointed out, this kind of guess for earthquake occurrence cannot be said to be a prediction. It is only a kind of regionalization. A true prediction of earthquake should be based on an actually-observed physical symptom or precursor. This section is aimed at reviewing geophysical and geochemical precursors of various disciplines so far reported. A brief account on anomalous behavior of animals preceding an earthquake will also be presented.

The author (RIKITAKE, 1975 b) analysed a data-set of 282 earthquake precursors reported by 1974. Another set of precursor data amounting to 109 in number was later added (RIKITAKE, 1978 b). It appears that the speed of data acquisition is being very much accelerated because of the promotion of earthquake prediction programs mainly in China, Japan, the U.S.A., and the U.S.S.R. It would not be unreasonable to think that a fairly comprehensive set of precursor data can shortly be provided, 10 years later say, should the programs on earthquake prediction be developed at a pace as rapid as the present one.

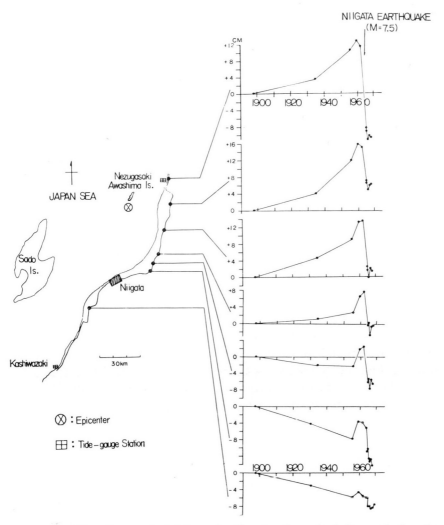

FIG. 11. Changes in height at levelling bench marks before and after the 1964 Niigata earthquake (DAMBARA, 1973).

7.1 Land deformation

One of the most reliable precursors of a coming earthquake is sometimes detected by means of repetition of geodetic survey. Figure 11 is the now-famous land uplift which preceded the 1964 Niigata earthquake of magnitude 7.5. It is noticeable that bench marks for levelling surveys along a levelling route passing through Niigata City on the coast of the Japan Sea indicated an upheaval since around 1955, largely deviating from the trend of gradual secular movement since 1900. The earthquake occurred when the anomalous uplift was suspended. Should the uplift have been a premonitory effect of the earthquake, the precursor time, that is the time between the onset of the effect and the main shock, is estimated as 8 years or thereabout.

On the day of the Tonankai earthquake ($M=8.0$, 1944), a levelling survey was underway around Kakegawa City, Shizuoka Prefecture about 150 km distant from the epicenter. It was noticed by the surveyer that the closing error of the survey was ridiculously large. SATO (1970) pointed out that a premonitory tilting of the order of 1 second of arc had been taking place at least several hours preceding the earthquake. The precursory tilting took place in a direction the same as that accompanied by the main shock. It is also reported that the bubble of the level became so restless that the setting of the instrument became difficult probably several minutes prior to the shock.

Other examples of premonitory land deformation, both horizontal and vertical, have been reported as summarized by RIKITAKE (1976 a).

At the moment, two anomalous land uplifts are known. The one in California has already been described in Subsection 2. 2. Another one has been found in the Izu Peninsula, Japan (see Fig. 8) as can be seen in Fig. 12 (GEOGRAPHICAL SURVEY INSTITUTE, 1976, 1977 a, b). The uplift seems to have expanded a little to the west in 1977. Since the 1974 Izu Hanto-Oki earthquake ($M=6.9$), that occurred at the southern-most tip of the peninsula, the seismicity has been high in the peninsula which occasionally experienced swarm activities involving a few shocks of magnitude 5 or over.

An earthquake of magnitude 7.0 occurred under the sea between the peninsula and Oshima Island accompanying an earthquake fault extending to the east coast of the peninsula. Although a land subsidence of 17 cm in height took place probably in association with the shock on the east coast of the peninsula, the anomalous uplift to the north of the fault does not seem to have been affected by the earthquake.

It is not at all clear how the uplift is correlated to the present and future seismicity. There is evidence, however, that the strain accumulation around the Tanna fault, which moved at the time of the 1930 North Izu earthquake ($M=7.0$), is large (GEOGRAPHICAL SURVEY INSTITUTE, 1974). As was pointed out by RIKITAKE (1975 a), the rate of strain accumulation is so large there that the strain would reach the mean ulti-

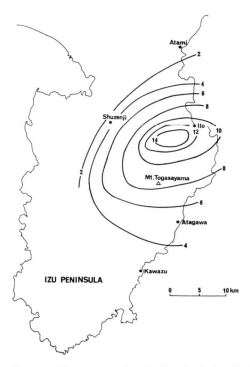

Fig. 12. Land uplift in cm in Izu Peninsula during 1967 and 1969 to 1976. The contours are obtained by combining the levelling surveys in 1967, 1969, and 1976 (Geographical Survey Institute, 1976).

mate strain as discussed in Subsection 6. 2 by 1980. On the other hand, a Sacks-Evertson strainmeter (Sacks *et al.*, 1971) at Ajiro near Atami City to the east of the Tanna fault and to the north of the anomalously uplifted area has been recording and anomalous compression since the end of the year 1977 (Japan Meteorological Agency, 1978, see Subsection 7. 3).

Although nothing certain is known about the correlation between the seismicity, land uplift and strain accumulation predominating over the Izu Peninsula, it is suspected that the crustal stress at the northeastern corner of the Philippine Sea plate is getting high in recent years. As the Tokai area adjacent to the Izu Peninsula is under threat of a great earthquake as stated in Sections 4 and 5, special attention should be paid to the crustal activity in the Izu Peninsula.

7. 2   Change in sea level

There are a number of reports that a severe earthquake occurred following a sudden retreat of sea water in Japan's history. At the time of the 1802 Sado earthquake, of which the magnitude is estimated as 6.6, for example, it was reported that a fairly strong earthquake occurred at about 10 o'clock in the morning when an anomalous sea retreat was observed at Ogi port of the Sado Island off the Japan Sea coast in Central Japan. While the local people feared that a tsunami might hit there, the main shock struck the island at about 2 o'clock in the afternoon causing collapse of houses. Such a sea retreat is nothing but a local land-uplift observed relative to the sea surface.

Present-day monitoring of sea level is made by a tide gauge which enables us to conduct continuous recording. Telemetering of sea level data from tide gauge stations in an earthquake-threatened area to a central monitoring office in Tokyo is in operation in Japan.

Changes in sea level as observed by a tide gauge are usually contaminated by noise due to meteorological and oceanographical origins. The sea level is in fact affected by atmospheric pressure, wind, water temperature, ocean currents and so on. The usual procedure of eliminating such noise is to make the difference in height of sea level between two stations, several tens of kilometers distant from each other. It has been known, however, that an error of several centimeters cannot be avoided even if such an elimination procedure has been applied to raw data.

In recent years, search for precursors in terms of sea level change has been made notably by WYSS (1975, 1976 a, b). He reported on sea level changes amounting to 3 to 6 cm forerunning the 1960 Antofagasta, Chile ($M=6.8$), 1966 Peru ($M=7.5$) and 1961 and 1968 Hyuganada ($M=7.0$ and 7.5) earthquakes. The precursor times are 2190, 1830, 1280 days, respectively. Other recent data are summarized in RIKITAKE (1978 b).

7. 3   Tilt and strain

A great number of premonitory changes in ground tilt and horizontal strain as observed with tiltmeters and strainmeters have been reported by many authors (e.g. RIKITAKE, 1976 a). Judging from the hitherto-accumulated data, tilt and strain observation is doubtless useful for detecting earthquake precursors. Although no description of the data presented so far (RIKITAKE, 1976 a) is repeated here, the author should like to point out that a few precursory tilting took place a few days and also several months prior to the Haicheng earthquake ($M=$ 7.3, 1975). Such changes would seem to have provided a basis for the prediction of that earthquake.

The extensive array of bore-hole tiltmeters set up along the San Andreas fault as mentioned in Subsection 2. 2 appears to be useful for earthquake prediction. Prior to the occurrences of an earthquake of magnitude 3 to 5, precursory changes in tilt direction have often been reported.

One of the most outstanding successes in monitoring precursory strain change has been achieved by an array of bore-hole strainmeters as mentioned in Item (4) of Subsection 2. 1.   When a shock of magnitude 7.0 occurred nearby Izu Peninsula and Oshima Island on Jan. 14, 1978, it was noticed that a strainmeter at Irozaki station situated at the southern-most extremity of the peninsula had been indicating an anomalous compression since Dec. 3, 1977 as can be seen in Fig. 13. The trend of compression changed its sign about 4 days prior to the main shock. Such an anomalous change in volume strain might be an earthquake precursor.

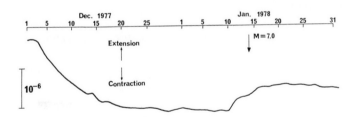

FIG. 13. Anomalous changes in volume strain as observed by a bore-hole volume strainmeter at Irozaki located at the southern extremity of Izu Peninsula (JAPAN METEOROLOGICAL AGENCY, 1978).

The record at Ajiro station, 55 km north of Irozaki, is also anomalous. Starting around Dec. 20, 1977, the volume strain there began to indicate a remarkable contraction with a ridiculously large strain rate amounting to 2×10^{-7}/day. Should the contraction continue over a period of a few hundred days with the same rate, the strain would reach its ultimate value. In contrast to tiltmeters and strainmeters set up in an underground vault, the bore-hole strainmeter used by JMA does not have a long history of observation. It is therefore difficult to evaluate the strain changes as mentioned above. However, the records obtained at Irozaki on the occasion of the 1978 earthquake and other stations on different occasions strongly suggest that the recorded changes in volume strain may have something to do with possible earthquake precursors.   At the moment, the JMA strainmeters set up at 12 stations along the Pacific coast in Central Japan provide the most reliable means for imminent prediction of the suspected Tokai earthquake.

7. 4   Tidal strain

That the amplitude of earth tidal constituents such as M_2, O_1, S_2 and the like may be modified before an earthquake has long been suggested (e.g. RIKITAKE, 1976 a).   Fairly convincing observations of premonitory changes in tidal amplitude have been reported in recent years.

LATYNINA and RIZAEVA (1976) reported a 6% decrease in M_2 amplitude 30 days prior to a magnitude 4.5 earthquake (1967) near Dushambe, Middle Asia.   MIKUMO et al. (1977) also reported a 15% increase in tidal admittance about 11 months before the Central Gifu earthquake ($M=6.6$, 1969) on the basis of observation at Kamitakara Observatory about 60 km distant from the epicenter.   It may be that the tidal strain can be regarded as one of the disciplines useful for earthquake prrediction.

7. 5 Foreshock, anomalous seismicity, b-value, microseismicity, source mechanism, and hypocentral migration

In this section, the author proposes to summarize recent findings in disciplines of earthquake prediction related to seismic activity. Foreshock data are steadily increasing.   The importance of foreshocks on earthquake prediction was emphasized on the occasion of the Haicheng earthquake ($M=7.3$, 1975), while the Tangshang earthquake ($M=7.8$, 1976) could not be foretold partly because no outstanding foreshocks had been observed.   It appears to the author that the precursor time, namely the time-interval between the very beginning of foreshocks and the main shock, does not in any way depend upon the magnitude of the main shock.

The Haicheng earthquake is characterized by the fact that a considerable increase in number of small earthquakes was observed in Liaoning Province during the 1.5-year period previous to the main shock and that an enormous increase in number as well as in magnitude of foreshocks in the Haicheng area was followed by a sudden decrease in foreshock activity that took place several hours prior to the main shock.

It is said that the earthquake warning was actually issued by the local authorities when the abrupt decrease in foreshock activity was recognized.

Small shocks had been observed since a few days prior to the January 14, 1978 earthquake ($M=7.0$) that occurred near Izu Peninsula and Oshima Island. The seismic activity tended to increase in the morning of the very day of the main shock occurrence. Such an activity was suspended quite suddenly at about 10 o'clock, so that JMA issued information about a possible occurrence of a slightly damaging earthquake to the local people. The main shock actually occurred at 12:24 on that day though its magnitude was far larger than expected. It may well be said, therefore, that a sudden decrease in seismic activity sometimes provides a forerunner of a large earthquake as has previously been discussed (e.g. RIKITAKE, 1976 a).

It has been well known that the Gutenberg-Richter formula which is represented by

$$\log {}_{10}N = a - bM \tag{4}$$

holds good for seismic activities, where N is the number of earthquakes of magnitude M that occur during a certain period and a and b are constants. Usually, b takes on a value around 1 for normal seismic activity.

As has often been pointed out notably by Suyehiro (SUYEHIRO et al., 1964; SUYEHIRO, 1966, 1969) and others, b-value for foreshocks is considerably lower than its normal value. Values as low as 0.5–0.6 have been reported for foreshocks. The b-value provisionally estimated for the foreshock activity of the aforementioned 1978 Izu earthquake takes on such a small value.

It is interesting to note that the b-value for microearthquake activity decreased from 1.0 to 0.8 about 730 days prior to the magnitude 6.2 earthquake that occurred in the southeast Akita Prefecture, Japan in 1970 (HASEGAWA et al., 1975). A summary of precursory change in b-value can be found in RIKITAKE (1976 b).

One of the most interesting findings relevant to seismic activity in recent years is the anomalous seismicity that seems to precede a large earthquake as discussed by SEKIYA (1976) and EVISON (1977 a, b). These authors pointed out that a swarm-like seismic activity sometimes took place in an area which later becomes the epicentral area of a major earthquake. The larger the magnitude of the main shock is, the longer is the lead time of the anomalous activity. Sekiya presented 10 examples of this kind of seismic activity, while 11 examples have been put forward by Evison. The anomalous seismic activity as stated here may be a kind of foreshock activity in a broad sense. But the present activity is discriminated from usual foreshocks by the magnitude dependence of precursor time.

Only one example of premonitory change in source mechanism was reported in these several years (GUPTA, 1975). Meanwhile, precursory migration of earthquake focus has drawn the attention of seismologists (BUFE et al., 1974; GUPTA, 1975; ENGDAHL and KISSLINGER, 1977). It has been reported by BUFE et al., (1974) that the depth of microearthquake became deep before a moderately large earthquake that occurred at the Stone Canyon area in California. ENGDAHL and KISSLINGER (1977) reported on the gradual migration of the foci of foreshocks converging to that of the main shock as observed in the Aleutian Islands area.

7. 6   Change in seismic wave velocities

Many data of premonitory changes in P wave velocity (V_P) and S wave velocity (V_S) have been reported notably from China in recent years. The number of data was doubled in these 2 to 3 years' time (see RIKITAKE, 1978 b). These data suggest the feasibility of using changes in V_P, V_S or V_P/V_S ratio as a possible means for earthquake prediction on the condition that the accuracy of observation is good enough.

It is important, however, to point out that much of ambiguous and even negative results emerged from the studies of seismic waves produced by man-made explosion, as intensive researches on this subject developed, notably in California. As it is tendious to present here all the references, the author hopes that those who are particularly interested in the topic may refer to CHOU and CROSSON (1978) in which most of the existing references can be found.

Judging from the recent results as mentioned in the above, it is clear that we do not necessarily expect to observe changes in seismic wave velocities that are presumed to be induced by a dilatancy generation in the earth's crust (e.g. SCHOLZ et al., 1973). Although it was at one time believed that the V_P/V_S ratio observation might provide a very powerful means of earthquake prediction, it now turns out that there are classes of earthquakes that accompany no change in the V_P/V_S ratio.

7. 7   Geomagnetism, earth-currents, resistivity and conductivity anomaly

In addition to the precursory seismomagnetic effects as described in RIKITAKE (1976 a), a number of geomagnetic data which indicate possible changes forerunning an earthquake have recently been put forward.   Except the 1.5 gamma change associated with the earthquake on Nov. 28, 1974 of magnitude 5.2 that occurred near Hollister, California (SMITH and JOHNSTON, 1976), those changes in association with the Yakutat ($M=7.9$) and Sitka ($M=7.1$), Alaska, earthquakes that occurred respectively in 1958 and 1972, and the Haicheng earthquake ($M=7.3$, 1975) amounted to 20 gammas or so although the overall accuracy of observation is not clearly known.

WILLIAMS *et al.* (1977) detected a magnetic change exceeding 10 gammas in relation to the southern California uplift as mentioned in Subsection 2. 2.   The change took place near the junction of the San Andreas and San Jacinto faults.   A 4 gamma decrease in the total geomagnetic intensity was also observed at a station which is situated immediately above a magnitude 4.4 earthquake that occurred during a period between the two magnetic surveys.

Premonitory changes in earth-currents have often been reported by many researchers.   In spite of much experience in earth-current observation after major earthquakes, however, the author has never observed an outstanding precursory change in earth-currents associated with aftershock activity.   Much of classical as well as modern earth-current precursors are summarized by YAMAZAKI (1977).

Although it is almost impossible to perform a geophysically-significant observation of earth-currents in Japan because of stray electric currents leaking into the ground from d.c.-operated railways, it appears to the author that much stress is put on each-current observation as a possible means of earthquake prediction in the U.S.S.R. and China where man-made noise could be fairly low.

In China, earth-current observations by amateur volunteers with simple home-made instruments are popular.   Figure 14 shows an outstanding change in earth-potential forerunning the Haicheng earthquake ($M=7.3$, 1975) as observed at Yingkou, a few tens of kilometers distant from the epicenter (ZHU, 1976).   The electrode distance for observation is reported as 60 m.

Anomalous earth-current changes of various types which preceded the same earthquake have been reported, and popularized among geophysicists in western countries by an American seismology delegation

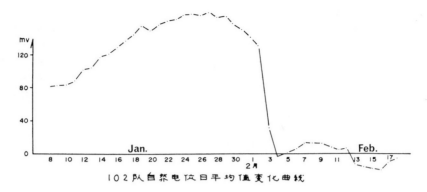

102队自然电位日平均值变化曲线

FIG. 14. Changes in earth-potential forerunning the Haicheng earthquake as observed at Yingkou, a few tens of kilometers distant from the epicenter (ZHU, 1976).

to China (RALEIGH *et al.*, 1977). According to the delegation, fluctuations in earth-potential with a period of a few minutes were observed at a number of observation sites shortly before the main shock of the Haicheng eathquake. It is really hard to evaluate the geophysical meaning of these amateur observations because they are sometimes perturbed by heavy rain fall, and other unknown causes.

That the underground resistivity sometimes decreases preceding an earthquake was first reported in the Garm area, Tadzhik Republic, the U.S.S.R. (BARSUKOV, 1972). Electric pulses of 100 amperes are sent into the ground, and the voltage between a pair of electrodes located at a distance of several kilometers is measured. The ratio of the voltage to the electric current provides a quantity proportional to the ground resistivity. As the electric currents are believed to penetrate effectively to a depth of focal region in this case, the observed decrease in resistivity may reflect some changes in the physical properties at the focal region preceding an earthquake. As are summarized by RIKITAKE (1976 a) and YAMAZAKI (1977), similar changes of a precursory nature in resistivity have been reported in China and the U.S.A.

A resistivity variometer of unusually high sensitivity has been in operation at Aburatsubo, a near-shore station about 60 km south of Tokyo. As the detail of the variometer has been reported elsewhere (YAMAZAKI, 1975, 1977), no mention of its construction will be given here. The author should, however, like to point out that coseismic resistivity steps of the order of 10^{-5}–10^{-4} in rate of resistivity change are observed for moderate to large earthquakes depending upon the earthquake magnitude and the epicentral distance, and that many of them are associated with a premonitory change having a precursor time of a few hours.

The variometer recorded a very clear premonitory effect starting 4 hours prior to the 1974 Izu Hanto-Oki earthquake ($M=6.9$), the epicentral distance being 100 km in this case.   No clear-cut premonitory effect was observed, however, on the occasion of the 1978 earthquake ($M=7.0$) that occurred near Izu Peninsula and Oshima Island in spite of the fact that the epicentral distance amounted to only 60 km this time. It is not entirely known why the mode of occurrence of a precursor is different from earthquake to earthquake.

That the amplitude ratio of the vertical intensity to the horizontal one of short-period geomagnetic variations undergoes a considerable change prior to a large earthquake has sometimes been pointed out (YANAGIHARA, 1972; MIYAKOSHI, 1975; YANAGIHARA and NAGANO, 1976). The present author, who examined the magnetograms obtained at Sitka, Alaska, Magnetic Observatory before and after the Sitka earthquake ($M=7.1$, 1972), found a similar change (RIKITAKE, 1979).   Such a premonitory effect is suspected to be caused by anomalous changes in underground electric conductivity.

7. 8   Radon, underground water and oil flow

Changes in water level, temperature, chemical composition, radon content and the like of underground water have been reported in relation to premonitory effects of an earthquake (e.g. RIKITAKE, 1976 a).   Even fluctuations of oil flow from wells in the Gulf of Suez seem to provide pre-earthquake signals (ARIEH and MERZER, 1974).

The Chinese put much stress on radon observation (GROUP OF HYDRO-CHEMISTRY, 1975; RALEIGH et al., 1977).   Although they claim that changes in radon content indicate a symptom of a coming earthquake, it is sometimes hard to distinguish the precursory signals just by looking at published curves.   Other reports on radon precursors, especially in the U.S.S.R., are summarized by RIKITAKE (1976 a).

For the first time in Japan, a precursor-like change in radon content was reported by WAKITA and ASADA (1978) who made observations at a point in Izu Peninsula, some 30 km distant from the epicenter of the 1978 earthquake ($M=7.0$) that occurred near the peninsula and Oshima Island.   As can be seen in Fig. 15, the radon content indicated a remarkable decrease since November, 1977.   The trend has reversed in

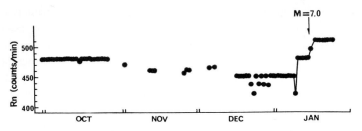

FIG. 15.   Anomalous changes in radon content as observed at a station in Izu Peninsula, the epicentral distance being some 30 km (WAKITA and ASADA, 1978).

the beginning of January, 1978 turning to a rapid increase several days before the main shock.

7. 9   Animal behavior

We have many legends, folklores and the like that cattle, dogs, cats, rats, mice, snakes, birds, fish, insects, earth worms, and other species behave abnormally before a large earthquake. It has now become widely known that the Chinese Seismological Bureau encourages local people to observe anomalous behavior of animals (ALLEN *et al.*, 1975; RALEIGH *et al.*, 1977; ZHU, 1976). It is even reported that anomalous animal behavior sometimes plays an important role on issuing earthquake prediction information to the public in China.

Classical reports on earthquake-related animal behavior are some-times fantastic. But modern reports in association with the 1975 Haicheng earthquake ($M=7.3$) in China, the 1976 Friuli earthquake ($M=6.7$) in northeastern Italy, the 1977 Rumania earthquake ($M=7.2$), the 1978 earthquake near Oshima Island ($M=7.0$) in Japan and so on are so concrete and realistic that one can hardly discredit them. Moreover, many reports about similar behaviors of common species come from mutually independent and widely separated countries.

Under the circumstances, scientific efforts toward clarifying the relation between animal behavior and earthquake occurrence have recently started at least in Japan and the U.S.A. The present author (RIKITAKE, 1978 d) analysed the existing 157 data on abnormal animal behavior prior to an earthquake. It turns out that the distribution of precursor time for animal behavior, should it be a precursor at all, indicates two peaks at about 2–3 hour range and at about 0.5 day. The latter seems to cover the time range for which very few geophysical and geo-

chemical precursors are observed.   An analysis of the 129 data in association with the 1978 earthquake near Oshima Island results in a similar
conclusion (RIKITAKE, 1980 a, b).

Although the nature of animal precursors is not at all clear, there
is a possibility that animals are very sensitive to some kind of stimula-
tions which are hard to detect by usual physical and chemical means.
The author thinks that study of animal behavior should be strengthened
although he believes that the major part of earthquake prediction must
rely on geophysical symptoms.   It is a good thing that people can obtain
knowledge about earthquake and related phenomenon through watching
animal behavior.

8. Classification of Precursor

8. 1   *Three kinds of precursors*

The nature of earthquake precursors as presented in the last
section is different from discipline to discipline.   For instance, precursor times seem to be distributed over a wide range from a few minutes
to several thousand days depending upon incompletely-understood
nature of precursors.   In order to make use of these precursor data for
forming a theory of earthquake prediction, it is therefore required to
bring the nature of precursors to light as clearly as we can.

The present author (RIKITAKE, 1975, 1976 a, 1978 b) attempted to
classify the precursors so far reported according to the dependence of
precursor time on earthquake magnitude.   As are presented in Table 2,
it was possible to collect precursor data for 19 disciplines as stated in the
last chapter with the total number amounting to 391.   The number
of precursors is increasing very rapidly even while the author is preparing the manuscript of this article, so that the most recent ones cannot
be included in the following statistics.

According to a previous study (RIKITAKE, 1975 b), it was shown
that no systematic dependence of precursor time (T) upon earthquake
magnitude (M) is seen for disciplines foreshock (f) and tilt and strain
(t).   First of all, therefore, the relationship between $\log_{10} T$ (T is measured
in units of days) and M is plotted in a graph as shown in Fig. 16 for all
the disciplines except f and t.

We can now see in the figure that there are two classes of precursors.
The first one, called the *precursor of the first kind*, are scattered in such
a manner that the precursor time is longer as the magnitude of main

TABLE 2.   Number of precursors.

Discipline	Abbreviation	Number of data
Land deformation	l	30
Tilt and strain	t	89
Tidal strain	t′	2
Foreshock	f	83
Anomalous seismicity	f′	14
b-value	b	12
Microseismicity	m	4
Source mechanism	s	7
Hypocentral migration	h	4
Fault creep anomaly	c	2
V_P/V_S	v	50
V_P and V_S	w	19
Geomagnetism	g	6
Earth-currents	e	17
Resistivity	r	32
Conductivity anomaly	c′	3
Radon	i	12
Underground water	u	2
Oil flow	o	3
Total		391

shock becomes larger.   On the other hand, it seems likely that some precursors are scattered around $\log_{10} T = -1$ irrespective of magnitude. We may call this class of precursor the *precursor of the second kind*.

The best fitting by a straight line for the $\log_{10} T - M$ relationship of the precursors of the first kind is obtained as

$$\log_{10} T = 0.60M - 1.01 \tag{5}$$

by making use of the least squares method.   The constants involved in Eq. (5) are slightly different from those in the previous analysis (RIKITAKE, 1975 b, 1976 a) in which only 89 data were used in contrast to the 163 data for the present one.

It is perhaps necessary to mention that anomalous tiltings as observed by water-tube and bore-hole tiltmeters approximately fit in Eq. (5) although all the data belonging to discipline t are excluded from the plots in Fig. 16 and determination of the constants in Eq. (5).

The present author (RIKITAKE, 1978 b) made a further analysis of precursors for disciplines f and t.   Figure 17 shows a histogram of precursor times from which the precursors that belong to the first kind are omitted.   Only the data for earthquakes having a magnitude equal to

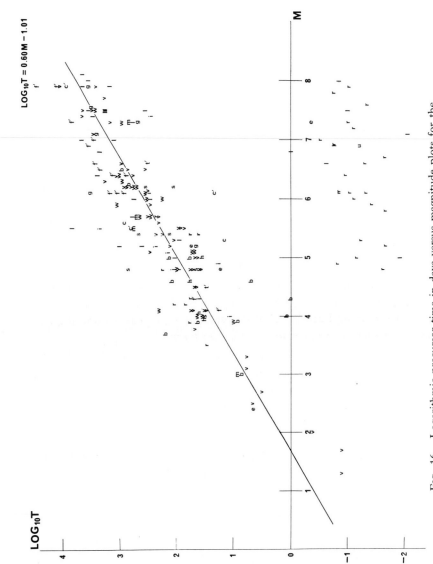

FIG. 16.   Logarithmic precursor time in days versus magnitude plots for the precursor data excluding those for disciplines t and f.   Each alphabet indicates a discipline as listed in Table II (RIKITAKE, 1978b).

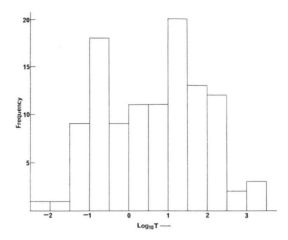

FIG. 17. Histogram of logarithmic precursor time $\log_{10} T$ (T is measured in units of days) for precursors of the third kind. No data for an earthquake of which the magnitude is smaller than 6 are included (RIKITAKE, 1978b).

or larger than 6 are used. It is clearly seen that there are two peaks in the histogram. The peak around $\log_{10} T = -1$ coincides with the precursors of the second kind. Another type having a peak around $\log_{10} T = 1$ may be called the *precursor of the third kind*.

8. 2 *Physical processes related to precursors*

The analysis of precursors in the last subsection is solely based on empirical data, so that nothing about the physical processes that give rise to them preceding an earthquake can be said from the above studies alone.

It is suggestive, however, that a relation such as Eq. (5) holds good for the precursors of the first kind. As was discussed by SCHOLZ et al. (1973) and others, for instance, it is possible to deduce a relationship such as Eq. (5) provided a diffusion process is assumed in a focal region.

In this context, a hypothesis assuming dilatancy in the focal region is of some interest. According to SCHOLZ et al. (1973) and others, many microcracks would be formed in the earth's crust when the crustal stress becomes very high. In that case, the increase in volume would cause an anomalous uplift at the earth's surface. At the same time, a decrease in V_P may well be expected in such a porous medium. If one assumes that underground water diffuses into such a dilatant region, lowering of electric resistivity and change in water level may occur.   The water passing through newly-formed cracks may contain much radon.   Although no quantitative discussion is possible at the present stage of investigation, most of the precursors of the first kind may be accounted for provided a dilatancy model is assumed.

Nothing is known about the cause of the precursors of the second kind.   But it may not be unreasonable to assume that a preliminary rupture had already started in the focal region when a precursor of this kind is observed.

The physical cause of the precursors of the third kind, the mean precursor time of which is estimated as 4.5 days, is even more obscure. We may only suppose that the crustal stress increases to such an extent that foreshocks and premonitory land deformations take place.

9. Algorithm of Earthquake Prediction

What are mentioned in the last section may lead us to a practical approach to an earthquake prediction.   It is of course important to have a long-term prediction or regionalization of earthquake occurrence by what are discussed in Sections 5 and 6.   Some other information such as seismicity gap and active fault (e.g. RIKITAKE, 1976 a) is also useful for identifying the possible site of a coming earthquake.

Summarizing all these preparatory knowledge, we first guess the most likely place where a large earthquake is expected to occur in the near future.   If we strengthen observations of various kinds over such an area, we may well expect to observe precursors of the first kind.

DAMBARA (1966) put forward an empirical relation between earthquake magnitude M and the mean radius of crustal deformation r measured in units of kilometers as

$$M = 1.96 \log_{10} r + 4.45. \tag{6}$$

The author thinks that a relation such as given by Eq. (6) may hold good even for precursory changes such as land deformation and other precursors of the first kind because anomalies mus tbe detected over the entire focal region.

Whenever one detects an anomalous symptom which may possibly be a precursor of the first kind, it is important to conduct a survey by which we can see the spatial extent of the area over which the anomaly is extended.   If the mean radius of the anomalous area can be obtained, equation (6) approximately specifies the magnitude of the coming earthquake.

The magnitude thus inferred may then be used for estimating the precursor time from Eq. (5). It should be borne in mind, however, that equations (5) and (6) are only empirical formulas of which the constants involved are subjected to errors. Strictly speaking, therefore, the determination of magnitude and precursor time is necessarily probabilistic.

All through the above procedures we see that there are instances for which the place, magnitude and occurrence time of an earthquake can roughly be inferred. As precursors of a destructive earthquake of magnitude 6–7 appear with a precursor time of several months to a few years, it is possible to strengthen observations for earthquake prediction over the targeted area before the shock actually hits there.

If a highly-dense array of observation instruments of various disciplines can be set up over the earthquake-threatened area by the supposed epoch of earthquake occurrence, the author believes that there is a high probability that precursors of the second and third kinds may be detected by the array. In case signals that are likely to be the precursors of the second kind are caught, it must be thought that an imminent prediction, that should be conveyed to authorities for disaster prevention, can be achieved.

The algorithm of earthquake prediction as stated above presents attractive features. Actually, the Haicheng earthquake seems to have been imminently predicted after long-, medium-, and short-term predictions. Under the circumstances there is good reason to suppose that what are described in this section provide a general guide-line of practicable approach to earthquake prediction.

We should pay attention, however, to the fact that the nature of earthquake precursors as indicated in Figs. 16 and 17 are obtained as averaged characteristics with respect to the whole earthquake data from various parts of the world. As the occurrence mode of earthquakes may possibly be different from region to region, it is feared that application of the general tendency to a particular earthquake may lead to a considerable deviation. It is even likely that the disciplines of precursor that are associated with earthquakes in any one area may differ from earthquake to earthquake.

One of the drawbacks of earthquake prediction observation is the presence of possible noise to precursors. Even if one observes an anomalous signal for one of the disciplines of earthquake prediction, it is not known in theory whether the signal is a precursor or not until the main shock occurs.   It is sometimes misleading, therefore, to talk about earthquake occurrence on the basis of a signal of single discipline.   It is the author's belief that an imminent prediction can only be made when anomalies are detected for a number of disciplines with the background that long- or medium-term prediction has already been made on the basis of precursors of the first kind.

10. How to Issue an Earthquake Warning?

10. 1 Prediction Council for the Tokai area

It was mentioned in the introduction that a Prediction Council for the Tokai earthquake, that is feared to occur sooner of later, has been set up since 1977.   It may probably be of interest especially for non-Japanese readers to know how the Council was formed and how it is proposed to make a prediction.

Towards the end of the year 1976, a young seismologist then at the University of Tokyo, who, relying on a newly-found historical document, claimed that the epicentral area of the 1854 Tokai earthquake must have covered the Suruga Bay.   This was contrary to the widely-accepted view that the epicenter of the quake was located off the coast in the Enshunada Sea (see Fig. 5).   As long as the present author is aware, he presented no study on the occurrence time of the quake, nevertheless, it was reported in newspapers, weekly magazines for sensationalism and the like that he claimed that the shock may occur " even tomorrow." As the fault model for the coming earthquake presented by him extends even beneath Shizuoka City, a panic-like unrest was induced among the people living in Shizuoka Prefecture.   It appears to the present author that the present-day seismology is not sufficiently developed that we can determine the fault model of a yet unoccurred earthquake.

The seismologist frequently appeared on TV and publicized his idea.   Even NHK (Japanese equivalent to BBC of London) presented TV programs apparently approving his view.   He even proposed to set up a local center for preventing possible damage due to the expected earthquake.

T. Asada, a professor of seismology of the Faculty of Science, University of Tokyo was called before the Committee of Budget, the House of Conucilors, and testified that the danger of having a great earthquake is approaching, and that there is a seismologist who has been say-ing that such a quake may occur " even tomorrow." Although he did not say that the shock would soon occur, the fear among the local inhabitants increased even more.

That the Tokai area is under the threat of a coming earthquake has recently been pointed out by many Japanese seismologists. Actually in 1973, the present author testified before a committee of the House of Representatives that the Tokai area is under threat of a large earthquake. The CCEP nominated the area to the " area of intensive observation " in the following year.

The author thinks that no new evidence of the occurrence time of the next Tokai earthquake has been put forward by the above-mentioned seismologist, who, as long as the present author knows, has as yet published no paper in international journals with a reputable refereeing system.

The rumor that such a large earthquake is soon forthcoming excited much debate on earthquake prediction at the congress and cabinet level. As a result, intensification of an observation system has become to be urgently planned even though CCEP expressed its view that there is no indication that the earthquake would occur in the Tokai area in the immediate future. Ironically enough, the bold utterance of a young seismologist, no matter whether it be true or not, led to the long-hoped development of a prediction-oriented observation system in the Tokai area.

All the members of CCEP are working on part-time basis. GSI, the host institution of CCEP, is closed on Sundays. No formal legislation has been made about CCEP. Under the circumstances, it is clear that CCEP cannot be responsible for imminent prediction although what has been done by CCEP for long-time prediction should be appreciated.

In reply to the social demand arising in association with the debate on the threatening Tokai earthquake, the Prediction Council was formed in 1977 as already mentioned in the Introduction. Although the council is a subsystem of CCEP, it is hosted by JMA to which signals of volume strainmeters buried in the Tokai and South Kanto areas are telemetered on an on-line real-time basis. Many data of other disciplines observed by various organizations are also telemetered to JMA. Someone at JMA's Seismology Division is always watching the recorders even during the night. In case the record should indicate an anomalous change exceeding the prescribed limit, the chairman of the council will be called.   He is always carrying a wireless alarm instrument, the so-called " pocket-bell."

Whenever the chairman thinks that the anomaly may have something to do with earthquake occurrence, the council will be summoned immediately, and it will make a decision whether or not an earthquake is impending.   The council members being university professors, however, they cannot take the full responsibility of predicting the earthquake because of their routine duties.   In spite of such structural weakness, it should be appreciated that a 24-hour alert system for a large earthquake has been launched.

Dramatic intensification of a watching system in the Tokai area is now under planning.   In addition to the existing volume strainmeter system, output of seismographs including the sea-battom ones extending off the Pacific coast by 115 km or so, tide gauges, tiltmeters, underground water monitoring instruments and the like will be telemetered to the University of Nagoya, Earthquake Research Institute, GSI, National Research Center for Disaster Prevention and Geological Survey. Branched signals will also be transferred to JMA from the respective institutions providing imminent data for the Prediction Council.   Facsimile instruments installed at these institutions will also be used for data exchange.   The brief outline of such a watching system to be completed by 1979 is shown in Fig. 18.

The author gathered from his American colleagues that a prediction council has been set up in USGS.   But no detail of its activity has as yet reached him.   It is also said that the Governor of California State has his own evaluating committee for earthquake information.   It is the author's understanding that, whenever someone puts forward any earthquake prediction information, the committee will meet and judge whether the information is credible or not.

10. 2   From prediction to warning

When the Japan's Prediction Council was formed, there was serious discussions about how the earthquake prediction information should be made public provided a fairly reliable prediction of the suspected Tokai earthquake is achieved.   It was the final consensus at that time that prediction should be made purely on scientific basis not taking its social consequences into account, and that the information may be made public, if necessary, after administrative consideration by government authorities.   In a country such as Japan where many people are living

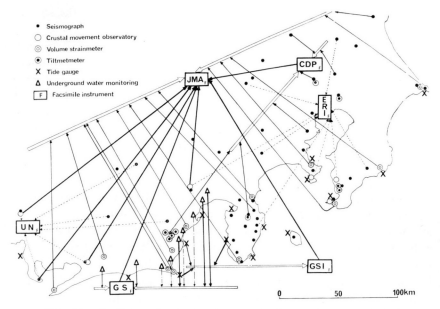

- ● Seismograph
- ○ Crustal movement observatory
- ◎ Volume strainmeter
- ◉ Tiltmetmeter
- X Tide gauge
- △ Underground water monitoring
- [F] Facsimile instrument

FIG. 18.   A diagram indicating how geophysical and geochemical elements are sent to JMA for the porpose of 24 hours' watch for possible occurrence of the suspected Tokai earthquake.   The abbreviations are as follows: CDP, National Research Center for Disaster Prevention; ERI, Earthquake Research Institute, University of Tokyo; GS, Geological Survey, Japan; GSI, Geographical Survey Institute; JMA, Japan Meteorological Agency; UN, Geophysical Institute, University of Nagoya.

on heavy industries and tourism, closure of their businesses due to earthquake prediction information, even if it is only temporary, is liable to cause much economical loss.   If the scientists, who are responsible for prediction information, are afraid of such an influence of prediction information, no unbiased judgement can be made, so that the first point of the above consensus should be emphasized in actual prediction work.

The Prediction Council was asked by the government officials, who are responsible for prevention of earthquake hazards, to specify earthquake prediction information in a proper manner which can be applied to actual issuance of earthquake warning.   Government officials do not know how to act if the prediction is purely academic as usually mentioned in terms of confidence limit or something of that kind.   It is for example required to classify the information into 3 classes according to time windows covering the dangerous period during which earthquake occurrence is expected with a high probability; i.e. A (an earthquake would occur within several hours' time), B (within a few days' time) and C (within a period that is not exactly known but not very long).

In addition to the above classification, the government officials are anxious to know the credibility of the information hopefully classified into the following 3 ratings; X (highly reliable), Y (slightly reliable) and Z (possible). It is, then, their proposal to take porper action according to the 9 segments given by the combination of (A, B, C) for time window and (X, Y, Z) for credibility of information.

Although the author does not know much about the likely action to be taken by the Japanese government in case of the warning concerning the expected Tokai earthquake, the following may be the possible measures to be taken in such an emergency. If the information is set forth with the combination of A and X, for example, an action to stop the world-famous Shinkansen (bullet train) running through the Tokai area will be taken. Freeways running through the area will be closed, too. Firebrigades, polices, self-defense forces, hospitals, and other governmental agencies will immediately stand by. If the rating of credibility is slightly lower or the time window is longer, the preparations to be made will be somewhat limited. For instance, the Shinkansen will be operated at a maximum speed of 70 km/hour instead of the routine maximum speed of 200 km/hour.

The above strategy responding to an earthquake warning may sound sufficient. However, it is no easy matter to classify earthquake information as stated above in the framework of the time window (A, B, C) and the credibility rating (X, Y, Z). Only a tentative approach to such classification and rating has been proposed by the present author (RIKITAKE, 1978 c). The time windows are specified on the basis of the period during which the cumulative probability of short-range precursors such as precursors of the second and third kinds and animal behavior increases from 30 to 80%. Such probabilities can be estimated from Weibull distribution analyses of each precursor. It has provisionally been shown that the windows corresponding to A, B, and C are specified as 1–4 hours, 3 hours–4 days, and 1.5–28 days respectively for the second kind, for animal behavior and for the third kind precursors, respectively.

To cancel earthquake prediction information is as difficult as to issue one.   Since no clear-cut criterion for cancelling has as yet been put forward, it will be assumed that the signal was false if an earthquake did not occur within a certain time even if the probability exceeded a certain limit, 80% say.

No difinite standard has been provided for rating the reliability of earthquake prediction information.   It is the present author's idea that such ratings may be achieved by a combination of numbers of disciplines for which an anomaly is detected and the number of observation stations where we observe anomalies although nothing in detail is presented here (see RIKITAKE, 1978 c).

10. 3   Special legislation

As earthquake prediction is becoming a matter of actuality, it turns out that many problems have to be solved in relation to actual issuance of earthquake warning on the basis of earthquake prediction information.   With regard to the Tokai area, the Shinkansen would first of all have to be stopped as already mentioned in the last section if a strong earthquake is shortly impending.   The freeway connecting Nagoya to Tokyo should also be closed.   The nuclear power station situated right in the center of the threatened area would be requested to take measures for safety.   Along the Suruga Bay coast, we have many factories including oil refinery complexes and other chemical industry works which would not only be easily set on fire but would also cause extensive environmental contamination if damaged.   Some measures must be taken for preventing disaster from these heavy industries including a complete shutdown of operation.

Daily life of local citizens would also be very much affected.   People would tend to run away with their motorcars causing much traffic tumult.   They would also rush to banks for cashing and to supermarkets for buying foods and daily necessaries.   It may even be possible that they would become looters.   Police and self-defense forces would spend much energy in dealing with them.   It would probably be advised by local governments to close all shops and stores.

It is tedious to talk about such anticipated confusion in case of an earthquake warning.   What are written in the preceding paragraph in this section are based on the author's simple guess which is necessarily fragmentary because he is a non-specialist in the field of sociology and socio-psychology.   But it is obvious that two important points arise even from such a crude thinking.   First of all, it is not entirely clear who

has the right to order necessary arrangements such as stopping the Shinkansen, shutting down heavy industries and so on.   In the next place,
nothing has been decided about compensation of economical losses associated with an earthquake warning.

The above two points cannot be handled in the framework of the
existing law.   In the course of the debate on the danger of the Tokai
earthquake, it has become well recognized that some kind of new legislation must be set forth in order to deal with the forthcoming disaster.
A law which can be applied to an area threatened by an extremely large
earthquake had been on the carpet at the 1978 ordinary session of the
Japanese Diet.   After much debate, the " Large-scale Earthquake Countermeasures Act " has passed the Diet, and was enacted in December,
1978.   Although the author is not in a position to mention the details of
the legislation, it is his understanding that it is the prime minister who
declares the possibility of having a great earthquake on the basis of
earthquake prediction information which reaches him through the director of JMA.   It will be made possible to use the self-defense force
at this stage.   The prime minister will be temporarily enpowered to
restrict certain personal rights in the targeted area.   But the Japanese
government will not be liable for compensating economical losses due
to the declaration of earthquake danger.   As such a law must not become a kind of martial law and should not empower the prime minister
too much in a democratic country, many problems will be necessarily
left unsolved in the actual application of the law.   It should be appreciated, however, that one step has been taken on a nation-wide scale
toward preventing earthquake hazards on the basis of possible prediction.

Nothing concrete is likely to be included in the law about strengthening the system of earthquake prediction.   But such strengthening is
absolutely necessary for achieving a timely prediction not only for the
Tokai area but also for entire Japan.

The author believes that the time for establishing an organization
tentatively called the " Earthquake Prediction Center," which is responsible for earthquake prediction and warning, is now approaching.   The
center, where well-trained specialists are watching observed data day
and night, will take over the job of CCEP and Prediction Council.   All
the data taken by JMA, GSI and so on should be telemetered to the
center mostly on real-time basis.   The center must have its own mobile
parties to be sent to earthquake-threatened area immediately.   The

director of the center must have the right to ask various organizations to make necessary observations aiming at earthquake prediction, wherever necessary. Such extraordinary work should be funded by the center.

There are varieties of opinion concerning such a central organization for earthquake prediction and warning. Some may say that it is premature to set up such a center because earthquake prediction is still in a developing stage. The author feels, however, that some central organization which works on a full-time basis should nevertheless be set up in order to meet the social demand for earthquake prediction that covers all parts of Japan.

10.4 Reaction of the public against an earthquake warning

The fact that tremendous social unrest arose in the Tokai area when it was rumored that a great earthquake would impend suggests that the public would be terribly panicked if an earthquake warning is suddenly issued by a credible organization. It is therefore important that earthquake prediction information should be released with much care and at the right time. Such a timing may possibly be inferred provided the public reaction against a credible earthquake warning is known.

Socioeconomic study of possible consequences of earthquake prediction is advanced particularly in the U.S.A. As the development of American investigations has been reviewed elsewhere (e.g. RIKITAKE, 1978 a), only essentials of the results will be quoted here. The readers who are particularly concerned with this aspect of earthquake prediction are referred to publications by NATIONAL ACADEMY OF SCIENCES (1975), HAAS et al. (1975), HAAS and MILETI (1976, 1977), PANEL ON EARTHQUAKE PREDICTION OF THE COMMITTEE ON SEISMOLOGY (1976), and WEISBECKER et al. (1977).

On the assumption that a highly credible prediction of a magnitude 7.3 earthquake was made by USGS a few years prior to the actual earthquake occurrence, reaction of the public against such a release of earthquake prediction information was studied through interviewing and questionnaries with important people in two Californian cities where earthquakes are a standing menace. Responsible people at city councils, police, firebrigade, hospital, waterworks, sewage, road, traffic, gas, electricity, telephone, school, and other public services cooperated in the work. Even retail dealers, realtors, developpers, bankers, insurers, and the like were involved.

All through such studies, the following points have been clarified by HAAS *et al.* (1975). Local people would make a rush for buying earthquake insurance. As the needs would exceed the capacity of insurance companies, an insurance system subsidized by the government would be requested to start. Property values would drop considerably. Speculative realtors would try to buy land at a bargain hoping to sell it at higher price after the quake. Mortgage would become unavailable. Businesses would move out permanently or temporarily. Construction would halt. Overall economic activity would very much decline. As a result, tax income of local government would considerably decrease in spite of the fact that much money is needed for increasing police force, firebrigade capability and so forth. A considerable decrease in population would also take place as the predicted time approaches. It is concluded that the overall economic damage to local community would in some cases be comparable to actual damage caused by the earthquake.

Public reaction toward an earthquake warning has also been studied by taking into account the fact that the prediction information would become more accurate as the earthquake approaches (HAAS and MILETI, 1976, 1977). It is instructive that the timing for evacuation of inhabitants living downstream a dam of water reservoir, of vulnerable things such as computers, of hospitals and prisons, and the like has been suggested.

It is clear that public reaction of this kind differs from country to country. In Japan, for instance, very few people think of evacuation even if an earthquake warning is issued because they are living on businesses closely connected to local community. The reaction also depends on national traits and economic potential, so that specific studies must be made for each country.

It is in the next place important to guess what would happen if the earthquake alarm were false; and a false warning is easily expected at the present stage of earthquake prediction. The author gathered from Chinese colleagues that an alarm for evacuation was issued about 40 days before the Haicheng earthquake, that local inhabitants spent two nights in a field camp in freezing weather, and that people complained of the false warning. If it happens that an earthquake alarm is false in a country like the U.S.A. and Japan, the federal and local governments would fall in a difficult situation because of requests of local people for compensating economic and spiritual damage arising from the false alarm.

It is instructive to read the reports by WEISBECKER *et al.* (1977) in which possible results of successful and unsuccessful predictions are described. Even if an earthquake prediction is successful in a scientific sense, it is possible that the prediction would give rise to much social confusion and economic loss.

The author nevertheless feels that an earthquake prediction information should be made public in the name of humanity, whenever possible. Mass deaths and injuries can only be avoided by such a warning.

REFERENCES

ALLEN, C. R., M. G. BONILLA, W. F. BRACE, M. BULLOCK, R. W. CLOUGH, R. M. HAMILTON, R. HOFHEINZ, Jr., C. KISSLINGER, L. KNOPOFF, M. PARK, F. PRESS, C. B. RALEIGH, and L. R. SYKES, Earthquake Research in China, *EOS (Trans. Am. Geophys. Union)*, **56**, 838–882, 1975.

ARIEH, E. and A. M. MERZER, Fluctuations in oil flow before and after earthquakes, *Nature*, **247**, 534–535, 1974.

BARSUKOV, O. M., Variation of electric resistivity of mountain rocks connected with tectonic causes, *Tectonophysics*, **14**, 273–277, 1972.

BUFE, C. G., J. H. PFLUKE, and R. L. WESSON, Premonitory vertical migration of microearthquakes in Central California—Evidence of dilatancy biasing?, *Geophys. Res. Lett.*, **1**, 221–224, 1974.

CASTLE, R. O., J. P. CHURCH, and M. R. ELLIOT, Aseismic uplift in southern California, *Science*, **192**, 251–253, 1976.

CHOU, C. W. and R. S. CROSSON, Search for time-dependent seismic travel times from mining explosions near Centralia, Washington, *Geophys. Res. Lett.*, **5**, 97–100, 1978.

DAMBARA, T., Vertical movements of the earth's crust in relation to the Matsushiro earthquake, *J. Geod. Soc. Japan*, **12**, 18–45, 1966 (in Japanese).

DAMBARA, T., Crustal movements before, at and after the Niigata earthquake, *Rep. Coord. Comm. Earthq. Predict.*, **9**, 93–96, 1973 (in Japanese).

ENGDAHL, E. R. and C. KISSLINGER, Seismological precursors to a magnitude 5 earthquake in the central Aleutian Islands, *J. Phys. Earth*, **25**, Suppl., S243–S250, 1977.

EVISON, F. F., Fluctuations of seismicity before major earthquakes, *Nature*, **266**, 710–712, 1977a.

EVISON, F. F., The precursory earthquake swarm, *Phys. Earth Planet. Inter.*, **15**, 19–23, 1977b.

GEOGRAPHICAL SURVEY INSTITUTE, Recent crustal movement in South Kanto district, 4, *Rep. Coord. Comm. Earthq. Predict.*, **8**, 32–34, 1972 (in Japanese).

GEOGRAPHICAL SURVEY INSTITUTE, Precise strain measurements in Yamakita and Tanna regions, *Rep. Coord. Comm. Earthq. Predict.*, **11**, 94–95, 1974 (in Japanese).

GEOGRAPHICAL SURVEY INSTITUTE, Crustal deformation in the central part of Izu Peninsula, *Rep. Coord. Comm. Earthq. Predict.*, **16**, 82–87, 1976 (in Japanese).

GEOGRAPHICAL SURVEY INSTITUTE, Crustal deformation in the central part of Izu Peninsula, *Rep. Coord. Comm. Earthq. Predict.*, **17**, 59–64, 1977a (in Japanese).

GEOGRAPHICAL SURVEY INSTITUTE, Crustal deformation in the central part of Izu Peninsula, *Rep. Coord. Comm. Earthq. Predict.*, **18**, 56–60, 1977b (in Japanese).

GROUP OF HYDRO-CHEMISTRY, the Seismological Brigade of Hobei Province, Studies on forecasting earthquakes in the light of the abnormal variations of Rn concentration in ground water, *Acta Geophys. Sinica*, **18**, 279–283, 1975 (in Chinese).

GUPTA, I. N., Premonitory seismic-wave phenomena before earthquakes near Fairview Peak, Nevada, *Bull. Seismol. Soc. Am.*, **65**, 425–437, 1975.

HAAS, J. E. and D. S. MILETI, Consequences of earthquake prediction on other adjustments to earthquakes, Paper presented at Australian Academy of Science Symposium on National Hazards in Australia, Section 3: Community and Individual Responses to National Hazards, Canberra, Australia, 26–29 May, 1976.

HAAS, J. E. and D. S. MILETI, *Socioeconomic Impact of Earthquake Prediction on Government, Business and Community*, Institute of Behavioral Science, University of Colorado, 1977.

HAAS, J. E., D. S. MILETI, and J. R. HUTTON, Use of earthquake prediction information, Paper presented at XVI General Assembly of the International Union of Geodesy and Geophysics, International Association of Seismology and Physics of the Earth's Interior, Symposium No. 4; Geophysical Phenomena Preceding, Accompanying and Following Earthquakes, Grenoble, France, 2 September, 1975.

HASEGAWA, A., T. HASEGAWA, and S. HORI, Premonitory variation in seismic velocity related to the Southeastern Akita earthquake of 1970, *J. Phys. Earth*, **23**, 189–203, 1975.

JAPAN METEOROLOGICAL AGENCY, Reported to the February, 1978 meeting of the Coord. Comm. Earthq. Predict., 1978.

KAKIMI, T., H. SATO, K. TSUMURA, and M. ISHIDA, Seismicity, crustal movement and neotectonics in Kanto district, in *Symposium on Earthquake Prediction Research*, edited by Z. Suzuki, pp. 21–45, Sub-committee on Earthquake Prediction, National Committee for Geodesy and Geophysics,

Science Council of Japan and Seismological Society, Japan, 1977 (in Japanese).

LATYNINA, L. A. and S. D. RIZAEVA, On tidal-strain variations before earthquakes, *Tectonophysics*, **31**, 121–127, 1976.

MIKUMO, T., M. KATO, H. DOI, Y. WADA, and T. TANAKA, Possibility of temporal variations in earth tidal strain amplitudes associated with major earthquakes, *J. Phys. Earth*, **25**, Suppl., S123–S136, 1977.

MIYAKOSHI, J., Secular variation of Parkinson vectors in a seismically active region of Middle Asia, *J. Fac. Gen. Educ., Tottori Univ.*, **8**, 209–218, 1975.

NATIONAL ACADEMY OF SCIENCES, *Earthquake Prediction and Public Policy*, National Academy of Sciences, Washington, D.C., 1975.

PANEL ON EARTHQUAKE PREDICTION OF THE COMMITTEE ON SEISMOLOGY, *Predicting Earthquakes—A Scientific and Technical Evaluation with Implications for Society*, National Academy of Sciences, Washington, D.C., 1976.

PRESS, F., H. BENIOFF, R. A. FROSCH, D. T. GRIGGS, J. HANDIN, R. E. HANSON, H. H. HESS, G. W. HOUSNER, W. H. MUNK, E. OROWAN, L. C. PAKISER, Jr., G. SUTTON, and D. TOCHER, *Earthquake Prediction: a Proposal for a Ten Year Program of Research*, Office Sci. Technol., Washington, D.C., 1965.

RALEIGH, B., G. BENNET, H. CRAIG, T. HANKS, P. MOLNAR, A. NUR, J. SAVAGE, C. SCHOLZ, R. TURNER, and F. WU, Prediction of the Haicheng earthquake, *EOS (Trans. Am. Geophys. Union)*, **58**, 236–272, 1977.

RIKITAKE, T., Statistics of ultimate strain of the earth's crust and probability of earthquake occurrence, *Tectonophysics*, **26**, 1–21, 1975a.

RIKITAKE, T., Earthquake precursors, *Bull. Seismol. Soc. Am.*, **65**, 1133–1162, 1975b.

RIKITAKE, T., *Earthquake Prediction*, Elsevier, Amsterdam, 1976a.

RIKITAKE, T., Recurrence of great earthquakes at subduction zones, *Tectonophysics*, **35**, 335–362, 1976b.

RIKITAKE, T., Probability of a great earthquake to recur off the Pacific coast of Central Japan, *Tectonophysics*, **42**, T43–T51, 1977.

RIKITAKE, T., Earthquake prediction and warning, *Interdisciplinary Sci. Rev.*, **3**, 58–70, 1978a.

RIKITAKE, T., Classification of earthquake precursors, *Tectonophysics*, **54**, 293–309, 1978b.

RIKITAKE, T., Classification of earthquake prediction information for practical use, *Tectonophysics*, **46**, 175–185, 1978c.

RIKITAKE, T., Biosystem behaviour as an earthquake precursor, *Tectonophysics*, **51**, 1–20, 1978d.

RIKITAKE, T., Changes in the direction of magnetic vector of short-period geomagnetic variations before the 1972 Sitka, Alaska, earthquake, *J. Geomag. Geoelectr.*, **31**, 441–448, 1979.

RIKITAKE, T., Anomalous animal behavior preceding the 1978 earthquake of magnitude 7.0 that occurred near Izu-Oshima island, Japan, this volume, 1981a.

RIKITAKE, T., Precursors to the 1978 near Izu-Oshima island, Japan, earthquake of magnitude 7.0, this volume, 1981b.

SACKS, I. S., S. SUYEHIRO, D. W. EVERTSON, and Y. YAMAGISHI, Sacks-Evertson strainmeter, its installation in Japan and some preliminary results concerning strain steps, Pap. Meteorol. Geophys., 22, 195–208, 1971.

SATO, H., Crustal movements accompanying the Tonankai earthquake in 1944, J. Geod. Soc. Japan, 15, 177–180, 1970.

SCHOLZ, C. H., L. R. SYKES, and Y. P. AGGARWAL, Earthquake prediction: A physical basis, Science, 181, 803–810, 1973.

SEKIYA, H., The seismicity preceding earthquakes and its significance to earthquake prediction, Zisin (J. Seismol. Soc. Japan), Ser. 2, 29, 299–311, 1976 (in Japanese).

SMITH, B. E. and M. J. S. JOHNSTON, A tectonomagnetic effect observed before a magnitude 5.2 earthquake near Hollister, California, J. Geophys. Res., 81, 3556–3560, 1976.

SUYEHIRO, S., Difference between aftershocks and foreshocks in the relationship of magnitude to frequency of occurrence for the great Chilean earthquake of 1960, Bull. Seismol. Soc. Am., 56, 185–200, 1966.

SUYEHIRO, S., Difference in the relationship of magnitude to frequency of occurrence between aftershocks and foreshocks of an earthquake of magnitude 5.1 in Central Japan, Pap. Meteorol. Geophys., 20, 175–187, 1969.

SUYEHIRO, S., T. ASADA, and M. OHTAKE, Foreshocks and aftershocks accompanying a perceptible earthquake in Central Japan, Pap. Meteorol. Geophys., 15, 71–88, 1964.

TAKAGI, A. and A. HASEGAWA, Seismic activity in the northeastern Japan island arc system and study of earthquake prediction, in Symposium on Earthquake Prediction Research, edited by Z. Suzuki, pp. 15–20, Subcommittee on Earthquake Prediction, National Committee for Geodesy and Geophysics, Science Council of Japan and Seismological Society, Japan, 1977 (in Japanese).

TSUBOI, C., Investigation on the deformation of the earth's crust found by precise geodetic means, Jap. J. Astron. Geophys., 10, 93–248, 1933.

TSUBOI, C., On seismic activities in and near Japan, in Contribution in Geophysics, in Honor of Beno Gutenberg, edited by H. Benioff, M. Ewing, B. F. Howell, and F. Press, pp. 87–112, Pergamon Press, London, 1958.

TSUBOI, C., K. WADATI, and T. HAGIWARA, Prediction of Earthquakes— Progress to Date and Plans for further Development, Rep. Earthquake Prediction Res. Group, Japan. Earthquake Res. Inst., Univ. Tokyo, Tokyo, 1962.

UMINO, N. and HASEGAWA, On the two-layered structure of deep seismic plane in northeastern Japan arc, Zisin (J. Seismol. Soc. Japan), Ser. 2, 28,

125–139, 1975 (in Japanese).

UTSU, T., Space-time pattern of large earthquakes occurring off the Pacific coast of the Japanese Islands, *J. Phys. Earth*, **22**, 325–342, 1974.

WAKITA, H. and T. ASADA, Reported to the February, 1978 meeting of the Coord. Comm. Earthquake Prediction, 1978.

WEISBECKER, L. R., W. C. STONEMAN, S. E. ACKERMAN, R. K. ARNOLD, P. M. HALTON, S. C. IVY, W. H. KAUTZ, C. A. KROLL, S. LEVY, R. B. MICKLEY, P. D. MILLER, C. T. RAINEY, and J. E. VAN ZANDT, *Earthquake Prediction, Uncertainty, and Policies for the Future*, Stanford Research Institute, Menlo Park, California, 1977.

WILLIAMS, F., M. J. S. JOHNSTON, R. O. CASTLE, and S. H. WOOD, Local magnetic field measurements in southern California—Comparison with leveling data and moderate magnitude earthquakes, *EOS (Trans. Am. Geophys. Union)*, **58**, 1122, 1977.

WYSS, M., A search for precursors to the Sitka, 1972, earthquake: sea level, magnetic field, and P-residuals, *Pure Appl. Geophys.*, **113**, 297–309, 1975.

WYSS, M., Local changes of sea level before large earthquakes in South America, *Bull. Seismol. Soc. Am.*, **66**, 903–914, 1976a.

WYSS, M., Local sea level changes before and after the Hyuganada, Japan, earthquakes of 1961 and 1968, *J. Geophys. Res.*, **81**, 5315–5321, 1976b.

YAMAZAKI, Y., Precursory and coseismic resistivity changes, *Pure Appl. Geophys.*, **113**, 219–227, 1975.

YAMAZAKI, Y., Tectonoelectricity, *Geophys. Surv.*, **3**, 123–142, 1977.

YANAGIHARA, K., Secular variation of the electrical conductivity anomaly in the central part of Japan, *Mem. Kakioka Mag. Obs.*, **15**, 1–12, 1972.

YANAGIHARA, K. and T. NAGANO, The change of transfer function in the Central Japan Anomaly with special reference to earthquake occurrences, *J. Geomag. Geoelectr.*, **28**, 157–163, 1976.

ZHU, F. M., Prediction, warning and disaster prevention related to the Haicheng earthquake of magnitude 7.3, in *Proceedings of the Lectures by the Seismological Delegation of the People's Republic of China*, pp. 15–26, Seismol. Soc. Japan, 1976 (in Japanese).

Note added in proof

In conformity with the "Large-scale Earthquake Countermeasures Act", 170 cities, towns and villages in the Shizuoka and neighboring prefectures were designated by the Prime Minister as the "Area under Intensified Measures against Earthquake Disaster" on Aug. 7, 1979. Accordingly, the Prediction Council has become officially named the "Prediction Council for the Area under Intensified Measures against Earthquake Disaster". The Council is now an advisory organ to the Director of JMA.

Should the Prime Minister hear an imminent earthquake prediction information from the Council via the Director of JMA, he would issue an "Earthquake Warnings Statement" upon consultation with the Cabinet, and then various measures described in the text would be put into practice.

PRECURSORS TO THE 1978 NEAR IZU-OSHIMA ISLAND, JAPAN, EARTHQUAKE OF MAGNITUDE 7.0

Tsuneji RIKITAKE

Department of Applied Physics, Faculty of Science, Tokyo Institute of Technology, Ookayama, Meguro-ku, Tokyo, Japan

Abstract Geophysical precursors such as foreshocks, volume strain, geomagnetic field intensity, amplitude of short-period geomagnetic variation, earth potential, radon content and level of underground water are analysed in association with the 1978 earthquake of magnitude 7.0 that occurred near the Izu-Oshima Island about 100 km south of Tokyo. The mean precursor time for the present earthquake occurs at a time range of scores of days in contrast to several days for the mean value of the existing data.   It may be that the mean precursor time for a particular earthquake differs considerably from that for the data including many earthquakes of various kinds.   Anomalous animal behaviour, should it be a precursor at all, indicates a much shorter precursor time amounting to 0.5 day or so.

1. Introduction

An earthquake of magnitude (M) 7.0 occurred undernearth the sea between Izu-Oshima Island and Izu Peninsula, Japan at 12.24 local time on January 14, 1978. The epicenter is located at 34°46′N and 139°15′E as can be seen in the insert of Fig. 2. The Izu area has been seismically active since the 1974 Izu-Hanto Oki earthquake $(M=6.9)$ that occurred near the southernmost extremity of the peninsula. A crustal uplift has been noticed to have taken place in the central and northern part of the peninsula (see Fig. 2), while swarm-like seismic activities have been continually occurring notably on the southern flank of the uplift.

In the Tokai area to the west of the Izu area, there is evidence of much accumulation of crustal strain, and so it is feared that the probability of a great earthquake such as the 1954 one of magnitude 8.4 to recur there takes on a high value (RIKITAKE, 1977).

Under the circumstances, the Izu area and its surroundings have become one of the most important target areas in Japan over which various geophysical instruments for monitoring possible signals useful for earthquake prediction are installed. Although no warning of the Near Izu-Oshima Island earthquake was actually issued except the so-called " earthquake information ", which suggested possible occurrence of a slightly damaging earthquake, as issued by the Japan Meteorological Agency (JMA) about 1.5 hours preceding the main shock, it appears that precursory effects have been detected by observations of a number of different disciplines of geophysical elements although most of them had to be recognized after the earthquake because very few instruments are operated on the on-line real-time basis.

It may be said, however, that the precursor-like phenomena experienced in relation to the present earthquake encourage Japanese seismologists. Sould we have a well-distributed and well-organized observation network that works essentially on the on-line real-time basis, it would not be unreasonable to expect that some symptoms of occurrence of an earthquake of magnitude 7 or over can be observed preceding the shock.

2. Seismic Activity

According to JMA, it is observed that small earthquakes began to occur somewhere between Izu-Oshima Island and Izu Peninsula since January 12. This area is famous for swarm earthquakes; only the magnitude of the present swarm earthquakes seemed to be a little larger than that for usual activity.

The number of small earthquakes tended to increase at about 8h local time in the morning of January 14. The activity becomes so high at around 09.30 that the exact numbers of shocks are hardly counted. The number of shocks seemed to decrease suddenly around 10.30 as can be seen in Fig. 1. It has been known that such a sudden decrease in microseismicity sometimes provides an earthquake precursor, one of

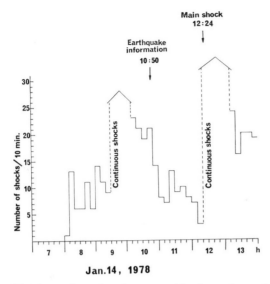

FIG. 1. Number of earthquakes per 10 min. observed at Izu-Oshima Weather Station on Jan. 14, 1978, the day of the Near Izu-Oshima Island earthquake of magnitude 7.0 (After Japan Meteorological Agency).

the best examples being the case for the 1975 Haicheng earthquake of magnitude 7.3 (RALEIGH *et al.*, 1977).

An "earthquake information" is then issued from JMA. Such information is essentially different from earthquake prediction. It is mostly concerned with the seismic activity now prevailing with some guess for future activity. It should be appreciated, however, that the possibility of having a slightly damaging earthquake is suggested by JMA this time. As can also be seen in Fig. 1, the number of shocks continued to decrease until the main shock took place.

Frankly speaking, no seismologist, including the present writer, thought that an earthquake having a magnitude as large as 7 would occur in the epicentral area of the present earthquake. Only earthquakes having a magnitude 6 or slightly over at maximum have been recorded in the area in history.

3. Crustal Movement

An outstanding uplift has been found in the central and northern part of Izu Peninsula in recent years. The center of the uplift is located

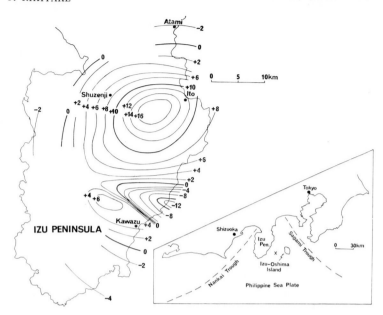

FIG. 2. The vertical movement in cm over Izu Peninsula during a period from 1967–1969 to 1978 (after the earthquake) (After Geographical Survey Institute).

at some 10 km to the west of Ito City as shown in Fig. 2. The maximum uplift, that occurred during a period of 10 years, is estimated as 15 cm or thereabout. Figure 2 shows the changes in height as obtained by comparing the 1978 post-earthquake survey to that for 1967–1969. A subsidence probably associated with the fault appeared at the time of the present earthquake is found to the south of the crustal uplift. However, there is no evidence that the uplift tends ot vanish in association with the present earthquake. It is rather suspected that the uplift seems likely to expand westwards.

A swarm of earthquakes occurred around the uplifted area, especially to the south of it, in 1975–1976 including a magnitude 5.4 earthquake. A fault, that is called the Tanna fault, appeared in the northern part of Izu Peninsula running in an almost north-south direction at the time of the North Izu earthquake ($M=7.0$, 1930). It has been pointed out that crustal strain is accumulating so rapidly around the fault that the probability of an earthquake to recur there would become fairly high by the end of this century (RIKITAKE, 1975). Although nothing is known about the relationship between the uplift, that takes place to-

ward the southernmost portion of the fault, and the future seismic activity related to the fault, an eye should be cautiously kept on the crustal
activity around the uplift and fault.

JMA is operating an array of bore-hole volume strainmeters (SACKS
et al., 1971) over the Tokai, Izu, and other areas. The signals from
these instruments are telemetered to the central office of JMA on the
on-line real-time basis in such a way that 24-hour watching of anomalous
crustal movement can be made.

One of the strainmeters set up at Irozaki at the southernmost point
of Izu Peninsula, which had been indicating no change over a long
period of time, tended to record contraction since December 3, 1977.
The strain change switched over to expansion about 4 days prior to
the earthquake as shown in Fig. 3. These turning points may be regarded as the commencement of a precursory signal, and the precursor
times are listed in Table 1.

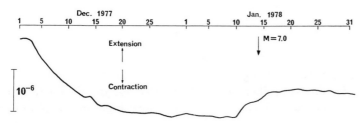

FIG. 3. Changes in volume strain at Irozaki (After Japan Meteorological
Agency).

Another strainmeter at Ajiro, about 55 km north of Irozaki, tended
to record marked contraction since December 20, 1977 with a rate as
high as 2×10^{-7}/day. Such a contraction tendency continues even
after the earthquake. As it appears that the strainmeter is affected to
some extent by the pumping up of hot-spring water at a nearby well,
evaluation of such an enormous strain rate is difficult to make.

No outstanding precursory changes were observed by the tiltmeters
and strainmeters at Aburatsubo Crustal Movement Observatory, only
60 km distant from the epicenter, in contrast to the volume strainmeters
as mentioned above.

4. Gravity, Geomagnetism, Telluric Currents, and Resistivity

A well-organized gravity surveys have frequently been conducted
by the Earthquake Research Institute (ERI) over the uplift in Izu Pe-

ninsula (HAGIWARA *et al.*, 1976).   As a result, a marked gravity change amounting to a few tens of microgal has been found.   The change seems to be accounted for by the so-called free-air one associated with the ground uplift, so that no evidence for an intrusion of high-density magma is supported.

A series of magnetic surveys conducted by the Tokyo Institute of Technology over the western peninsula brings out that 5–10 gamma changes in the geomagnetic total intensity took place during a period including the earthquake.   The changes seem to recover during a few months' period after the earthquake.

The Kakioka Magnetic Observatory group made a continuous observation of the geomagnetic total intensity at Matsuzaki on the southwest coast of the peninsula.   Comparing the results to those at a station in the eastern part of the peninsula, they reported on an anomalous change amounting 5 gammas or thereabout that occurred during a period of several days around November 10, 1977.

A three-component geomagnetic variometer has been operated by ERI at a station near the center of the anomalous uplift.   The ratio of the amplitude of short-period geomagnetic variations to that at a standard observatory about 130 km distant increased for the horizontal intensity by 5% in November, 1977, while that for the declination indicates a slight decrease.   According to an observation of earth potential at the same spot, a gradual decrease amounting to 30 mV/km has been observed since November 10.   Such a trend reversed its direction of change some time in December.

The resistivity variometer installed at Aburatsubo recorded a precursory change 4 hours forerunning the 1974 Izu-Hanto Oki earthquake.   The variometer does not record such an outstanding precursor in spite of the fact that the epicentral distance is much smaller this time. A large coseismic step of resistivity change is observed.

5.   Radon Content and Level of Underground Water

Figure 4 shows the changes in radon content at a well near the uplift center as reported by the Geophysical Institute, University of Tokyo.   The observers pointed out that the decrease started in the beginning of November and also that the increase since January 10 may have something to do with a premonitory effect.

Fairly many stations for obervation of radon content are provided

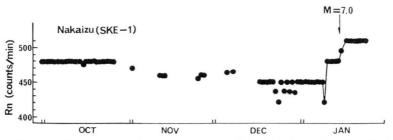

FIG. 4.   Changes in radon content at a well in Nakaizu (After Geophysical Institute, University of Tokyo).

by the Geological Survey of Japan over the Tokai-Izu area.   As most of them are monitored once a week, no detailed information about precursors is available even though there are a number of instances for which a decrease in content is detected before the earthquake.

That the water level of a well, which is located in the middle of the peninsula at an epicentral distance of 35 km, tended to rise since April, 1977 showing a rise of about 2 m by the middle of December, and that a sudden fall of water level amounting to 1 m took place 30 days before the occurrence of the present earthquake is reported by ERI.   The depth of the well is 600 m.   A 6 m lowering of the level is observed associated with the earthquake, and the level has been gradually recovering since the quake.

6. Summary of Geophysical Precursors

The precursory effects as described in the preceding sections are summarized in Table 1.   Although the number of data amounts to only 14 in total, a histogram of logarithmic precursor time T measured in units of days is presented in Fig. 5.

According to the writer's previous study (RIKITAKE, 1978b), earthquake precursors seem to be classified into the following classes.   Precursor time of precursors of the first kind depends on magnitude of the main shock.   Precursors of the second kind are characterized by the precursor time around $\log_{10} T = -1$ regardless of the magnitude of main shock.   In addition to the precursors of the above two kinds, it seems likely that there are precursors of the third kind which have a fairly broad spectrum.   Many of foreshock, ground tilting and other precursors seem to belong to this class.   It is noted in the previous paper, which dealt with all the existing data, that the mean precursor time of

TABLE 1.   Geophysical precursors.

Discipline	Precursor time (day)	Epicentral distance (km)	Observer	Observation point
Foreshocks	2		JMA	
Enormous increase in number of foreshocks	0.183		JMA	
Sudden decrease in foreshock activity	0.135		JMA	
Change in volume strain	42.5	32	JMA	Irozaki
Reversal of the above change	4	32	JMA	Irozaki
Change in volume strain	34.5	40	JMA	Ajiro
Change in geomagnetic field	64.5	38	JMA	Matsuzaki
Change in amplitude of short-period geomagnetic variation	69.5	30	ERI	Nakaizu
Change in earth potential	64.5	30	ERI	Nakaizu
Reversal of the above change	17	30	ERI	Nakaizu
Decrease in radon content	65.5	30	GI*	Nakaizu
Increase in radon content	7	30	GI	Nakaizu
Rise of water level	288.5	35	ERI	Funabara
Fall of water level	29.5	35	ERI	Funabara

*   GI means Geophysical Institute, University of Tokyo.

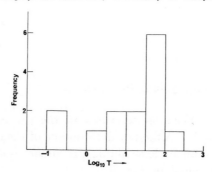

FIG. 5.   Histogram of $\log_{10} T$ for the geophysical precursors. T is measured in units of days.

the third kind precursor occurs around $\log_{10} T = -1$ as long as earthquakes of which magnitude is equal to or larger than 6 are concerned.

It is not at all clear whether a precursor of the first kind existed before the present earthquake.   The anomalous uplift may have something to do with the precursor concerned, but not exact time is known for its appearance.   As the uplift is still there even after the earthquake, it is tentatively assumed that the uplift is not related to a precursor of the calss.

The general feature of the histogram of logarithmic precursor time in Fig. 5 resembles that of the third kind precursor (see Fig. 3 in RIKI-TAKE, 1978b). It is noticeable, however, that the maximum frequency of $\log_{10} T$ occurs around 1.5 in contrast to 1.0 in the previous paper. The mean precursor time is estimated as scores of days for the present earthquake, the value being considerably shifted from 4.5 days obtained in the previous paper. As for the precursor of the second class, we could have only 2 reports this time.

It may be that the occurrence mode of precursor may differ from earthquake to earthquake even if precursors should exist at all. It is quite possible that the precursor times for a particular earthquake are considerably different from the mean value derived from a data set including all sorts of earthquakes.

It seems also highly likely that the disciplines of precursor, that are associated with an earthquake, are different from earthquake to earthquake although the reason why it is so is not known at the present stage of investigation.

7. Anomalous Animal Behaviour

Anomalous animal behaviour preceding the present earthquake is investigated by the present writer mostly by means of postal reply card questionnaire. As the report will appear elsewhere (RIKITAKE, 1980), only the conclusion is mentioned here very briefly. The frequency versus logarithmic precursor time in days is reproduced in Fig. 6, and the mean precursor time is estimated as 0.56 day, a value agreeing well with 0.4 day in the previous paper that analysed all the existing animal data (RIKITAKE, 1978a). The standard deviation of the present analysis is fairly small as can be observed from Fig. 6.

The data include anomalous behaviours of rats, moles, cats, dogs, raccoon dogs, pigs, rabbits, squirrels, deer, cows, crows, macaws, budgerigars, hens, pheasants, mynas, nightingales, ducks, sparrows, horse mackerel, catfish, yellowtails, goldfish, carp, crabs, gray mullets, daces, snakes, vipers, frogs, cockroaches, earthworms, alligators and so on. It is ascertained that the smaller the epicentral distance is, and the larger the seismic intensity is, the larger is the number of anomaly reports.

It appears to the writer that no evidence which rules out the possibility of anomalous animal behaviour to be a kind of earthquake precursor is brought out by the study. It is interesting to note that the

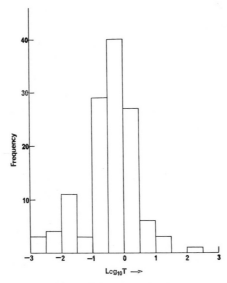

Fig. 6. Histogram of $\log_{10} T$ for the animal precursor data. T is measured in units of days.

mean precursor time of animal behaviour, should it be a precursor at all, is significantly different from that of the geophysical precursors.

8. Concluding Remarks

In association with the January 14, 1978, earthquake of magnitude 7.0 that occurred near Izu-Oshima Island about 100 km south of To- kyo, it seems likely that a number of precursors were observed although most of them were detected by close examination of data after the earth- quake. It is encouraging for Japanese seismologists, however, that the present experience suggests the possibility of predicting an earthquake provided that well-organized observation arrays are available on the on- line real-time basis and that someone is always watching the telemetered records.

It is quite surprising, however, that the mean precursor time is as long as scores of days for this earthquake in contrast to several days as estimated for the existing precursors of this class. It is no easy matter, therefore, to guess the occurrence time of a particular earthquake al- though an algorithm of earthquake prediction can be provided in a sta- tistical sense.

It should finally be remarked that much confusion occurred in conveying earthquake information to the public in association with the present earthquake. The "earthquake information" sent out from JMA seemed to reach local governments. Being close to the noon on a Saturday, the information received does not seem to have been conveyed to local citizens in a proper way. On January 18, the government of Shizuoka Prefecture issued information about aftershocks, which was only a repetition of the announcement made by the Coordinating Committee on Earthquake Prediction held a few days ago. Some local people, who watched at the news on TV, listen to the radio, or informed by soneone-else, misunderstood that a very large earthquake would soon hit. Those panicked people ran home by a car, went to supermarkets for buying daily necessaries, rang relatives up by telephone so many times that telephone lines failed to manage, and were involved in other unnecessary businesses.

Judging from the above experiences, it must be emphasized that administrators must know how local citizens react against prediction-oriented earthquake information and that the information should be presented with a proper timing.

In conclusion, it should be mentioned that almost all the precursor data presented here are taken from the reports by respective organizations to the Coordinating Committee for Earthquake Prediction. They will soon be published in a report of the committee. The writer is thankful to the organizations for the data used in this paper.

REFERENCES

GEOGRAPHICAL SURVEY INSTITUTE, Crustal deformation in the central part of Izu Peninsula, *Rep. Coord. Comm. Earthq. Prediction*, **18**, 56–60, 1977 (in Japanese).

HAGIWARA, Y., H. TAJIMA, S. IZUTUYA, and H. HANADA, Gravity changes in the eastern part of Izu Peninsula during the period of 1975–1976, *J. Geod. Soc. Japan*, **22**, 201–209, 1976 (in Japanese).

RALEIGH, B., G. BENNET, H. GRAIG, T. HANKS, P. MOLNAR, A. NUR, J. SAVAGE, C. SCHOLZ, R. TURNER, and F. WU, Prediction of the Haicheng earthquake, *EOS (Trans. Am. Geophys. Union)*, **58**, 236–272, 1977.

RIKITAKE, T., Statistics of ultimate strain of the earth's crust and probability of earthquake occurrence, *Tectonophysics*, **26**, 1–21, 1975.

RIKITAKE, T., Probability of a great earthquake to recur off the Pacific coast of Central Japan, *Tectonophysics*, **42**, T43–T51, 1977.

RIKITAKE, T., Biosystem behaviour as an earthquake precursor, *Tectonophysics*, **51**, 1–20, 1978a.

RIKITAKE, T., Classification of earthquake precursor., *Tectonophysics*, **54**, 293–309, 1978b.

RIKITAKE, T., Anomalous animal behaviour preceding the 1978 earthquake of magnitude 7.0 that occurred near Izu-Oshima Island, Japan, this volume, 1981.

SACKS, I. S., S. SUYEHIRO, D. W. EVERTSON, and Y. YAMAGISHI, Sacks-Evertson strainmeter, its installation in Japan and some preliminary results concerning strain steps, *Pap. Meteorol. Geophys.*, **22**, 195–208, 1971.

ANOMALOUS ANIMAL BEHAVIOUR PRECEDING THE 1978 EARTHQUAKE OF MAGNITUDE 7.0 THAT OCCURRED NEAR IZU-OSHIMA ISLAND, JAPAN

Tsuneji RIKITAKE

Department of Applied Physics, Faculty of Science, Tokyo Institute of Technology, Ookayama, Meguro-ku, Tokyo, Japan

Abstract   Anomalous animal behaviour preceding the January 14, 1978, earthquake of magnitude 7.0 that occurred near Izu-Oshima Island, Japan is studied mostly by means of reply card questionnaire.   As a result of the analysis of 129 reports, the mean value of precursor time is estimated as 0.56 day which approximately agrees with that of the previous study based on a worldwide data set for many earthquakes.   No outstanding difference in precursor time between animal species is found.   No evidence which rules out the possibility of aniaml behaviour to be a probable earthquake precursor is deduced.

1. Introduction

An carthquake of magnitude (M) 7.0 occurred near Izu-Oshima Island and Izu Peninsula at 12.24 local time on January 14, 1978 causing considerable damage mostly on the peninsula.   The epicenter is located at 34.8°N and 139.3°E according to the Japan Meteorological Agency as shown in Fig. 1.

As one of the members of a research group on reaction of biosystems against environmental stimulation probably related to a coming earthquake led by Professor Y.   Suyehiro, a famous ichthyologist, the writer has been preparing reply cards for questionnaire about anomalous animal behaviour preceding an earthquake.   Soon after the present earthquake, these cards were sent out to schools, town halls, zoos, fishery institutions and the like in the strongly-shaken areas, and the writer could collect many a report on animal behaviour preceding the earthquake with the cooperation of many people.

The precursor time data thus collected are subjected to a Weibull distribution analysis. That the frequency versus precursor time of anomalous animal behaviour graph indicates a fairly sharp peak with a mean value amounting to 0.56 day will be shown in a later section.

2. Data

Postal reply cards amounting to 290 in total number are dispatched mostly to organizations within a distance of 100 km from the epicenter, and 186 cards were recovered.   Of these, 119 cards replied that nothing anomalous had been noticed.   Some cards described anomalies after the earthquake, or only ambiguous descriptions were mentioned on some cards.   These have to be excluded from the present study.   But the writer could gather fairly clear reports from 60 cards.   In a number of cases, many reports were sent to the writer in addition to reply cards. Some data are also added with the aid of mass communication media. As a result, the writer could collect 129 reports on anomalous animal behaviour preceding the earthquake.   According to RIKITAKE (1978a), who examined all the available data on anomalous animal behaviour forerunning an earthquake, only 157 examples could be obtained from worldwide existing sources.   It is rather surprising, therefore, that data as many as 129 in number are collected for a single earthquake.

The data thus collected are listed in the Appendix separately for mammal, bird, fish, and snake, worm and so on.   The tables in the Appendix include the observation point, epicentral distance, precursor time and simplified description of the animal behaviour observed.   No details of observation site are given in the table because they mean almost nothing for non-Japanese readers.   No names of reporter are listed as well.   These points can be found in the Japanese version of this report (RIKITAKE, 1978b).

3. Spots Where Anomalous Animal Behaviour Is Observed

Figure 1 shows the spots where one or more anomalies of animal behaviour are observed.   Although it cannot be said that the reply cards are distributed quite uniformly, it is markedly noticed in the

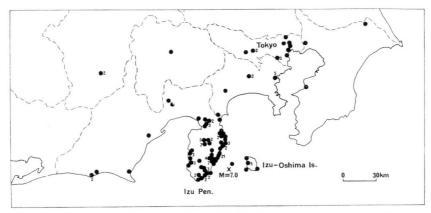

FIG. 1.   Spots where anomalous animal behaviour are observed.   The numerals indicate the number of reports at one spot.   No numeral is attached when we have only one report.   The epicenter is shown by a cross.

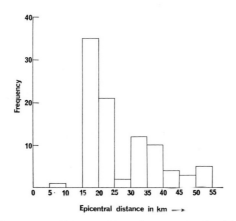

FIG. 2.   Histogram of the numbers of reported animal behaviour as arranged according to the epicentral distance.

figure that many reports are obtained from the strongly-shaken and consequently damaged areas.   These areas are naturally close to the epicenter.   In Fig. 2, the histogram of the numbers of report for each 5-km range of epicentral distance up to 55 km is presented.

4.   Precursor Time

Precursor times are sometimes described by terms such as " the morning of the very day " or " the last night ".   In such cases, no ex-

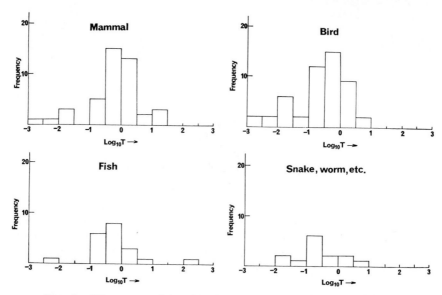

FIG. 3. Histograms of logarithmic precursor time in units of days for the 4 groups of animal.

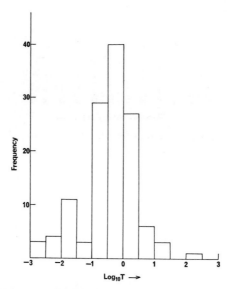

FIG. 4. Histogram of logarithmic precursor time in units of days for the whole animal data.

act estimate of precursor time is possible.   It is therefore assumed in
the statistics that the anomalous behaviour is observed at 6 a.m. on the
day of earthquake for the former case and at 6 p.m. on the yesterday
for the latter case.

In Fig. 3 are shown the histogram of logarithmic precursor tine
T measured in days for the four groups of animal, i.e. mammal, bird,
fish, and snake, worm and the like, respectively.   It appears to the
writer that no marked difference in the distribution of precursor time
between the respective groups is observed in the histograms in Fig. 3,
so that the following discussion about the precursor time of animal behaviour will rely on the whole data of which the histogram is shown in
Fig. 4.

Looking at Fig. 4, it is markedly noticed that the distribution indicates a sharp peak around $\log_{10} T = 0$.   It is clear, therefore, that most
anomalies of animal behaviour occur with a precursor time ranging 0.1–
3 days as far as the present data are concerned.

5. Weibull Distribution Analysis of Precursor Time Data

A Weibull distribution analysis is applied to the precursor time
data shown in Fig. 4.   No details of the analysis is needed to be described here because the analysis is made in the exactly same way as
that in the previous paper (RIKITAKE, 1978a).

It is assumed that logarithmic precursor time is governed by a
distribution such as

$$\lambda(s) = Ks^m \qquad (s = \log_{10} T + c) \qquad (1)$$

where λ is the probability for a logarithmic precursor time to fall in
an interval between s and $s + \Delta s$. Δs is much smaller than s. c is a constant which may be chosen in such a way as to make actual calculation
easy. $c = 3.0$ is taken here.

Denoting the cumulative probability for an earthquake to occur
during $0–s$ by $F(s)$, we define a function $R(s)$ by

$$R(s) = 1 - F(s) \qquad (2)$$

It is evident from (1) that $R(s)$ is given by

$$R(s) = \exp\left[-\int_0^s \lambda(s)ds\right] = \exp\left[-Ks^{m+1}/(m+1)\right] \qquad (3)$$

On taking the double logarithm of $1/R$, we obtain from (3)

$$\log_e \log_e (1/R) = \log_e \left(\frac{K}{m+1} \right) + (m+1) \log_e s. \qquad (4)$$

As R can be estimated from the actually observed data in a way which was presented in the previous paper (RIKITAKE, 1978a), the $\log_e \log_e (1/R)$ versus $\log_e s$ relation is readily obtained as shown in Fig. 5. $\varDelta s = 0.5$ is assumed in the actual plotting.

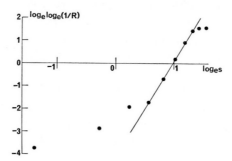

FIG. 5. $\log_e \log_e (1/R)$ versus $\log_e s$ relationship.

It seems likely from Fig. 5 that the distribution of precursor time consists of two different distributions respectively for smaller and larger precursor times. The plots for smaller precursor times are scattered because the data are rather scanty. However, the parameters of Weibull distribution, i.e. K and m, for longer precursor times can be determined fairly accurately by the straight line approximation as shown in Fig. 5.

The parameters thus determined enable us to estimate the mean value of logarithmic precursor time $E(s)$ and its standard deviation σ as follows;

$$E[s] = 2.43 \qquad (E[\log_{10} T] = -0.57)$$
$$\sigma = 0.16$$

so that the mean precursor time becomes

$$E[T] = 0.56 \text{ day.}$$

6. Conclusion

The above analysis of precursory anomalies of animal behaviour leads to the following conclusions:

1. The smaller the epicentral distance is, and the larger the seismic intensity is, the larger is the number of anomaly reports.

2.   The mean precursor time of anomalous animal behaviour is estimated as 0.56 day which agrees with the result of the previous analysis of worldwide data as obtained by RIKITAKE (1978a).   However, the standard deviation is much smaller this time.

3.   Although another peak of precursor time distribution seems likely to be found around several tens of minutes, its reality is not established because of scanty data.

4.   No outstanding difference in precursor time between animal species is found.

Summarizing the above points, it may be said that the point that animal behaviours can sometimes provide an earthquake precursor cannot be relued out.

Acknowledgements

Some portion of the expense for the present study was defrayed from the funds given to the group led by Suyehiro.   The writer is grateful to the Ministry of Education for the support.

APPENDIX

Precursor data 1—Mammal

Observation spot	Epicentral distance (km)	Precursor time	Description
Ito	20	3 weeks	7 rats (*rattus norvegicus* forma *albinus*) come out from a cage.
Ichikawa	140	15 d	Moles began to make a fuss in the garden.
Yokohama	90	2 weeks	A hamster does not come out from its nest and docs not eat.
Yugashima	32	1 week	Rats disappear from a house.
Inatori	16	1 week	Moles make a fuss biting trees in the garden.
Tokyo	120	3 d	Rats make a fuss violently on the ceiling.
Tokyo	110	3 d	Cats make a fuss in the night.
Tokyo	110	3 d	Rats (*rattus noevegicus*) walk on the street.
Mishima	50	3 d	Dogs howl unusually.
Ito	20	2–3 d	Rats, which used to come out every night, disappear.
Inatori	16	2–3 d	Rats disappear from a pigpen.
Atsugi	85	2 d	Dogs bark noisily all the night.

Precursor data 1—Mammal (continued)

Observation spot	Epicentral distance (km)	Precursor time	Description
Atsugi	85	2 d	Rats come out.
Motomachi, Izu-Oshima Is.	17	2 d	Rats make a fuss beneath the floor of a new house.
Inatori	16	2 d	Rats disappear.
Inatori	16	2 d	A dog howls and does not come out from the kennel.
Inatori	16	2 d	A raccoon dog kept in a house barks.
Nishina	38	1.5 d	Many molehills are found in the field.
Kamo	41	1 d	Many rats run away.
Yugashima	36	1 d	Pigs get excited.
Ito	32	1 d	A house dog shivers in bed.
Shimoda	21	1 d	Dogs howl.
Inatori	16	1 d	Dogs bark.
Inatori	16	1 d	A rabbit goes wild until it dies.
Inatori	16	1 d	Cats disappear.
Nirayama	46	18 h 24 m	Dogs get active.
Nirayama	46	18 h 24 m	A cat is missing for one week.
Nirayama	46	18 h 24 m	Squirrels move about actively and do not eat.
Ito	31	last night	A puppy unusually barks against the sea.
Kawazu	19	last night	Rats run about so noisily that one cannot sleep.
Kawazu	19	last night	A scream probably of a dog is heard.
Inatori	16	last night	Dogs howl sadly.
Motomachi, Izu-Oshima Is.	17	16 h 24 m	10 pigs make a fuss.
Iida	160	morning of the day	4 deer sit in the center of a play ground and do not move.
Yugashima	36	morning of the day	A dog goes wild and trys to run away.
Ito	24	6 h 24 m	10 rats run away from a drainpipe.
Inatori	16	morning of the day	Dogs do not bark.
Motomachi, Izu-Oshima Is.	17	3 h	8 cows cry out.
Kamo	40	24 m	Dogs bark.
Nakaizu	35	24 m	Dogs make a fuss.
Nakaizu	33	24 m	Cats make a fuss.
Inatori	16	10 m	House dogs bark out.
Ookawa	20	2 m	Dogs shiver.

Precursor data 2—Bird

Observation spot	Epicentral distance (km)	Precursor time	Description
Yugashima	32	1 week	Crows disappear.
Shimoda	24	1 week	50 birds cry out for 3 hours.
Nirayama	44	2 d 21 h	Macaws go wild.
Tokyo	120	2–3 d	A budgerigar gets restless and does not enter a cage.
Hiekawa	28	2–3 d	Hens in a poultry yard make a fuss. Egg production decreases considerably.
Shimoda	23	2–3 d	Crows as many as 100 fly away.
Yugawara	50	2 d	Crows cry in several strange ways for 1 hour.
Inatori	16	2 d	Japanese white-eyes stop to cry.
Inatori	16	2 d	Crows, which used to come for picking hens' eggs, go away.
Sahara	180	1 d 3 h	A budgerigar cries noisily.
Oimatsuda	75	1 d 3 h	Countless sea-birds appear in flocks.
Kumagaya	170	1 d	A bird of the pheasant family (*Chrysolophus pictus*) begins to cry.
Tokyo	120	1 d	Many birds of various kinds make a noise.
Tokyo	110	1 d	Many birds make a fuss for 30 min.
Minamiizu	30	1 d	Crows disappear.
Ito	24	1 d	Pheasants make a fuss.
Sawada	20	1 d	Many crows cry out.
Mitakahama	18	1 d	A bird flies away from a cage.
Hama	18	1 d	A myna is terribly noisy.
Inatori	16	1 d	Pheasants, larks and the like cry out.
Inatori	16	1 d	10 hens jump up on a tree and do not eat.
Matsuzaki	30	22 h	30 crows make a fuss.
Kisarazu	105	20 h 24 m	More than 100 crows fly.
Maebashi	190	last night	A macaw goes wild in a cage.
Yugashima	36	last night	A Japanese nightingale does not sleep in the night.
Ito	24	6 h 24 m	Pheasants make a fuss.
Motomachi, Izu-Oshima Is.	17	6 h 24 m	Several pheasants and 10 hens cry out.
Ito	31	morning of the day	50 buntings get together.
Ito	28	morning of the day	Wild pheasants make a noise.
Yatsu	19	morning of the day	Birds disappear.

Precursor data 2—Bird (continued)

Observation spot	Epicentral distance (km)	Precursor time	Description
Inatori	16	morning of the day	Crows fly in flocks.
Matsuzaki	38	4 h 24 m	Ducks, which are usually found on the river, disappear.
Matsuzaki	38	4 h 24 m	2 pheasants get together.
Shimoda	22	4 h 24 m	A macaw gets out of the cage.
Shimoda	23	3 h 30 m	Many crows walk around on the ground.
Matsuzaki	38	3 h 9 m	2 pheasants cry out crazily for 7–8 sec.
Yokohama	90	3 h	Many crows and sparrows make a noise.
Yokohama	90	2 h 30 m	Many crows appear quite unusually.
Yatsu	19	2 h	Pheasants go wild terribly.
Shimoda	21	1 h 24 m	Crows make a fuss and fly away.
Ito	30	30 m	Unusual sounds of bird cry.
Kofu	122	24 m	400 pheasants cry out.
Kamo	40	24 m	Crows disappear.
Nakaizu	35	24 m	Macaws make a fuss.
Shimoda	20	24 m	A myna and 100 crows make a terrible fuss.
Kumagaya	170	20 m	A bird of the pheasant family (*Chrysolophus pictus*) suddenly cries out.
Inatori	15	10 m	Sparrows fly away in flocks.
Kawazu	23	4 m	50–60 pheasants cry out.
Fujimiya	88	2 m	6 pheasants fly about and cry.
Kannami	50	1–2 m	4 macaws and 2 paddy birds go wild and make noise.
Tokyo	130	5–10 s	Pheasants cry out.

Precursor data 3—Fish

Observation spot	Epicentral distance (km)	Precursor time	Description
Ito	22	4 months	A deep-sea fish (*Trachipterus ishikawai*) is caught by a drift net.
Yatsu	19	14 d	Great catch of horse mackerel.
Oarai	220	4 d	Catfish are caught unusually.
Tokyo	130	3 d	Catfish in breeding tanks move unusually.
Ito	28	3 d	Yellowtails are fished 20 days earlier than usual date of year.
Tokyo	130	2 d	Catfish in breeding tanks move unusually.

Precursor data 3—Fish (continued)

Observation spot	Epicentral distance (km)	Precursor time	Description
Mikuni	320	1 d 1 h	11 catfish make a fuss.
Tokyo	120	1 d	Tropical fish move anomalously.
Tokyo	110	1 d	Goldfish swim around. This is unusual for the season.
Shimoda	24	1 d	Big catch of a kind of fish (*Girella punctata*).
Habu, Izu-Oshima Is.	23	1 d	Several carp swim around restlessly.
Inatori	16	1 d	Goldfish float up.
Yahazudashi fishing field	8	1 d	Catch of a kind of fish (*Beryx splendens*) suddenly ceases.
Matsuzaki	38	22 h	20–30 crabs float up.
Ito	24	6 h 24 m	Goldfish jump out from a water tank.
Kamo	19	morning of the day	Gray mullets disappear in flocks.
Off Izu-Oshima Is.	16	dawn of the day	No catch of fish of mackerel family (*Seriola anreovittata*) despite big catch until yesterday.
Minamiizu	31	4 h 24 m	Several carp jump up.
Minamiizu	31	4 h	3 among 50 carp float and jump up.
Shimoda	23	2–3 h	A dace in a breeding pond makes a fuss.
Kannami	50	10 m	4 carp jump.

Precursor data 4—Snake, insect, worm, etc.

Observation spot	Epicentral distance (km)	Precursor time	Description
Yatsu	19	4–5 d	A snake comes out and dies.
Shimoda	23	2–3 d	Several vipers come out.
Toyooka	130	2 d	A grass snake appears on the pavement.
Yokohama	90	1 d	A snake appears.
Inatori	16	1 d	Frogs and cockroaches come out.
Ryuyo	130	morning of the day	Many earthworms run away from the breeding nest.
Asaba	120	morning of the day	The same as the above.
Hamaoka	100	morning of the day	The same as the above.
Shizuoka City	80	morning of the day	The same as the above.

Precursor data 4—Snake, insect, worm, etc. (continued)

Observation spot	Epicentral distance (km)	Precursor time	Description
Yatsu	19	morning of the day	A snake appear.
Fujimiya	88	5 h 30 m	Many earthworms come out from the breeding nest.
Atagawa	17	1 h	Alligators in a zoo cry out. They seldom cry.
Atami	50	34 m	10 earthworms are found on a staircase of a school.
Irozaki	32	24 m	6 snakes come out from a stone wall.

REFERENCES

RIKITAKE, T., Biosystem behaviour as an earthquake precursor, *Tectonophysics*, **51**, 1–20, 1978a.

RIKITAKE, T., The 1978 earthquake that occurred near the Izu-Oshima Island and precursory animal behaviour, *Rep. Coord. Comm. Earthq. Predict.*, 1978b **20**, 67–76 (in Japanese).

RECENT EARTHQUAKE PREDICTION RESEARCH IN THE UNITED STATES

Max Wyss

Cooperative Institute for Research in Environmental Sciences, University of Colorado/NOAA, Boulder, Colorado, U.S.A.

Abstract Results of field observations pertaining to the problem of earthquake prediction by United States investigators during the period 1975 through mid-1978 are summarized. Slow but steady progress has been made toward understanding earthquake preparatory processes; however, many basic geophysical and geological problems need to be solved before routine predictions become possible. The most outstanding recent discoveries were those of the southern California uplift and the absence of extensive velocity anomalies before California earthquakes with magnitudes from 4.0 to 5.5. In southern California an area of about 400 km by 150 km was uplifted rapidly by approximately 30 cm during the 1960's and subsidence appears to be in progress now. Many investigators do not believe that this phenomenon is an earthquake precursor. The funding level of the national earthquake prediction research was increased substantially in 1978. This boost has strongly accelerated the rate of progress in the field of earthquake prediction studies. However, it is important to remember that geological processes are very slow compared to a human life span, and therefore we will need years and decades of careful observations before we can understand the preparatory process of major earthquakes.

1. Introduction

The purpose of this paper is to give a brief overview of recent earthquake prediction research published by U.S. scientists. Excluded in this summary are laboratory experiments and the details of velocity anomalies because these topics are covered by Martin (*Current Research in Earthquake Prediction II*) and Aggarwal (*Current Research in Earthquake Prediction II*). Perhaps the most important, and one hopes the most objective, part of such a report is the reference list. Even here subjectivity enters in the choice of what fields of research are to be considered closely enough related to the problem of prediction to be included. The list of about 200 references contains only work which is directly applicable to the problem of understanding the preparatory process of earthquakes and to predicting them.

Summaries on the state of U.S. earthquake prediction research were last compiled in 1975 (KISSLINGER and WYSS, 1975; BOLT and WANG, 1975). Therefore, I chose to define " recent " as work published from 1975 to mid-1978. Wherever possible I have tried to avoid reference to earlier work and to studies by investigators other than from the United States. Unless there were no other sources available, I have also refrained from quoting abstracts from meetings because one can rarely find out enough from abstracts to be able to use the information in one's own work.

Examining this reference list, one finds that three areas of studies dominate, suggesting that research is concentrated in these fields. Topics on seismicity, crustal deformation, and velocity changes each account for approximately $25\pm5\%$ of the references. Fields like modelling (theory), electric and magnetic methods, and animal behavior account for approximately $7\pm3\%$ each, and studies on ground water level, gas emanations, and miscellaneous topics make up the rest.

The publications on prediction related topics account for only a small fraction of the total seismological and geophysical publications. In the *Bulletin of the Seismological Society of America* the prediction papers have climbed from about 5% in 1975 to 12% in 1978, averaging roughly 7% during this period. In other journals the fraction is even smaller; in the *Journal of Geophysical Research* and in the *Geophysical Journal* they account for approximately 4% and less than 1%, respectively. It may be surprising that these numbers are so small. The reason probably is that it takes a while for a national research program to get started and to produce results.

The formal U.S. earthquake prediction program came into existence in 1974. Before that time basic research on processes related to the earthquake source was funded by the National Science Foundation within the geophysics program. Up to 1973 few researchers pursued the topic of prediction in earnest. When the U.S. earthquake prediction program was initiated, the U.S. Geological Survey was given responsibility for it. The total funds allocated were approximately $9 M per year for earthquake hazard, prediction and fundamental studies. About $2.5 M of these funds were spent to support research outside the U.S. Geological Survey. Of both of the above figures roughly half was used on prediction research. The funding remained constant through 1977 except for an additional one-time allocation of $2 M for the study of the southern California uplift. In 1978 funding was increased: A total of approximately $30 M per year is now spent for research on hazard, prediction and basic problems. Of this amount about $10 M support research outside of the Geological Survey, and again roughly half of these amounts is spent on earthquake prediction research.

In addition the National Science Foundation has expanded its support of basic research aimed at understanding of earthquake processes from $3.5 M per year before 1975 to $6.3 M per year in 1978 as a part of the U.S. national program of earthquake hazard mitigation. The engineering oriented part of this program is administered by the National Science Foundation's Applied Science and Research Application division (formerly RANN), which spent $18 M during 1978 on problems regarding the siting and design of structures, as well as sociological and policy questions arising in connection with earthquake hazards.

In this summary the emphasis is on research applied directly to the problem of earthquake prediction. Much of the work quoted was carried out or supported by the U.S. Geological Survey. A summary of studies supported by this program was compiled by HAMILTON (1978).

From the funding levels given above one can see that the United States prediction program is now starting to accelerate. The years 1974–77 were an initiation period with emphasis on research fields which seemed promising based on earlier work in the USSR and Japan. We will soon see a flood of research articles on the topic of earthquake prediction, and hopefully some great new discoveries will be made.

The discussion below is limited to a few areas in which a substantial amount of work has been done recently or which seem especially promising. These are velocity changes, seismicity patterns, crustal deformations, stress estimates, electric and magnetic methods, ground water levels, gas emanations, and animal behavior. Omitting other topics does not mean that I estimate other work or its potential to be inferior. Clearly theoretical work and modelling will ultimately be of great importance. The reference list includes all types of field work, and I hope that it is fairly complete.

2. Velocity Changes

When velocity ratio precursors were confirmed in New York State (AGGARWAL *et al.*, 1975), many people expected that this would be the number one parameter solving the problem of earthquake prediction. Several reasons contributed to these high hopes. First, the New York State data indicated that very strong changes (about 13% for V_p/V_s) could be observed even for small main shocks ($M \approx 3$). Second, the anomaly extended throughout a crustal volume with dimensions exceeding the source volume by a factor of 5 or so. And third, in laboratory results such velocity changes were observed at stresses near failure.

It seemed that it should be easy to observe anomalies with the above characteristics. Therefore, it was surprising when many searches for velocity precursors were unsuccessful. The disappointment was such that some researchers suggested that no velocity anomalies ever existed. However, the negative results do not as yet rule out the possible existence of velocity precursors in general, as it seemed to some, because of the following facts: (a) Almost all of the negative data were gathered in California and they applied to strike-slip or normal faulting. (b) Most of the studies involved relatively small main shocks ($M \leq 5.5$). (c) Some of the data did not involve main shocks, but were general velocity surveys (e.g. KANAMORI and HADLEY, 1975; ROBINSON and IYER, 1976; PEAKE *et al.*, 1977). (d) Some of the data were not sufficiently tight in time and space to actually rule out velocity precursors as clearly as one would wish (e.g. MCEVILLY and JOHNSTON, 1974). However, there is no doubt that the data available to date tell us that, in the strike-slip tectonism of California, velocity precursors are small in degree and dimensions if they exist at all.

Table 1 summarizes the work on velocity measurements from 1975 to the present as far as I am aware of it. I thought it useful to divide the list into three groups: (1) The top seven references represent " positive " results, which means that the authors stated that they found a precursory velocity anomaly. (2) The second group of papers include those where the experiment was " negative ", that is, no anomaly was found before specific main shocks. (3) The last three articles in Table 1 report constant velocities at times when no specific main shocks occurred.

The top group of " positive " research reports are ordered by degree to which they appear convincing to me. I am of the opinion that the first two articles (AGGARWAL *et al.*, 1975; A. C. JOHNSTON, 1978) are clearly the strongest ones. In both cases the data are documented and the resolutions as reflected by the reading accuracy and standard deviation were excellent (Table 1). The data presented by AGGARWAL *et al.* (1975) are especially strong because the V_p/V_s results were backed by explosion residuals, and because several stations registered the anomaly which was apparently decreasing with distance from the main event. A. C. JOHNSTON's (1978) anomaly is also strong because large numbers of data points define it (see Fig. 1), and because the totally independent data set of local seismicity indicates the onset of a period of quiescence approximately at the time when the travel times change (WYSS *et al.*, 1978).

The next two data sets in Table 1 (WESSON *et al.*, 1977; WYSS, 1975a) are less strong because the data used are of lesser quality. However, the data have been explicitly documented so they can be inspected. WESSON *et al.* (1977) present a reworked and improved data set originally used by ROBINSON *et al.* (1974). The new data still show the same and unique anomaly; however, the authors state that it might be possible that the anomaly was due to errors in locations and selection of arrival time. LINDH *et al.* (1978a) attacked the conclusions reached by WESSON *et al.* (1977), examining the same as well as new data. Clearly some late arrivals could not be explained away by LINDH *et al.* (1978a). Furthermore, CRAMER and KOVACH (1975) observed two delayed teleseismic residuals for the same station and time period. In my opinion the evidence indicates that a travel time delay probably did exist before the $M=5.0$ Bear Valley earthquake of 24 February 1972.

Teleseismic P-residuals as used by WYSS (1975a) are only reliable if close by reference stations are available. As we now know the P-wave residual anomaly at Matsushiro (WYSS and HOLCOMB, 1973) is probably not real, therefore one wonders whether the Garm anomaly could also have been due to non-random source distributions. This is not likely because shallow events were excluded, and because relative residuals suggest that the anomaly is real. Also the case is strengthened by the appearance of anomalies for independent parameters at approximately the same time.

The next two papers in the top group of Table 1 (GUPTA, 1975a, b;

TABLE 1.   Studies of velocities

Authors		Area studied	Wave type	Sources
AGGARWAL *et al.*	1975	New York	V_p/V_s, P	Local eqs, local expl.
JOHNSTON, A. C.	1978	Hawaii	P	Deep teleseisms
WESSON *et al.*	1977	California	P	Local eqs.
WYSS	1975	Garm	P	Deep teleseisms
GUPTA	1975	Nevada	S	Local eqs.
TALWANI	1978	S. Carolina	V_p/V_s	Local eqs.
CRAMER and KOVACH	1975	California	P	Teleseisms
MCEVILLY and JOHNSON	1974	California	P	Local expl.
BOORE *et al.*	1975	California	P	{Local expl., nucl. expl. {Deep teleseisms
ENGDAHL	1975	Aleutians	P	Deep teleseisms
KANAMORI and FUIS	1976	California	P	{Local expl. {Teleseisms
CRAMER	1976	California	P	
CRAMER *et al.*	1977	California	P	{Nuclear expl. {Deep teleseisms
BOLT	1977	California	P	Nuclear expl.
STEPPE *et al.*	1977	California	P	Local eqs.
MURDOCK	1978	Wyoming	P	Nuclear expl.
LINDH *et al.*	1978	California	P	Local eqs.
KANAMORI and HADLEY	1975	California	P	Local expl.
ROBINSON and IYER	1976	California	P	Nuclear expl.
PEAKE *et al.*	1977	California	P	Local expl.
RAIKES	1978	California	P	Teleseismic

TALWANI, 1978) present velocity precursors whose reliability cannot be checked because the data are not shown in detail in these papers. It is, therefore, difficult to form an opinion on the existence of anomalies in these cases.

The " negative " evidence (second group in Table 1) carries the clear message that moderate size strike-slip and normal faulting earthquakes do not have large or extensive velocity precursors. The table gives the approximate distance from the ray path to the nearest point of the aftershock area, and in parenthesis the distance to the epicenter. In the only " negative " example in a thrust environment (ENGDAHL, 1975) the distance is considerable. It is interesting to note that A. C. JOHNSTON (1978) observed that no anomaly existed at a station within the aftershock area 15 km from the main shock epicenter, while the station that did as a function of time.

Accuracy reading (sec)	σ	Travel time delay (sec)	Velocity ratio change	Magnitude	Distance (km)	Faulting type
0.01		0.13	13%	3±	4	Thrust
0.02	0.05	0.2		7.2	0 (3.5)	Horizontal
0.05	~0.1	0.3		5.0	0 (3)	Str.-slip
0.1	~1.0	0.4		5.7	20 (25)	Thrust
				4.0	~10	Normal
			13%	2.3±	~3	Str.-slip
	0.07	0.15		5.0	0 (3)	
0.1		2%		5±	10	Str.-slip
	0.15	<0.2		5.8	0.5 (18)	Str.-slip
0.05		<0.3		6.8	50	Thrust
0.01		<0.1		5±	~5	Str.-slip / Normal
		<0.1		5.1	2.5	Str.-slip
		<0.05		5.8	5 (10)	Normal
		<0.02		5.8	5 (10)	Normal
				4.5±	3	Str.-slip
		<0.1		6.0	8	Normal
0.05		<0.1		5.0	0 (3)	Str.-slip
0.01		≤3%		No specific earthquake	NA	NA
		<0.1			NA	NA
0.004		≤0.004			NA	NA
0.05	0.05	<9%			NA	NA

show the anomaly was located at 3.5 km from the epicenter. The rupture length of this quake was about 40 km. The top two examples in Table 1 therefore provide a strong contrast for the dimension ratio of the anomaly to the source volume: for the small thrust event in New York the ratio was 5 or 10; for the large " horizontal normal " fault in Hawaii, it was 1/5 to 1/10. The difference in dimension ratio is a factor of 25 to 100.

If one assumes that the strike-slip ruptures of the California type have the same dimension ratio as the Hawaii example (the smallest known ratio of anomaly to source), one would expect the 10 km long earthquakes of Table 1 to have had anomaly dimensions of about 2 km around the main shock epicenter. Clearly this could not have been detected in any of the " negative " cases. However, for great earth-

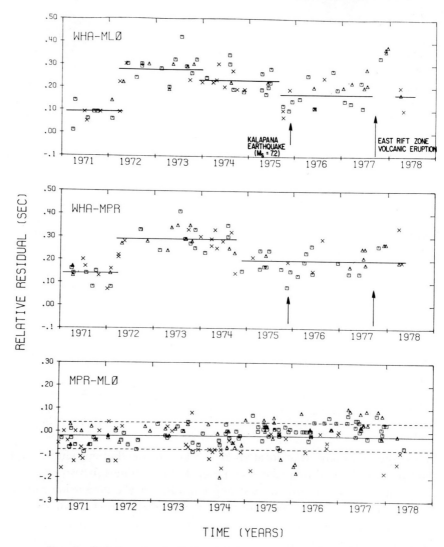

FIG. 1.   Relative teleseismic P-residuals versus time at Hawaiian stations (from
A. C. JOHNSTON, 1979).   WHA is located at 3.5 km from the $M_s = 7.2$ Kalapana earthquake of November 1975.   The reference stations used are located
at 16 and 44 km distance.   Deep ($h > 500$ km) Tonga-Fiji sources are used;
□ north, △ central, × south Tonga-Fiji.   On the seismic records 1 cm equals
1 sec.   Increased relative residuals are interpreted as indicating a travel time
delay.   Continuous lines are averages; dashed lines show ± one standard
deviation.

quakes in California, the ones for which prediction really matters, one would expect to observe a velocity precursor on several stations.   Under the assumption of similarity to Hawaii a 300 km rupture would have an anomaly in a volume of 30 to 60 km radius, which would make the methods useful for helping to predict large quakes even in California. These statements should, of course, be taken as a hypothesis only, which has to be confirmed by observations before it can be further developed.

The vibroseismic technique to measure travel times to accuracies of 10^{-4} sec is a promising but not fully explored tool (e.g. McEVILLY, 1977; CLYMER and McEVILLY, 1978).   Once this technique is perfected, the resolution of travel time may be improved by two to three orders of magnitude over the data quoted in Table 1.   This technique is more expensive than other travel time measurements mainly because it involves an active signal generator.   However, if small travel time delays can be measured before California earthquakes, the expense would be well worth it.

In summary, my view on velocity precursors is the following: Attitudes towards this parameter have gone through a cycle of exaggerated hopes to exaggerated negativism.   I believe that at least two velocity precursors, possibly six, were found by U.S. investigators.   However, many " negative " results and one " positive " result show that velocity anomalies for some earthquakes must be very small in extent and possibly may not be detectable for many events.   In future research it will be essential that the authors document their velocity precursors in fullest detail so that informed discussions of the data are possible.

3. Seismicity

Locating seismic events is of course the most basic of all studies dealing with the assessment of earthquake danger.   Seismic risk maps depend strongly on the documented seismicity.   The deployment of the Worldwide Standardized Seismic Network by the United States greatly improved the worldwide location capability.   Since 1963 the cutoff magnitude for complete location in most areas of the world lies between magnitudes 4.5 to 5.   Recently established dense networks of seismographs drop this level locally to magnitudes varying from 1 to 2. If these seismograph networks, and especially the WWSS network, are maintained over decades at their present high quality performance level, we may be able to gather long term seismicity data important for earthquake prediction in the future. Several types of studies which attempt to link seismicity patterns to following main shocks are described below.

A) *The identification of seismic gaps* along plate margins is generally considered to be a successful tool for specifying areas with increased seismic risk. It seems rare that aftershock areas of large earthquakes overlap, and several gaps were filled after they had been identified. An example of such a case in the United States is the Sitka earthquake of 1972 (SYKES, 1971).

At a recent conference, however, there was considerable debate about the usefulness of defining seismic gaps. One of the unsatisfactory aspects of seismic gaps is the fact that their identification does not constitute an earthquake prediction, because although location and approximate size are specified, the occurrence time can be estimated with much less certainty from the generally poorly known recurrence times. It can happen that a gap attracts so much attention that some people are left with the feeling that an earthquake is " imminent " in a particular gap. If the feeling of imminence prevails, the effect is the same as if a prediction based on facts had been made, and the consequences can be undesirable or even disastrous (GARZA and LOMNITZ, 1979).

The identification of seismic gaps is important in order to designate areas for intensive research. For example, southern Alaska and the Caribbean have been instrumented because of seismic gaps, but no one has claimed that predictions were issued for these areas.

McCANN *et al.* (1979) (Fig. 2) recently completed circumpacific seismic gap maps and summarized the seismicity work of many years (e.g. KELLEHER *et al.*, 1973). Figure 2 shows portions of plate boundaries which are relatively safe because stress has been released by a large quake within the last 30 years, or large ruptures never occur because creep accommodates a major portion of the plate motions. The latter areas could be called permanent gaps. Relatively mature gaps (category 1 in Fig. 2) are those where the last large quake occurred more than 100 years ago. With an average slip rate along plate boundaries of about 6 cm/year, displacements of more than 6 m could have accumulated in this time, which would correspond to a stress drop in excess of about 24 bars (assumed width = 100 km), which is sufficient for a great earthquake to occur in such locations. McCANN *et al.* point out that recurrence times in some areas seem to be fairly short, so that category 2 was

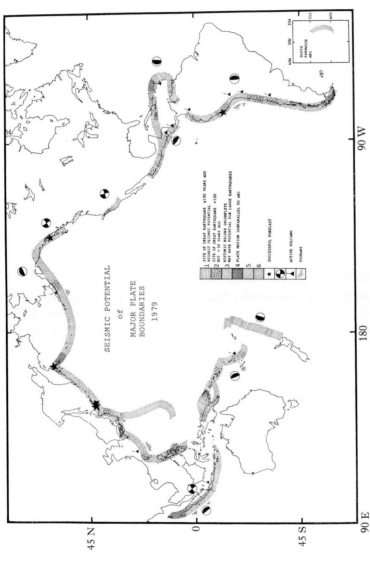

FIG. 2.   Seismic gaps and their relative state of maturity (from McCann *et al.*, 1979).   The segments of plate boundaries are differentiated into six categories: (1) the highest seismic potential is assigned to regions where the last large quake occurred more than 100 years ago; (2) the last large shock happened more than 30 but less than 100 years ago; (3) incomplete history of large quakes; (4) portions of subduction zones with mostly strike-slip motion; (5) no history of great earthquakes; (6) the last large earthquake happened less than 30 years ago, which is the category with the lowest quake potential.

assigned to plate boundaries where the last large shock occurred more than 30 but less than 100 years ago.

In each individual case one can debate just how mature a gap is, or how safe a part of a plate boundary is, depending on the local tectonic setting. The question of safety appears to be a difficult one, because two zones of large shallow earthquakes may exist parallel to each other along some plate boundaries (DEWEY and SPENCE, 1979). Therefore the closure of a gap by one large earthquake in these areas might not mean that the gap is now safe, but it could be followed by another large quake parallel to the first one. Examples of such plate boundaries are Peru (DEWEY and SPENCE, 1979) and the Kuriles (Fedotov, personal communication).

B) *Long term seismicity patterns* within gaps have received considerable attention recently. The thesis is that background seismicity within gaps may hold a clue for the prediction of the occurrence time, and perhaps the size of the expected main shock.

KELLEHER and SAVINO (1975) found that relative quiescence of the background seismicity preceded main shocks for many years in some rupture areas, except for the epicentral zone where they noticed increased activity. More recently periods of precursory quiescence were reported for earthquakes with magnitudes between 4.5 and 8.7 (OHTAKE et al., 1977b; ISHIDA and KANAMORI, 1977; ENGDAHL and KISSLINGER, 1977; KHATTRI and WYSS, 1978; WYSS and HABERMANN, 1979; OHTAKE et al., 1978; WYSS et al., 1978). These periods of quiescence are long term precursors in the sense that they approximately agree with the precursor time magnitude relation by SCHOLZ et al. (1973).

Many authors have not attempted to quantitatively measure the degree of relative quiescence. KHATTRI and WYSS (1978) and WYSS et al. (1978) reported seismicity rate decreases to less than 0.6 of normal levels. These periods of quiescence often contain precursory clusters (BRADY, 1976; ISHIDA and KANAMORI, 1977; ENGDAHL and KISSLINGER, 1977; McNALLY, 1978; KEILIS-BOROK et al., 1978; WYSS et al., 1978). These are groups of earthquakes without main shocks which are concentrated in space and time, and which are sometimes located at the hypocenter of the main shock they precede. Clusters appear to be long term precursors which lend themselves to detection by algorithm.

Many seismologists now believe that seismicity pattern studies hold promise for aiding in the earthquake prediction problem. However, we expect that much work will be needed until these patterns are understood so that they can be used for prognosis. A question difficult to answer is: How do we know that a given seismicity fluctuation is connected to the precursory process of a given earthquake? KEILIS-BOROK *et al.* (1978) propose that seismic clusters at distances of several times the source dimension of the main shock (more than 100 km for $M \approx 6.5$ quakes) qualify as precursors, arguing that sensitive spots in the crust, and different ones for each quake, could be activated by changing stress fields during the preparatory process. This hypothesis seems tenuous to others who can only believe a connection exists if the fluctuations (quiescence as well as cluster) are closely tied to the future rupture and hypocentral volume.

A very difficult problem will be the identification of clusters as precursory before the fact. Observations of changing focal mechanisms (LINDH *et al.*, 1978b), wave forms (ISHIDA and KANAMORI, 1978) and P to S amplitude ratios (WYSS *et al.*, 1978) give hope that identification of precursory clusters may be possible without too many false alarms.

C) *Precursory migration patterns* have not been studied in great detail, but it appears that in some cases such patterns can be found on a regional tectonic scale (KELLEHER *et al.*, 1973; DEWEY, 1976; SCHOLZ, 1977; TOKSÖZ *et al.*, 1978; DELSEMME and SMITH, 1979) and in other cases on a local scale (e.g. ENGDAHL and KISSLINGER, 1977). No specific hypothesis has been formulated as to how a precursory migration pattern can be defined as such and how it should be used for prediction.

D) *Foreshocks* appear to hold promise for refining an existing earthquake alert. JONES and MOLNAR (1976) have shown that more than half of the large quakes have foreshocks of sufficient magnitude to be detected with the standard worldwide seismic coverage. One would therefore expect that foreshocks could be detected by local networks, installed in mature gaps, before a high percentage of large earthquakes. The difficulty will be to differentiate foreshocks from long term precursory clusters and background seismicity. As this may not always be possible, a high rate of false alarms may have to be accepted.

4. Crustal Deformations

Crustal deformations may be the most useful phenomenon for specific earthquake prediction, once we understand the preparatory process. This feeling is based on the idea that, whatever the nature of the earthquake preparatory process, it should be some type of mechanical process which would have to be accompanied by crustal movements. The surface expressions of these deformations would be the most direct observations of the preparatory phase, and they may be measured by several methods.

The Palmdale Bulge (Fig. 3) has attracted a great deal of attention in the United States. If the leveling surveys in southern California are accurate, an area of dimensions approximately $400 \text{ km} \times 150 \text{ km}$ was apparently uplifted by about 30 cm (CASTLE, 1978). This uplift episode has several remarkable properties: (1) it was not noticed until more than 10 years after most of the uplift had occurred; (2) most of the uplift accumulated very rapidly, within 1 ± 0.5 years; (3) the uplifted area has spread and contracted, i.e., subsidence has followed uplift in portions of the bulge; (4) the uplift has occurred aseismically; (5) it extends parallel to, but is confined mostly to one side of, the San Andreas fault. Even though we are learning many new facts about the geology of southern California by studying intensively the uplifted area, we do not know

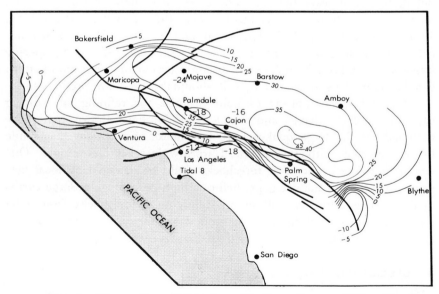

FIG. 3. The southern California uplift (from CASTLE, 1978). The flanks (south and north) are steep, the top is flat. The uplifted area is located off but parallel to the San Andreas fault. Contour lines indicate uplift in cm as labeled by the small numbers. Large numerals with minus signs indicate the amount of recent subsidence in cm.

whether this is a precursor to a future great earthquake in southern California.

Several arguments have been put forth against the precursor hypothesis for explaining the Palmdale Bulge. Based on geodetic data from the beginning of this century, CASTLE (1978) believes that the phenomenon occurs periodically and is followed by subsidence. THATCHER (1976) has proposed aseismic thrusting as the origin, because the bulge is offset from the San Andreas fault and has a steep gradient to the south. RUNDLE (1978) also preferred the thrust over the dilatancy model. A significant observation is that V_p/V_s and P-residuals are normal and constant, respectively, within the resolution of the data, at the locations where these parameters have been measured (e.g. KANAMORI and HADLEY, 1975; RAIKES, 1978).

One can argue that the thrust model is not convincing, because it is in violation of the uplift data south of San Fernando (WYSS, 1977a), and because another thrust dipping to the south would be required to explain the steep gradient found recently at the northern margin of the Palmdale Bulge (see Fig.3). The facts which support the idea of a precursory, perhaps dilatancy, origin of the Palmdale Bulge are that the uplifted area appears to be associated in some way with a segment of the San Andreas fault which has not ruptured since 1857, and that strongly increased microseismicity, which may be of the cluster type, has been observed near Palmdale during the last year (MCNALLY et al., 1978). Also in the laboratory bulges very similar to the Palmdale Bulge develop alongside of large cracks propagating into rock samples before failure (SOBOLEV et al., 1978; SPETZLER et al., 1978).

A large portion of the U.S. budget for earthquake prediction is spent to obtain all kinds of geophysical observations in the Palmdale Bulge area in the hope of finding out more about the past and possible future of this remarkable phenomenon.

Below several methods to detect crustal deformations which have been used in connection with earthquake prediction by U.S. researchers are discussed.

A) *Sea level changes* as measured by tide gauges can resolve uplifts with the amplitude of the Palmdale Bulge if such an event occurs along coastlines. An example of a tide record at Acapulco during an earthquake of $M=6.4$ at a distance of less than 20 km was used to show that clearly the apparent coseismic step recorded by the continuous tide curve is resolved as a permanent level change by long term average sea level (WYSS, 1978).

Sea level variations due to oceanographical and meteorological effects can be reduced even more effectively by taking the difference between neighboring tide stations along the same coast. The standard deviation of differences in sea level are typically about 1 cm if the gauges are located no further than 200 km from each other. If a step function in relative sea level is found at a station located between tide gauges with constant relative sea level, one is justified in assuming that the origin of the sea level change was a land elevation change. Two such cases were found before large earthquakes and one change was apparently not associated with an earthquake (WYSS, 1976a, b). It is interesting to note that the uplift rates in these cases were as rapid as those at Palmdale, i.e., about 6 cm in 6 months (WYSS, 1977b).

Negative findings were that at 7 km and 40 km from the San Francisco (1906) and the Sitka (1972) large strike-slip earthquakes no substantial movements (>5 cm) could have occurred during several years before these events (WYSS, 1975b).

The strengths of mean sea level data are long term stability, low cost (BILHAM, 1977), and relatively good time resolution. The weaknesses are that such observations are confined to coastlines, that the sensitivity is poor (resolution several centimeters), and that only vertical movements can be detected.

B) *Geodesy* affords a much better resolution of vertical as well as horizontal position (resolution in the order of millimeters). Unfortunately, the measurements are costly and in most cases time consuming. For instance, the time necessary to complete a leveling line may be the same as the time in which 10 cm of uplift can occur. Also, because of the large cost, geodetic measurements are not repeated very often and therefore the time resolution is poor.

The spectacular discovery of the Palmdale Bulge was discussed above (CASTLE et al., 1976). Additional existing geodetic data sets are being re-examined now for California and other areas (e.g. REILINGER et al., 1977), and a large releveling effort is underway in southern California (CASTLE, 1978). In addition to leveling, an extensive geodimeter network exists along the San Andreas fault system (e.g. SAVAGE, 1975; SAVAGE et al., 1976; PRESCOTT and SAVAGE, 1976; SAVAGE and PRESCOTT, 1978b). The lengths, L, of the survey lines range from a few km to 30 km or more and are remeasured about once a year. The precision of these measurements is given by SAVAGE et al. (1976) as having a standard deviation of $\sigma=(3 + 2 \cdot 10^{-7} \cdot L^2)^{1/2}$ in millimeters.

No significant pre-, co-, or post-seismic deformations have been observed by this network except for the San Fernando, 1971, earthquake (e.g. SAVAGE et. al, 1976; SAVAGE and PRESCOTT, 1978a). This is not surprising because only earthquakes with $M \leq 5$, that is with source dimensions smaller than the network dimensions, have occurred in the area covered. Recent observations (SAVAGE et al., 1978) suggest that the mode of deformation in southern California is north-south compression, which contrasts with the shear deformation observed in central California.

It is expected that the data from this geodimeter network will answer the question of long term horizontal precursory deformations at the level of detection before the next large San Andreas fault earthquake. Deformation trends of the network stations have been established by high quality measurements spanning almost a decade.

C) *Gravity surveys* are expected to furnish important constraints for determining the mechanism and cause of uplifts. Gravity data predating some of the Palmdale uplift have a resolution of around 20 microgal (e.g. STRANGE and WESSELLS, 1976), which means that the uplift should be resolvable if groundwater effects can be accounted for. If substantial uplifts like the Palmdale Bulge are common, gravity surveys should be able to determine whether a Bouguer or a free air change took place, an observation which may help to discriminate between conventional and dilatancy uplift. An extensive gravity resurvey of the Palmdale Bulge was carried out and is currently being reduced (JACHENS, 1978). Modern instruments allow the resolution of about a microgal, over short baselines, if great care is used (LAMBERT and BEAUMONT, 1977), but systematic errors over long baselines can be substantially larger. However, accuracies are sufficient to detect the gravity effects due to a few centimeters of uplift.

D) *Tilt observations* are carried out on the largest scale of any geophysical observations after seismicity studies (MORTENSEN and JOHNSTON, 1975). Approximately 70 tiltmeters are operated in California and 3 in Alaska by the U.S. Geological Survey, and about 35 more have been given to University researchers throughout the United States. Many investigators have serious reservations about the instrument type and the mode of installation used. The short baseline bubble tiltmeters are installed at a few meters depth in loose sediments. The objections to this setup are that the influence of weather (rain, temperature, pressure, etc.) and other unknown near surface noise on such installations may be large and unpredictable. Long baseline instruments or installations at depth would probably be less sensitive to such effects, but these installations are more expensive. With near surface observations the secular tilt changes due to unknown sources may exceed 1 sec of arc over time periods comparable with the duration of an event like the southern California uplift, which, at its steepest flank, produced a tilt of about 2 arc sec angle. The uplift therefore would hardly have registered above the seasonal noise on shallow tiltmeters. This means, in some people's opinion, that short baseline tiltmeters in shallow installations are not the best instruments for observing secular crustal deformations, unless amplification effects near faults increase the signal but not the noise (BILHAM and BEAVAN, 1979). On the other hand, short term tilt changes may be defined satisfactorily by the California tiltmeter array. However, the lack of correlation of secular tilt measured by several short baseline bubble tiltmeters over periods of several months at the same site suggests that there are unknown noise sources with scales down to tens of meters or that this type of installation is not stable (WYATT and and BERGER, 1978; BILHAM and BEAVAN, 1979).

On the other hand, tiltmeters may be good instruments to record possible changes in the local response to tidal forces if the elastic or anelastic parameters of the crust change during the earthquake preparatory process (BEAUMONT and BERGER, 1974). Such changes in solid earth tides have not been observed, but many reports of secular tilt changes have been published. Because of the noise problems, many researchers are not convinced that specific tilt anomalies can be attributed to tectonic origins. A recent example of a tilt anomaly, which may have been a precursor (it was the largest departure on one component during 3 years of data) or which may have been due to rainfall, was shown by JOHNSTON et al. (1978a). At this point it is necessary to study carefully the problems of secular tilt and how to overcome the noise problem.

E) *Strain observations* over short baselines are affected even more than tilt measurements by secular noise. Several strain meters are operated along the central segment of the San Andreas fault with short term sensitivities around 10^{-8}. Most of them are located between 0.1 and 1.5 km from the fault trace and their data have been used to derive models for understanding fault creep events. No reports on earthquake precursory strain changes in California have been published.

A laser strain meter of 800 m baseline is operated in southern California at sensitivity levels near 10^{-7} for long term secular changes. This unusually stable instrument did not record any precursory change due to a nearby magnitude 4.9 earthquake even though a change in tilt on a USGS instrument was interpreted as a precursor (WYATT and BERGER, 1975). Two explanations for the discrepancy have been offered: Either the tilt change was due to unknown noise, or the fault zone, where the tilt instrument is located, magnified precursory deformations somehow. The detailed data of this case have not been published.

5. Stress Estimates

The most important parameter for issuing earthquake prognoses would be the state of stress in the hypocentral volume, if this information could be obtained. For instance if we could measure the state of stress in the crust at numerous points in a seismic gap we could learn rapidly at what stress level rupture is initiated and warnings could be issued when dangerous stress levels are reached. Unfortunately it is difficult and costly to measure stress directly at hypocentral depths. Most in situ stress measurements are conducted near the earth's surface based on the relaxation strain which is exhibited by rock samples when they are removed from the rock mass in which they were imbedded. The overcoring technique is used in tunnels and bore holes.

A) *In situ stress* work was started in order to determine stress at several hundred meters depth in California by the hydrofracture technique (ZOBACK *et al.*, 1977; ZOBACK, 1978a, b). The direction of maximum compression is approximately N15°E and the magnitude of the minimum and maximum horizontal stresses range from 114 to 225 bars at depths of 240 to 270 meters near the central segment of the San Andreas fault. At Livermore the minimum compressive stress was estimated to be an order of magnitude smaller, about 20 bars.

The great expense of the hydrofracture technique is due to the cost of drilling a deep hole. Unfortunately it may be that shallow holes cannot furnish the information sought (HAIMSON, 1978). The cost of drilling to 5 km depth, the hypocentral depth of major quakes in California, is considered prohibitive by many, because an enormous amount of research can be carried out at the surface for the same price. Besides the staggering cost, hydrofracture has the disadvantage that the state of stress is determined for a very small volume of rock only. It is not known over what size of crustal volumes the stress is approximately constant.   The multiple rupture nature of many large quakes indicates that the stress level, or source strength, varies strongly in crustal volumes of a few kilometer dimensions.   In spite of the drawbacks of the method, it may be that hydrofracture stress measurements in a few deep holes can bring us so much closer to understanding the problems of the earthquake preparatory and rupture process, that the expense may be warranted.

B) *Earthquake stress drops.*   Estimates of the shear stress causing crustal rupture can be obtained from the stress drops of earthquakes, if one assumes that either the stress is relaxed completely or that the stress drop is proportional to the total shear stress.   It is difficult to accept the idea that the total stress is measured directly by the stress drops because many earthquakes are multiple ruptures in which the radiation from crustal volumes with high and low energy density is mixed. Therefore the critical value of the largest stress concentration may easily escape determination.   However, it seems reasonable to accept the idea that regional and temporal differences of the average stress drop of many quakes are probably reflecting such differences in the total regional shear stress.

ARCHAMBEAU (1976, 1978) determined the ratio of the long-period to the short-period spectral amplitude from the ratio of m_b to M_s (body wave to surface wave magnitudes).   Using his earthquake source model he then estimates the stress drop from these spectral characteristics. Assuming that the stress drop is either proportional or equal to the total shear stress, ARCHAMBEAU (1978) has constructed stress contours for the circumpacific area, with special attention to the seismic gaps.   One of his high stress concentrations was an area where later the most recent earthquake off Japan (May 1978, $M=7.5$) occurred (Fig. 4).

The stress drop method certainly holds promise for identifying especially mature gaps.   It may, however, be difficult to observe changing stress conditions as a function of time, because earthquake sources may not be available for the most pertinent source volume and time interval because of the precursory seismicity quiescence.   For instance, WYSS and BRUNE (1971) using a similar method could not estimate the shear stress along most of the San Andreas gap of the 1857 rupture because of lack of seismicity.   Data were only available for the southernmost part of this gap, the San Bernardino–San Gorgonio area.   This region was identified as the one outstanding high stress area along the San Andreas (WYSS and BRUNE, 1971), and it is now also the area where

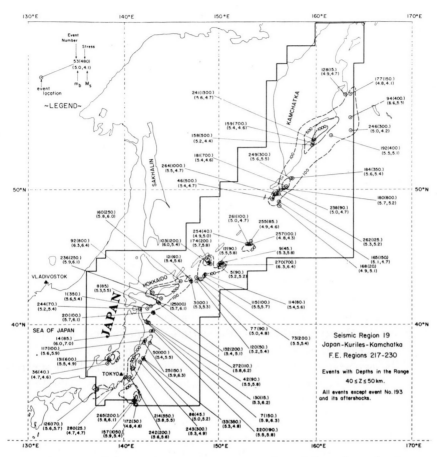

FIG. 4.   Minimum shear stress contours estimated from the ratio of m_b to M_s using the stress relaxation earthquakes source model (from ARCHAMBEAU, 1978).   Each event is identified by an event number; next to it is the stress-drop in bars (see legend, upper left corner).   Note that the earthquake of 12 June 78 with $M_s = 7.5$ was located at 38.4 °N/142.0 °E near the center of a high stress region.

the greatest uplift of the Palmdale Bulge was recorded.

C) *Directional changes of the principal stresses* may be revealed by rotations of fault plane solutions during the preparatory phase.   Some models of the preparatory process require stress rotation (e.g. MJACHKIN *et al.*, 1975; BRADY, 1976) and several observations of fault plane rotations have recently been reported by U.S. investigators (e.g. ENG-

DAHL and KISSLINGER, 1977; LINDH *et al.*, 1978b; ISHIDA and KANAMORI, 1978).

It may be that the precursory seismicity clusters discussed earlier can provide important data on stress rotations and stress level changes during the earthquake preparatory phase.

6. Magnetic Field Measurements

Local magnetic field changes have been reported in connection with earthquakes since the turn of the century. Some of these were coseismic and some preseismic changes. More recently magnetic anomalies caused by underground explosions have also been observed (HASBROUCK and ALLEN, 1972). Anomalies reported prior to the 1960's are suspect because modern instruments were not available at that time. Nevertheless, the hope of documenting tectomagnetic effects has persisted mainly because it is known that in the laboratory remanent magnetization and susceptibility of minerals and rocks change as a function of stress. It is therefore expected that tectonic stress build up and stress release will cause magnetization changes in crustal rocks. The resultant local perturbation of the earth's magnetic field is what one hopes to measure.

The expectation of measuring such field changes has been severely dampened by theoretical calculations which suggest that even for the largest earthquakes, tectonomagnetic changes could likely not exceed ten gammas. This result notwithstanding, anomalies of several gammas, up to tens of gammas, have been documented.

The U.S. effort is a substantial one with 21 observation sites where the magnetic field is recorded once per minute by proton precession magnetometers with a sensitivity of 0.25 gammas. In addition difference measurements of the magnetic field along profiles involving about 150 stations are carried out several times a year. Up to five years of a homogeneous baseline have been accumulated by now (M. J. S. JOHNSTON, 1978a).

Unfortunately almost all of this work is concentrated in California and up to recently has been carried out by one single research team. Plans now exist to expand the magnetometer observations into other tectonic environments and to involve additional researchers.

Up to now three anomalies of apparently tectonic origin have been reported from California. (1) The increase of the magnetic field in excess of 1 gamma at a station in the Garlock fault profile was apparently associated with a magnitude 4.3 earthquake which occurred in June 1975. It cannot be said for certain how much of this anomaly was pre-, co- or post-seismic. Some postseismic changes did take place during several months after the event (JOHNSTON *et al.*, 1975). The epicentral distance was about 3 km, the hypocentral depth was 8 km. (2) The Thanksgiving Day (1974) earthquake of $M = 5.2$ was apparently preceded by a bay-shaped magnetic field anomaly of 1.8 gamma amplitude at a permanently recording instrument at 11 km epicentral distance (Fig. 5). Two stations located only a few kilometers farther from the quake did not record anomalies. The increase in magnetic field began seven weeks before the quake, and the field returned to normal four weeks before Thanksgiving Day (SMITH and JOHNSTON, 1976). (3) Near the southern end of the 1906 San Francisco earthquake rupture M. J. S. JOHNSTON (1978a) reported a 2 gamma decrease of the magnetic field over five years.

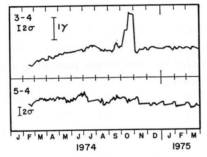

FIG. 5. Total magnetic field differences between neighboring observation points along the San Andreas fault (from SMITH and JOHNSTON, 1976). Top record shows the field difference between stations 3 and 4; the bottom curve is the difference between reference stations 4 and 5. An earthquake of $M_L = 5.2$ occurred on 20 November 1974 at about 11 km distance from station 3.

Improved techniques allow the reduction of geomagnetic field noise to values in the order of 0.02 gamma (WARE and BENDER, 1978; WARE, 1979) if sensitive narrow line rubidium magnetometers are used. This means that the magnetic field could be monitored with an order of magnitude better resolution than presently available in the California network, and very small tectonomagnetic effects could be measured if they exist.

Apparently no medium sized earthquake other than the ones mentioned have occurred within 10 km of the magnetometer sites during the observation period. That means that no " negative " observations (ones in which anomalies were not observed but would have been expected) have been reported. However, several creep events along the central San Andreas fault have not been associated with measurable magnetic field changes even at a few hundred meters distance (M. J. S. JOHNSTON, 1978a).

7. Resistivity and Selfpotential

Because rocks are essentially insulators their resistivity is very sensitive to the presence of pore fluid and to the number of pores and cracks per unit rock volume. Even at constant volume of pore fluid the resistivity can be dropped drastically by increasing the number and dimensions of cracks. It seems that a film of pore fluid on the crack walls is enough to act as a conductor. If these conductors are interconnected by increasing the crack density the resistivity decreases. WANG et al. (1975) observed pronounced resistivity changes in low confining stress stick-slip experiments in the laboratory. Under the same conditions no velocity changes could be measured. Therefore it is hoped that resistivity measurements may be a sensitive technique to detect premonitory changes.

Three research teams started to monitor resistivity in California during 1972, 1973, and 1975 respectively. FITTERMAN and MADDEN (1977) measured resistivity with an electrode spacing of 100 m across the central California San Andreas fault. The penetration depth was 30 to 50 meters. A 1 Hz signal was monitored continuously at a sensitivity of about 0.005%. MAZELLA and MORRISON (1974) measure deep resistivity with electrode spacing of 10 to 15 km across the same section of the San Andreas fault. The measurements were made about twice a month with standard deviation of about 2%. More recently this network was considerably expanded and measurements are made weekly (MORRISON et al., 1978). REDDY et al. (1976) used the magnetotelluric method to measure resistivity at three sites in southern California with a sensitivity of approximately 10%.

An apparent resistivity anomaly was reported by MAZELLA and MORRISON (1974). They found a 15% resistivity decrease during about a month before a magnitude 4.2 shock, with resistivity returning to normal and somewhat larger than normal values weeks before the event.

Unfortunately the anomaly occurred at the very beginning of the experiment. Because of this fact the authors cautioned that one should reserve judgment on this result and on the method until more anomalies are recorded. MORRISON et al. (1978) found subsequently that an $M_L=4.0$ earthquake in December 1977 was not preceded by any resistivity changes exceeding 2%. The December 1977 event occurred at the center of the extended array and continuous measurements during the time of the quake also showed that no short term resistivity change and no self-potential change occurred before this quake.

Another negative result was reported by FITTERMAN and MADDEN (1977) who found that fault creep events were not associated with resistivity changes at the 0.01% level.

CORWIN and MORRISON (1977) reported selfpotential changes before two San Andreas earthquakes. Ninety mV change across a 640 m dipole was observed 55 days before a magnitude 5 earthquake which occurred at a distance of 37 km, and a 4 mV anomaly started 110 hours before a magnitude 2.4 event at 2.5 km distance, and it was observed across a 300 m dipole.

It seems that judgment on the usefulness of the resistivity method should be reserved until an earthquake occurs within the San Andreas network which involves a substantial crustal volume (i.e. source dimensions of 10 km), and until more resistivity observations in tectonic environments other than the San Andreas strike-slip type become available.

8. Ground Water Level, Gas Pressure in Wells and Gas Emanation from Soil

A) *Ground water levels* have not been studied extensively in the U.S. It may be that an opportunity to obtain cheaply some interesting precursory information is overlooked. The data presented by KOVACH et al. (1975) are from a 0.15 km deep well in the central San Andreas fault zone. Observations started in 1971. The authors propose that decreases of a few centimeters, which are superimposed on seasonal fluctuations of 30 ± 10 cm, may be related to precursory processes of local earthquakes of magnitudes 2.5 to 5.0. LAMAR and MERIFIELD (1978) started a water level measuring program in October 1976 which includes 40 wells in the Palmdale area.

B) *The well head pressure* of gas producing wells increased about

10 days before an $M_s=6.75$ quake at approximately 20 km hypocentral distance (Wu, 1975). The wells are located in Taiwan and reach depths of around 1.5 km. Pressure increases of 15 to 60% were observed on three wells, and at no other time did " such rapid changes appear " (Wu, 1975). The data presented in Wu's paper appear to be strong, however it is unfortunate that the entire record of well pressure is not shown. I feel that all the data available should always be shown, because it is important to demonstrate to all critical readers that no similar changes occurred at any other time.

C) *Gas emanations* are studied on a moderate effort level in the U.S. The radon concentration was monitored since mid-1975 in 20 shallow dry holes along a 60 km segment of the central California San Andreas fault (KING, 1978a). Fluctuations with a period of one year are coherent over a large area. At some observation points this periodicity is not present, therefore KING (1978a) argues that the fluctuations may be precursory to two $M_L=4.3$ and 4.0 quakes rather than due to seasonal changes. With more data this problem may be resolved in the future. More recently several additional radon sampling networks have been established by a number of investigators (see KING, 1978b). The total number of stations at which radon is monitored is about 130 in California. In addition, radon fluctuation results were reported in connection with the Lake Jocasse earthquakes (TALWANI, 1978).

Several studies of other stable isotopes, notably helium, have also been started, but few results have been published to date (O'NEIL, 1977).

9. Animal Behavior before Earthquakes

It is an old and intriguing question whether or not animals can somehow sense the approach of an earthquake. The mythology and history of many cultures are laced with the belief that animals can have super powers with which they can control, or at least predict, the fate of individuals and nations. This illustrates to me that humans have a tendency, or even a need, to believe in these mysterious capabilities of animals. This may be the reason why some people without information will firmly believe in earthquake prediction by animals, and that some people will, after the fact, claim that their pet had acted abnormally before an earthquake.

An interesting story was told by a high official of a U.S. city which was affected by a strong earthquake: The night before the shock two police patrols independently radioed from their cars to the dispatcher that large numbers of rats scurried across the streets. This was considered a most unusual event, since this city was not supposed to have rats. I followed up on the case, because here was a report that should easily be documented by interviews with the police officers and dispatcher as well as by examining possible records of radio conversations. It turned out that no police could be found in the city in question who had any knowledge of the described incident or who had even heard of it. A newspaper reporter who also followed up on the story had the same results. I conclude from this incident that reports on strange animal behavior before earthquakes cannot be accepted as correct even if they appear to be extremely well-documented, unless the source is a trained observer who has conducted a controlled experiment.

After a recent $M_s=7.3$ earthquakes in Turkey, Toksöz et al. (1977) added to other field studies the questioning of residents about abnormal animal behavior. They found that physical precursors were observed by people but that no definite widespread unusual animal behavior was noted. Lott et al. (1978) on the contrary found that before an $M_L=4.9$ California earthquake unusual animal behavior was observed. While these studies were conducted carefully they suffered from the fact that they were carried out after the quake, which might have changed the importance which the observer attached to the deviatoric behavior.

A few controlled animal observation experiments are now underway. Continuing systematic ratings provide a long term baseline to establish the norm of behavior against which deviations will be measured in the future. Skiles and Lindberg (1978) reported that rodents at an observation station in the Mohave desert may have shown deviatoric behavior before an $M=3.5$ quake at 22 km distance. A network of volunteer observers was set up by Kautz (personal communication) in central California to report once a week, and in the event of unusual actions, on the behavior of domestic animals. No after the fact reports are collected in this study. A promising study which showed apparently deviatoric behavior of chimpansees the day before local quakes in central California (Kraemer et al., 1976) was not continued.

Clearly the first step to solve this problem will have to be to scientifically establish whether or not animals do act peculiarly before some earthquakes. If this is the case, it may be necessary to conduct laboratory experiments, controlling the environment and duplicating some precursory environmental changes to find out what it is the animals are sensing.   It may be that presently unknown precursory effects could be discovered in the process.

10. Discussion and Conclusions

In 1973 the feeling of some U.S. researchers was that in a few years substantial progress would be made toward understanding the earthquake preparatory process.   Now it appears that we will need a decade or more of careful data collecting, laboratory studies and theoretical work in order to document precursors and understand the physics of the preparatory process.

The most outstanding discovery during the last few years is probably the great uplift episode in southern California.   However, many researchers are of the opinion that this phenomenon is not an earthquake precursor, mostly because it is not centered on the San Andreas fault.   The Palmdale Bulge exemplifies the fact that large tectonic movements can occur rapidly and can go unnoticed even if geodetic surveys are made.   We clearly need better links between geodetic and tectonic studies.

The field of geodesy has probably one of the greatest potentials for development in the study of earthquake prediction.   Not only could we use existing data more extensively, but new and powerful techniques are being developed (NATIONAL RESEARCH COUNCIL, 1978).   The promises of the space techniques (like the satellite and very long baseline interferometry) with which one will be able to measure thousand kilometer baselines with a few centimeter accuracy have still not been fulfilled, but steady progress is being made (e.g. ONG et al., 1977).   Several new and promising methods are in the development and early implementation stage.   For instance, it has been proposed that satellites could carry a laser which could be reflected from an array of ground based corner cubes covering an area of interest (KUMAR, 1976).   This is essentially an inverted laser ranging technique.   It is hoped that by this method the relative position of points with a spacing of tens of kilometers could be obtained rapidly and frequently to an accuracy of a few centimeters.   An alternate approach which appears capable of achieving centimeter accuracy is the use of signals of the Global Positioning Systems satellites (MUELLER, 1978).   Further along in their development are the three wavelength generation of distance measuring instruments. The equipment described by HUGGETT and SLATER (1975) measures 10 km lines with accuracies of 10^{-7}, that of MOODY and LEVINE (1978) is designed for $5 \cdot 10^{-8}$ accuracy over lines of 50 km lengths, respectively. One of the reasons why these techniques may be very helpful in precursory deformation studies is that for large scale preparatory movements reference points in the relatively stable surrounding areas can be reached. An additional merit is the possibility of making continuous measurements in an observatory type mode of operation with these instruments. The superconduction gravimeter (WARBURTON and GOODKIND, 1978) also appears suitable for the detection of rapid crustal uplift by observatory observations, since its drift rate is low. Continuous geodetic monitoring may be important in order to detect short term precursory crustal movements if they exist.

Another important discovery during the last few years was that velocity anomalies are not widespread in the tectonic environment of California. The skepticism against velocity change data has grown so strong that only extremely well documented cases have a chance of being accepted as real.

Studies of rock bursts and in situ rock property changes as a function of stress changes in underground mines could further the understanding of pre-failure processes. Laboratory experiments are mostly confined to small rock samples which do not contain major cracks. However, most of the current theories modeling the earthquake preparatory process assume that cracks play a major role in this process. In some U.S. mines rock bursts reach magnitudes of approximately two. The scale (time and dimensions) is intermediate between laboratory experiments and the earthquakes one wishes to predict. Since these microearthquakes are triggered by the excavation process, one has an idea when and where future rock bursts will happen. The study of rock bursts could therefore be called a semi-controlled experiment. It is clear that underground mines with their complex systems of tunnels and shafts could afford an ideal laboratory to study the changes of all sorts of precursor parameters as a function of changes of the known stress tensor. (Because of the relatively small dimensions, the stresses could be obtained by overcoring and hydrofracturing at relatively low expense.) Except for some work by the U.S. Bureau of Mines, no such studies are presently underway or planned. It is surprising that this opportunity to study precursors and earthquake source processes has been overlooked so long, and continues to be neglected.

Knowledge of the distribution of stress in the crust at several kilometers depth would furnish important insight into the problems of earthquake source mechanisms and prediction if it can be acquired. Even though it is not impossible to obtain such data, it is not very likely that adequate measurements will be made, because they are too expensive.

The data from the electric and magnetic studies appear to be promising. But many methods appear promising as long as there are few data available, as was again demonstrated by the recent negative results of MORRISON et al. (1979). We simply have to wait until large amounts of high quality data are collected before assessing the potential of various methods which are only beginning to be explored, such as ground water level, gas emanations and other measurements.

What then is our outlook and what are our hopes of solving the problem of earthquake prediction? The problem is a difficult one, even without considering the sociological aspects. Main shocks of $M \leq 5$ occur relatively frequently (several a year in California) to give repeated opportunities to measure possible precursors. However, the dimension of these shocks (usually less than 10 km) and the dimensions of the precursor volume (unknown but probably about equal to or less than source dimensions in California) is smaller than most of the instrument networks deployed to measure precursors. Also the precursor times are short (less than three months). For these reasons it is very difficult to document a precursor well if it occurs; that is, one cannot expect to observe large numbers of anomalous data points on several observation stations surrounded by many close by reference stations which remain normal. Therefore the prospects of making significant advances by studying earthquakes with $M \leq 5$ are poor.

One can argue that it is the large and great earthquakes which matter most and therefore one should concentrate on learning how to predict these. Since the precursor times in these cases may be decades, it may take a half to a full century until we have well documented precursory anomalies for great quakes. Considering the life expectancy of the researchers and the funding programs we cannot wait that long to obtain results. Nevertheless we must lay the foundations by starting systematic observations which can be carried on by others over many decades. Only by planning and implementing long term experiments now, will we have data in the future to understand precursors to great earthquakes.   Individual researchers, funding agencies and the U.S. Congress are impatient and wish to get quick results, however it is important to realize that solving the earthquake prediction problem may take a long time and will require careful long range planning.

It seems that perhaps the needed results may be obtained reasonably soon (i.c. in a decade) from earthquakes in the range of $M = 6.5 \pm 0.5$. Anomalies before such quakes might have large enough dimensions and durations that they can be documented well.   A severe handicap for detailed near source observations is the ocean cover of most of the world's highly active areas, that is the subduction zones.   Since we are constrained to islands and continents with dense observation networks, it may take a long time before we catch several magnitude 6.5 earthquakes within our networks, and therefore it may be more than a decade before we get a reasonably good idea of what changes occur during the precursory process.

Predictions of earthquakes useful for the public can only be issued once we are able to measure synchronously changes of several crustal parameters, of which the patterns of each have previously been documented repeatedly.   A prediction only deserves this name if location, size and occurrence time are tightly specified, with uncertainties given for all three quantities.   Since precursor times of major earthquakes last for years (e.g. SCHOLZ et al., 1973), one may expect the uncertainty for occurrence time prediction to be in the order of a year, unless short term precursors can be used to refine an existing prediction.   Clearly it would be of great importance if short term precursors could be documented.   Not much progress has been made in finding short term changes.   Fresh en-echelon cracks were noticed in Cholame Valley along the surface trace of the San Andreas fault before the Parkfield earthquake of 1966 (ALLEN and SMITH, 1966).   This observation suggests that accelerated crustal movements must have preceded the earthquake by a few weeks.   In the case of this moderate size quake ($M_s = 6.4$) no foreshocks were observed by a mobile seismograph station eight days before the event, but a large ($M_L = 5.1$) foreshock occurred 17 minutes before it.   This foreshock did not vary significantly in its spectral radiation content from an aftershock of about the same magnitude (FILSON and McEVILLY, 1967).   This fact is discouraging, because one would hope to find some means of identifying the evidently often recorded foreshocks (JONES and MOLNAR, 1976).   The only other recently reported short term precursor was that of increased spring discharge before the November 1976 Turkey earthquake (TOKSÖZ *et al.*, 1977).

There is a great deal of gradation in working towards firming up a " prediction " and refining it, and therefore there is a need for clearly defined language. Now the word " prediction " is loosely used by many. For instance, some consider the definition of a seismic gap a " prediction ", whereas others would not like to apply the term unless the knowledge has been advanced and refined to the point where an evacuation order could be justified, that is, the occurrence time error margin is reduced to a week or so. Because of these gradations to which a " prediction " is specific, it would be useful if seismologists could agree on using a scale system for measuring the seriousness of an earthquake warning. For the sake of discussion, a scale system is proposed in Table 2. Such a classification may be useful for informing the public and elected officials responsible for precautionary measures. I am proposing to use one expression for all stages with a number measuring the seriousness, " earthquake alert stage 1 ", rather than different words for each stage because the public does not easily remember whether

TABLE 2. Gradation of predictions.

Term used	Conditions and observations necessary
Earthquake alert stage 1	An approximately defined area is estimated to be more likely than surrounding seismic areas to experience a future earthquake (e.g., seismic gap or at least one geophysical, geological or geodetic anomalous observation).
Earthquake alert stage 2	One or several crustal parameters show the biginning of a long to medium term pattern of change known to have occurred before some other earthquakes. At least one of the prediction elements (location, size, time) is still poorly defined (e.g. occurrence time uncertainty approximately equal to 50% of precursor time).
Earthquake alert stage 3	Changes of crustal parameters are observed which can be interpreted as indicating that the end of the long term preparatory process is near (e.g. return of anomalies to normal). The three prediction elements are fairly well defined (e.g. occurrence time uncertainty less than about 20% of precursor time).
Earthquake alert stage 4	In addition to conditions of stage 3, an anomaly is measured which can be interpreted as a short term precursor. Occurrence time uncertainty may range from hours to weeks.

alert was defined as more or less serious than warning, etc.

It can be a very sensitive matter to issue an earthquake prediction. It can happen that a " prediction " is not firmly supported by data, and should be called " earthquake alert, stage 1 " according to the proposed language, but the local public gains the impression that a short term earthquake prediction (stage 4, Table 2) has been issued. The risk of disasters by mistake (GARZA and LOMNITZ, 1979) could perhaps be avoided by adopting some language code such as suggested in Table 2, because the public in general will understand intuitively that there is no imminent danger if a statement is made that " an earthquake alert of stage 1 on a scale of 4 (with 4 the most dangerous) is in effect."

It is clear that the individual seismologist is faced with a difficult problem and a serious responsibility when he succeeds in obtaining information with regard to future earthquakes. It is also clear that it is no solution to bury this information in the drawers of our desks. Criticism of peers is one of the fundamentals of productive research into the unknown of our environment; censorship, however, is totally unacceptable to most researchers. Since it is important to avoid misunderstandings and wide publicity of low quality " predictions ", we should develop a set of terms and a system of peer evaluation (e.g. KISSLINGER, 1978) which would make it clear to the public what the merits and urgency of a given "prediction " are.

Even though not many spectacular discoveries were made during the last five years, we have made steady and significant progress toward understanding of earthquake source and preparatory processes. Instrumental capabilities are developing rapidly. We have made progress in designing field experiments and in dealing with noise problems. However, the main factors which will never allow us to make very rapid progress cannot be set aside: Earth processes are slow, therefore years of observation are needed; and furthermore we cannot fully design our experiments. Earthquakes will happen when and where they please rather than when and where we are ready for them to happen. Nevertheless, I believe that most seismologists are confident that we will make very substantial progress in understanding the earthquake preparatory process in the near future. Specific predictions of major earthquakes to within 10 days or so on a regular basis might not be possible for a long time to come, however we hope that predictions of major quakes to within one year or so may be possible in the not too distant future.

Acknowledgements

I thank P. Ward, C. Kisslinger, D. Hill, and J. Aggarwal for reading the manuscript and criticizing it.   I am also grateful for permission by the respective authors to use their figures and illustrations.   This work was supported by the National Science Foundation by grant number EAR 78–03633.

REFERENCES

ADVISORY GROUP ON EARTHQUAKE PREDICTION AND HAZARD MITIGATION, Earthquake prediction and hazard mitigation options for USGS and NSF programs, National Science Foundation and U.S. Geological Survey, 1976.

AGGARWAL, Y. P., Dilatancy mechanisms of earthquake precursors: A critical test, Ph. D. Thesis, Columbia University, 1975.

AGGARWAL, Y. P., L. R. SYKES, D. W. SIMPSON, and P. G. RICHARDS, Spatial and temporal variations in t_s/t_p and in P wave residuals at Blue Mountain Lake, New York: Application to earthquake prediction, *J. Geophys. Res.*, **80**, 718, 1975.

AKI, K., Origin of the gap: What initiates and stops a rupture propagation?, in U.S. Geological Survey, open file report, 78-943, 3, 1978.

ALLEN, C. R., Responsibilities in earthquake prediction, *Bull. Seismol. Soc. Am.*, **66**, 2069, 1976.

ALLEN, C. R. and S. W. SMITH, Parkfield earthquakes of June 27–29, 1966, Monterey and San Luis Obispo Counties, California—Preliminary report, *Bull. Seismol. Soc. Am.*, **56**, 966, 1966.

ANDERSON, D. L. and J. H. WHITCOMB, Time dependent seismology, *J. Geophys. Res.*, **80**, 1497, 1975.

ARCHAMBEAU, C. B., Earthquake prediction based on tectonic stress determination, *EOS, Trans. Am. Geophys. Union*, **57**, 290, 1976.

ARCHAMBEAU, C. B., Estimation of non-hydrostatic stress in the earth by seismic methods: Lithospheric stress levels along Pacific and Nazca plate subduction zones, in U.S. Geological Survey, open file report, 78-943, 47, 1978.

BEAUMONT, C. and J. BERGER, Earthquake prediction: Modification of the Earth tidal tilts and strains by dilatancy, *Geophys. J. R. Astr. Soc.*, **39**, 111, 1974.

BILHAM, R. G., A sea-level recorder for tectonic studies, *Geophys. J. R. Astr. Soc.*, **48**, 307, 1977.

BILHAM, R. G. and R. J. BEAVAN, Strains and tilts on crustal blocks, *Tectonophysics*, **52**, 121, 1979.

BILHAM, R. G. and R. J. BEAVAN, Tilt measurements on a small tropical island, in *Measurements of Ground Strain Phenomena Related to Earthquake Prediction, Proc. Conf. VII*, ed. J. Evernden, U. S. Geological Survey, open file report, 79-370, 1979.

BISCHKE, R. E., Secular horizontal displacements: A method for predicting great thrust earthquakes and for assessing earthquake risk, *J. Geophys. Res.*, **81**, 2511, 1976.

BOLT, B. A., Constancy of P-travel times from Nevada explosions to Oroville Dam station 1970–1976, *Bull. Seismol. Soc. Am.*, **67**, 6, 1977.

BOLT, B. A. and C. Y. WANG, The present status of earthquake prediction, *CRC Crital Reviews in Solid State Science*, 125, 1975.

BOOKER, J. R., Dilatancy and crustal uplift, *Pure Appl. Geophys.*, **113**, 119, 1975.

BOORE, D. M., T. V. McEVILLY, and A. LINDH, Quarry blast sources and earthquake prediction: The Parkfield, California, earthquake of June 28, 1966, *Pure Appl. Geophys.*, **113**, 293, 1975a.

BOORE, D. M., G. ALLAN, A. G. LINDH, T. V. McEVILLY, and W. M. TOL-MACHOFF, A search for travel-time changes associated with the Parkfield, California earthquake of 1966, *Bull. Seismol. Soc. Am.*, **65**, 1407, 1975b.

BRADY, B. T., Theory of earthquakes: IV. General implications of earthquake prediction, *Pure Appl. Geophys.*, **114**, 1031, 1976.

BRADY, B. T., Anomalous seismicity prior to rock bursts: Implications for earthquake prediction: *Pure Appl. Geophys.*, **115**, 357, 1977.

BRIGGS, P., F. PRESS, and Sh. A. GUBERMAN, Pattern recognition applied to earthquake epicenters in California and Nevada, *Geol. Soc. Am. Bull.*, **88**, 161, 1977.

BROWN, L. D., Prediction of large scale deformations in the solid earth, in *Geophysical Predictions, Studies in Geophysics*, p. 19, National Acad. Sciences, Washington, D. C., 1978.

BUFE, C. G., P.W. HARSH, and R. O. BURFORD, Steady-state seismic slip—a precise recurrence model, *Geophys. Res. Lett.*, **4**, 91, 1977.

CARR, M. J., Volcanic activity and great earthquakes at convergent plate margins, *Science*, **197**, 655, 1977.

CASTLE, R. O., J. P. CHRUCH, and M. R. ELLIOTT, Vertical crustal movements preceding and accompanying the San Fernando earthquake of February 9, 1971: a summary, *Tectonophysics*, **29**, 127, 1975.

CASTLE, R. O., J. P. CHURCH, and M. R. ELLIOTT, Aseismic uplift in southern California, *Science*, **192**, 251, 1976.

CASTLE, R. O., J. P. CHURCH, M. R. ELLIOTT, and J. C. SAVAGE, Preseismic and coseismic elevation changes in the epicentral region of the Point Magu earthquake of February 21, 1973, *Bull. Seismol. Soc. Am.*, **67**, 219, 1977.

CASTLE, R. O., Leveling surveys and the southern California uplift, *Earthq. Info. Bull.*, **10**, 88, 1978.

CHATTERJEE, A. K., A. K. MAL, and L. KNOPOFF, Elastic module of two-component systems, *J. Geophys. Res.*, **83**, 1785, 1978.

CHERRY, J. T., R. N. SCHOCK, and J. SWEET, A theoretical model of the dilatant behaviour of a brittle rock, *Pure Appl. Geophys.*, **113**, 183, 1975.

CHOU, C. W. and R. S. CROSSON, Search for time-dependent seismic travel times from mining explosions near Centralia, Washington, *Geophys. Res. Lett.*, **5**, 97, 1978.

CLYMER, R. and T. V. MCEVILLY, Measured changes in crustal P-wave travel times, *Earthq. Notes*, **49**, 60, 1978.

COHEN, S. C., Computer simulation of earthquakes, *J. Geophys. Res.*, **26**, 3781, 1977.

CORWIN, R. F. and H. F. MORRISON, Self-potential variations preceding earthquakes in central California, *Geophys. Res. Lett.*, **4**, 171, 1977.

CRAMER, C. H., Teleseismic residuals prior to the November 28, 1974, Thanksgiving Day earthquake near Hollister, California, *Bull. Seismol. Soc. Am.*, **66**, 1233, 1976.

CRAMER, C. H., and R. L. KOVACH, Time variations in teleseismic residuals prior to the magnitude 5.1 Bear Valley earthquake of February 24, 1972, *Pure Appl. Geophys.*, **113**, 281, 1975.

CRAMER, C. H., C. G. BUFE, and P. W. MORRISON, P-wave travel time variations before the August 1, 1975, Oroville, California earthquake, *Bull. Seismol. Soc. Am.*, **67**, 9, 1977.

CROSSON, R. S., Crustal structure modeling of earthquake data: A simultaneous least squares estimation of hypocenter and velocity parameters, *J. Geophys. Res.*, **81**, 3036, 1976.

DELSEMME, J. and A. T. SMITH, Analysis of seismic migration, *Pure Appl. Geophys.*, **117**, 1271, 1979.

DEWEY, J. W., Seismicity of Northern Anatolia, *Bull. Seismol. Soc. Am.*, **66**, 843, 1976.

DEWEY, J. W., and W. SPENCE, Seismic gaps in coastal Peru, *Pure Appl. Geophys.*, **117**, 1148, 1979.

DIETERICH, J. H., Preseismic fault slip and earthquake prediction, *J. Geophys. Res.*, **83**, 3940, 1978.

ENGDAHL, E. R., Effects of plate structure and dilatancy on relative teleseismic P-wave residuals, *Geophys. Res. Lett.*, **10**, 420, 1975.

ENGDAHL, E. R., Seismicity patterns precursory to larger earthquakes in the Adak region, in U.S. Geological Survey, open file report, 78-943, 163, 1978.

ENGDAHL, E. R., J. G. SINDORF, and R. A. EPPLEY, Interpretation of relative teleseismic P-wave residuals, *J. Geophys. Res.*, **36**, 5671, 1977.

ENGDAHL, E. R. and C. KISSLINGER, Seismological precursors to a magnitude 5 earthquake in the central Aleutian Islands, *J. Phys. Earth*, **25**, S23, 1977.

EVERNDEN, J. F., editor, *Proceedings of Conference I: Abnormal animal behavior prior to earthquakes*, U.S. Geological Survey, Office of Earthquake Studies, Menlo Park, California, 1976.

EVERNDEN, J. F., editor, *Proceedings of Conference II: Experimental studies of rock friction with application to earthquake prediction*, U.S. Geological Survey, Office of Earthquake Studies, Menlo Park, California, 1977.

EVERNDEN, J. F., editor, *Proceedings of Conference III: Fault mechanics and its relation to earthquake prediction*, U.S. Geological Survey, open file report, 78-380, 1978a.

EVERDEN, J. F., editor, *Proceedings of Conference VI: Defining seismic gaps and soon-to-break gaps*, U.S. Geological Survey, open file report, 78-943, 1978b.

EVERNDEN, J. F., editor, *Proceedings of Conference VII, Measurements of Ground Strain Phenomena Related to Earthquake Prediction*, U.S. Geological Survey, open file report, 79-370, 1979.

FILSON, J. and T. V. MCEVILLY, Love wave spectra and the mechanism of the 1966 Parkfield sequence, *Bull, Seismol. Soc. Am.*, **57**, 1245, 1967.

FITCH, T. J. and M. W. RYNN, Inversion of V_p/V_s in shallow source regions, *Geophys. J. R. Astr. Soc.*, **44**, 253, 1976.

FITTERMAN, D. V. and T. R. MADDEN, Resistivity observations during creep events at Melendy Ranch, California, *J. Geophys. Res.*, **82**, 5401, 1977.

FLEISCHER, R. L., and A. MOGRO-CAMPERO, Mapping of integrated Radon emanation for detection of long-distance migration of gases within the Earth: Techniques and principles, *J. Geophys. Res.*, **83**, 3539, 1978.

FLETCHER, J. B. and L. R. SYKES, Earthquakes related to hydraulic mining and natural seismic activity in western New York state, *J. Geophys. Res.*, **82**, 3767, 1977.

GARG, S. K., D. H. BROWNELL, and J. W. PRITCHETT, Dilatancy induced fluid migration and the velocity anomaly, *J. Geophys. Res.*, **82**, 855, 1977.

GARZA, T. and C. LOMNITZ, The Oaxaca gap: a case history, *Pure Appl. Geophys.*, **117**, 1187, 1979.

GELFAND, I. M., Sh. A. GUBERMAN, V. I. KEILIS-BOROK, L. KNOPOFF, F. PRESS, E. Y. RANZMAN, I. M. ROTWAIN, and A. M. SADOVSKY, Pattern recognition applied to earthquake epicenters in California, *Phys. Earth Planet Inter.*, **11**, 227, 1976.

GRIGGS, D. T., D. D. JACKSON, L. KNOPOFF, and R. L. SHREVE, Earthquake prediction: modeling the anomalous V_p/V_s source region, *Science*, **187**, 537, 1975.

GUPTA, I. N., Premonitory seismic-wave phenomena before earthquakes near Fairview Peak, Nevada, *Bull. Seismol. Soc. Am.*, **65**, 425, 1975a.

GUPTA, I. N., Premonitory changes in t_s/t_p due to anisotropy and migration of hypocenters, *Bull. Seismol. Soc. Am.*, **65**, 1487, 1975b.

HAIMSON, B. C., Near surface and deeper hydrofracturing stress measurements in the Waterloo Quartzite (abstract), *EOS, Trans. Am. Geophys. Union*, **59**, 327, 1978.

HAMILTON, R. M., Earthquake hazards reduction program—fiscal year 1978, studies supported by the U.S. Geological Survey, *Geol. Survey Circular*, **780**, 1–36, 1978.

HARRISON, J. C. and K. HERBST, Thermoelastic strains and tilts revisited, *Geophys. Res. Lett.*, **4**, 535, 1977.

HART, R. S. and H. KANAMORI, Search for compression before a deep earthquake, *Nature*, **253**, 333, 1975.

HASBROUCK, W. P. and J. H. ALLEN, Quasi-static magnetic field changes associated with the Cannikin nuclear explosion, *Bull. Seismol. Soc. Am.*, **62**, 1479, 1972.

HAYS, W. W., editor, *Proceedings of Conference V, Communicating Earthquake Hazard Reduction Information*, U.S. Geological Survey, open file report, 78-933, 1978.

HEALY, J. H., Recent highlights and future trends in research on earthquake prediction and control, *Reviews Geophy. Space Phys.*, **13**, 316, 1975.

HUGGETT, G. R. and L. E. SLATER, Precision electromagnetic distance measuring instruments for determining secular strain and fault movement, *Tectonophysics*, **29**, 19, 1975.

HUNTER, R. N., Earthquake prediction fact and falacy, *Earthq. Info. Bull.*, **8**, 24, 1976.

HUNTER, R. N. and J. S. DERR, Prediction monitoring and evaluation program: program report, *Earthq. Info. Bull.*, **10**, 93, 1978.

ISHIDA, M. and H. KANAMORI, The spatio-temporal variation of seismicity before the 1971 San Fernando earthquake, California, *Geophys. Res. Lett.*, **4**, 345, 1977.

ISHIDA, M. and H. KANAMORI, The foreshock activity of the 1971 San Fernando earthquake, California, *Bull. Seismol. Soc. Am.*, **68**, 1265, 1978.

JACHENS, R. C., Gravity measurements in studies of crustal deformation, *Earthq. Info. Bull.*, **10**, 120, 1978.

JOHNSTON, A. C., Localized compressional velocity decrease precursory to the Kalapana, Hawaii earthquake, *Science*, **199**, 882, 1978a.

JOHNSTON, A. C., An investigation of temporal P-velocity variation using travel time residuals at Hawaii and selected Circum-Pacific sites, Ph. D. thesis, 1979.

JOHNSTON, M. J. S., Tectonomagnetic effects, *Earthq. Info. Bull.*, **10**, 82, 1978a.

JOHNSTON, M. J. S., Local magnetic field variations and stress changes near a slip discontinuity on the San Andreas fault, *J. Geomag. Geoelectr.*, 1978b (in press).

JOHNSTON, M. J. S. and F. J. WILLIAMS, Tectonomagnetic anomaly during the southern California downwarp, to be submitted to *J. Geophys. Res.*, 1979.

JOHNSTON, M. J. S., A. C. JONES, and W. DAUL, Continuous strain measurements during and preceeding episodic creep on the San Andreas fault, *J. Geophys. Res.*, **82**, 5683, 1977.

JOHNSTON, M. J. S., B. E. SMITH, and R. MUELLER, Tectonomagnetic experiments and observations in western U.S.A., *J. Geomag. Geoelectr.*, **28**, 85, 1976.

JOHNSTON, M. J. S., A. C. JONES, W. DAUL, and C. E. MORTENSEN, Tilt near an earthquake ($M_L = 4.3$), Briones Hills, California, *Bull. Seismol. Soc. Am.*, **68**, 169, 1978.

JOHNSTON, M. J. S., G. D. MYREN, N. W. O'HARA, and J. H. RODGERS, A possible seismomagnetic observation on the Garlock fault, California, *Bull. Seismol. Soc. Am.*, **65**, 1129, 1975.

JOHNSTON, M. J. S., B. E. SMITH, J. R. JOHNSTON, and F. J. WILLIAMS, A search for tectonomagnetic effects in California and western Nevada, in Proceedings of the Conference on Tectonic Problems of the San Andreas Fault System, ed. R. L. Kovach and A. Nur, 225 pp., Stanford Univ., 1973.

JONES, L. and P. MOLNAR, Frequency of foreshocks, *Nature*, **262**, 677, 1976.

KAGAN, J. J. and L. KNOPOFF, Do epicenters migrate on the San Andreas fault?, *Nature*, **257**, 160, 1975.

KAGAN, J. J. and L. KNOPOFF, Earthquake risk prediction as a stochastic process, *Phys. Earth Planet Inter.*, **14**, 97, 1977.

KAGAN, J. J. and L. KNOPOFF, Statistical study of the occurrence of shallow earthquakes, in U.S. Geological Survey, open file report, 78-943, 267, 1978.

KANAMORI, H., Nature of seismic gaps and foreshocks, in U.S. Geological Survey, open file report, 78-943, 319, 1978.

KANAMORI, H. and G. FUIS, Variation of P-wave velocity before and after the Galway Lake earthquake ($M_L = 5.2$) and the Goat Mountain earthquakes ($M_L = 4.7$), 1975, in the Mojave desert, California, *Bull. Seismol. Soc. Am.*, **66**, 2017, 1976.

KANAMORI, H. and D. HADLEY, Crustal structure and temporal velocity changes in southern California, *Pure Appl. Geophys.*, **113**, 257, 1975.

KEILIS-BOROK, V., L. KNOPOFF, I. ROTWAIN, and T. SIDORENKO, Burst of aftershocks as long-term precursor of strong earthquakes, in U.S. Geological Survey, open file report, 78-943, 351, 1978.

KELLEHER, J. A. and J. SAVINO, Distribution of seismicity before large strike-slip and thrust-type earthquakes, *J. Geophys. Res.*, **80**, 260, 1975.

KELLEHER, J. A., L. SYKES, and J. OLIVER, Possible criteria for predicting earthquake locations and their application to major plate boundaries of the Pacific and the Caribbean, *J. Geophys. Res.*, **78**, 2547, 1973.

KENNETT, B. L. N. and R. S. SIMONS, An implosive precursor to the Colombia earthquake 1970 July 31, *Geophys. J. R. Astr. Soc.*, **44**, 471, 1976.

KHATTRI, K. and M. WYSS, Precursory variation of seismicity rate in the Assam area, India, *Geology*, **6**, 685, 1978.

KING, C. Y., Radon emanation on San Andreas fault, *Nature*, **271**, 516, 1978a.

KING, C. Y., Radon emanation on the San Andreas fault, *Earthq. Info. Bull.*, **10**, 136, 1978b.

KING, D. W., E. S. HUSEBYE, and R. A. W. HADDON, Processing of seismic precursor data, *Phys. Earth Planet. Inter.*, **12**, 128, 1976.

KISSLINGER, C., A code of ethics for earthquake predictions?, *EOS, Trans. Am. Geophys. Union*, **59**, 899, 1978.

KISSLINGER, C. and M. WYSS, U.S. National Report, I.U.G.S., 1975, Earthquake Prediction, *Reviews Geophys. Space phys.*, **13**, 298, 1975.

KOVACH, R. L., A. NUR, R. L. WESSON, and R. ROBINSON, Water level fluctuations and earthquakes on the San Andreas fault zone, *Geology*, **3**, 437, 1975.

KOSLOFF, D., Numerical simulations of tectonic processes in southern California, *Geophys. J. R. Astr. Soc.*, **51**, 487, 1977.

KRAEMER, H. C., B. E. SMITH, and S. LEVINE, An animal behavior model for short term earthquake prediction, in *Proceedings of Abnormal Animal Behavior Prior to Earthquakes*, ed. J. Evernden, Vol. 1, p. 213, U.S. Geol. Survey, 1976.

KUMAR, M., Monitoring of crustal movements in the San Andreas fault zone by a satellite-borne ranging system, Dept. Geodetic Sciences Report No. 243, Ohio State Univ., Columbus, Ohio, 1976.

LAMAR, D. L. and P. M. MERIFIELD, Water level monitoring in the area of the Palmdale uplift, southern California, *Earthq. Info. Bull.*, **10**, 144, 1978.

LAMBERT, A. and C. BEAUMONT, Nano variations in gravity due to seasonal ground water movements: implications for the gravitational detection of tectonic movements, *J. Geophys. Res.*, **82**, 297, 1977.

LINDH, A. G., D. A. LOCKNER, and W. H. K. LEE, Velocity anomalies: an alternative explanation, *Bull. Seismol. Soc. Am.*, **68**, 721, 1978a.

LINDH, A., G. FUIS, and C. MANTIS, Seismic amplitude measurements suggest foreshocks have different focal mechanism than aftershocks, *Science*, **201**, 56, 1978b.

LOTT, D. F., B. L. HART, and K. L. VEROSUB, Unusual animal behavior prior to a moderate earthquakes, 1979 (in preparation).

MAZZELLA, A. and H. F. MORRISON, Electrical resistivity variations associated with earthquakes on the San Andreas fault, *Science*, **185**, 1974.

McCANN, W., S. NISHENKO, L. R. SYKES, and J. KRAUSE, Seismic gaps and plate tectonics: Seismic potential for major boundaries, *Pure Appl. Geophys.*, **117**, 1082, 1979.

McEVILLY, T. V., Crustal velocity monitoring with a vibroseis system, Fall Meeting, Am. Geophys. Union, Dec. 1977.

McEVILLY, T. V. and L. R. JOHNSON, Earthquakes of strike-slip type in central California: evidence on the question of dilatancy, *Science*, **182**, 581,

1973.

MCEVILLY, T. V., D. M. BOORE, A. G. LINDH, and W. W. TOLMACHOFF, A search for travel-time changes associated with the Parkfield, California, earthquake of 1966, *Bull. Seismol. Soc. Am.*, **65**, 1407, 1975.

MCNALLY, K. C., H. KANAMORI, J. C. PECHMANN, and G. FUIS, Earthquake swarm along the San Andreas fault near Palmdale, southern California, 1976 to 1977, *Science*, **201**, 814, 1978.

MITCHELL, B. J., W. STAUDER, and C. C. CHENG, The New Madrid seismic zone as a laboratory for earthquake prediction, *J. Phys. Earth*, **25**, 543, 1977.

MJACHKIN, V. I., W. F. BRACE, G. A. SOBOLEV, and J. H. DIETERICH, Two models for earthquake forerunners, *Pure Appl. Geophys.*, **113**, 169, 1975.

MOODY, S. E. and J. LEVINE, Design of an extended range three wave-length distance measuring instrument, *Tectonophysics*, **52**, 77–82, 1979.

MORRISON, H. F., R. F. CORWIN, and M. CHANG, High accuracy determination of temporal variations of crustal resistivity, *Am. Geophys. Union Monograph*, **20**, 1977.

MORRISON, H. F., R. CORVIN, and R. FERNANDEZ, Earth resistivity self-potential variations and earthquakes; a negative result for $M=4.0$, *Geophys. Res. Lett.*, **6**, 139–142, 1979.

MORTENSEN, C. E. and M. J. S. JOHNSTON, The nature of surface tilt along 85 km of the San Andreas fault—preliminary results from a 14 instrument array, *Pure Appl. Geophys.*, **113**, 237, 1975.

MORTENSEN, C. E. and M. J. S. JOHNSTON, Anomalous tilt preceding the Hollister earthquake of November 28, 1974, *J. Geophys. Res.*, **20**, 3561, 1976.

MORTENSEN, C. E., R. C. LEE, and R. O. BOURFORD, Observations of creep-related tilt, strain, and water-level changes on the central San Andreas fault, *Bull. Seismol. Soc. Am.*, **67**, 641, 1977.

MUELLER, I. I., editor, *Proc. International Symposium on the Application of Geodesy to Geodynamics*, Ohio State Univ., Columbus, Ohio, 1978.

MURDOCK, J. N., Travel time spectrum stability in the two years preceding the Yellowstone, Wyoming, earthquake of June 30, 1975, *J. Geophys. Res.*, **83**, 1713, 1978.

MURPHY, A. J., W. MCCANN, and A. FRANKEL, Anomalous microearthquake activity in the north-east Caribbean: potential precursors to two moderate-sized earthquakes, abstract, *EOS, Trans. Am. Geophys. Union*, **59**, 811, 1978.

NATIONAL RESEARCH COUNCIL, Geodesy: trends and prospects, National Academy of Sciences, 86 pp., 1978.

NUR, A., A note on the constitutive law for dilatancy, *Pure Appl. Geophys.*, **113**, 197, 1975.

OHTAKE, M., T. MATUMOTO, and G. V. LATHAM, Temporal changes in seismicity preceding some shallow earthquakes in Mexico and Central America, *Bull. Int. Inst. Seismol. and Earthq. Eng.*, **15**, 105, 1977a.

OHTAKE, M., T. MATUMOTO, and G. V. LATHAM, Seismicity gap near Oaxaca, southern Mexico, as a probable precursor to a large earthquake, *Pure Appl. Geophys.*, **115**, 375, 1977b.

OHTAKE, M., G. V. LATHAM, and T. MATUMOTO, Patterns of seismicity preceding earthquakes in Central America, Mexico and California, in U.S. Geological Survey, open file report, 78-943, 585, 1978.

O'NEIL, J. R., Stable isotope variations in earthquake prediction, *Trans. Am. Geophys. Union*, **58**, 437, 1977.

ONG, K. M., P. E. MACDORAN, A. E. NIELL, G. M. RESCH, D. D. MORABITO, and T. G. LOCKHART, Radio interferometric geodetic monitoring in southern California, abstract, *EOS, Am. Geophys. Union*, **58**, 1121, 1977.

PATWARDHAN, A. S., R. B. KULKARNI, and D. TOCHER, A semi-markov model for characterizing recurrence of great earthquakes, in U.S. Geological Survey, open file report, 78-943, 635, 1978.

PEAKE, L. G., J. H. HEALY, and J. C. ROLLER, Time variance of seismic velocity from multiple explosive sources southeast of Hollister, California, *Bull. Seismol. Soc. Am.*, **67**, 1339, 1977.

PIERCE, E. T., Atmospheric electricity and earthquake prediction, *Geophys. Res. Lett.*, **3**, 185, 1976.

PLAFKER, G. and M. RUBIN, Uplift history and earthquake recurrence as deduced from marine terraces on Middleton Island, Alaska, in U.S. Geological Survey, open file report, 78-943, 687, 1978.

POEHLS, K. A. and D. D. JACKSON, Tectonomagnetic event detection using empirical transfer functions, *J. Geophys. Res.*, **83**, 4933, 1978.

PRESCOTT, W. H. and J. C. SAVAGE, Strain accumulation on the San Andreas fault near Palmdale, California, *J. Geophys. Res.*, **81**, 4901, 1976.

PRESCOTT, W. H., J. C. SAVAGE, and W. T. KINOSHITA, Strain accumulation rates in the western United States between 1970 and 1978, *J. Geophys. Res.*, **84**, 5423, 1979.

PRIESTLY, K. F., Possible premonitory strain changes associated with an earthquake swarm near Mina, Nevada, *Pure Appl. Geophys.*, **113**, 251, 1975.

RAIKES, S. A., The temporal variation of teleseismic P-residuals for stations in southern California, *Bull. Seismol. Soc. Am.*, **78**, 711, 1978.

REDDY, I. K., R. J. PHILLIPS, J. H. WHITCOMB, D. M. COLE, and R. A. TAYLOR, Monitoring of time dependent electrical resistivity by magnetotellurics, *J. Geomag. Geoelectr.*, **28**, 165, 1976.

REILINGER, R. E., G. P. GITRON, and L. D. BROWN, Recent vertical crustal movements from precise leveling data in southwestern Montana, western Yellowstone National Park, and the Snake River Plain, *J. Geophys. Res.*, **82**, 5349, 1977.

RICE, J. R., On the stability of dilatant hardening for saturated rock masses, *J. Geophys. Res.*, **80**, 1531, 1975.

ROBINSON, R. and H. M. IYER, Temporal and spatial variations of travel-time residuals in central California for Novaya Zemlya events, *Bull. Seismol. Soc. Am.*, **66**, 1733, 1976.

ROBINSON, R., R. L. WESSON, and W. E. ELLSWORTH, Variation of P-wave velocity before the Bear Valley, California, earthquake of 24 February 1972, *Science*, **184**, 1281, 1974.

RUNDLE, J. R., Gravity changes and the Palmdale uplift, *Geophys. Res. Lett.*, **5**, 41, 1978.

ROWLETT, H. and J. KELLEHER, Evolving seismic and tectonic patterns along the western margin of the Philippine Sea plate, *J. Geophys. Res.*, **81**, 3518, 1976.

SAVAGE, J. C., A possible bias in the California State geodimeter data, *J. Geophys. Res.*, **80**, 4078, 1975.

SAVAGE, J. C. and W. H. PRESCOTT, Geodolite measurements near the Briones Hills, California, earthquake swarm of January 8, 1977, *Bull. Seismol. Soc. Am.*, **68**, 1, 175–180, 1978a.

SAVAGE, J. C. and W. H. PRESCOTT, Geodimeter measurements and the southern California uplift, *Earthq. Info. Bull.*, **10**, 131, 1978b.

SAVAGE, J. C. and W. H. PRESCOTT, Geodimeter measurements of strain during the southern California uplift, *J. Geophys. Res.*, **84**, 171, 1979.

SAVAGE, J. C., M. A. SPEITH, and W. H. PRESCOTT, Preseismic and coseismic deformation associated with the Hollister, California, earthquake of November 28, 1974, *J. Geophys. Res.*, **81**, 3567, 1976.

SAVAGE, J. C., W. H. PRESCOTT, M. LISOWSKY, and N. KING, Measured uniaxial north-south regional contraction in southern California, *Science*, **202**, 883, 1978.

SBAR, M. L., T. ENGELDER, R. PLUMB, and S. MARSHAK, Stress pattern near the San Andreas fault, Palmdale, California from near surface in-situ measurements, *J. Geophys. Res.*, **84**, 156–164, 1979.

SCHOLZ, C. H., A physical interpretation of the Haicheng earthquake prediction, *Nature*, **267**, 121, 1977.

SCHOLZ, C. H. and T. KATO, The behavior of a convergent plate boundary: crustal deformation in the South Kanto District, Japan, *J. Geophys. Res.*, **83**, 783, 1978.

SCHOLZ, C. H., L. R. SYKES, and Y. P. A. AGGARWAL, Earthquake prediction: a physical basis, *Science*, **181**, 803, 1973.

SIEH, K. E., Prehistoric large earthquakes produced by slip on the San Andreas fault at Pallett Creek, California, *J. Geophys. Res.*, **83**, 3907, 1978.

SIEH, K. E., Central California foreshocks of the great 1857 earthquake, *Bull. Seismol. Soc. Am.*, **68**, 1731–1749, 1978.

SIMS, J. D., Determining earthquake recurrence intervals from deformational structure in lacustrine sediments, *Tectonophysics*, **29**, 141, 1975.

SKILES, D. D. and R. G. LINDBERG, Can animals predict earthquakes?: a search for correlation between changes in the activity patterns of two fossorial rodents and subsequent seismic events, in *Summary of Technical Reports*, ed. J. Evernden, U.S. Geological Survey, **5**, 249, 1978.

SLATER, L. E. and G. R. HUGGETT, A multiwavelength distance-measuring instrument for geophysical experiments, *J. Geophys. Res.*, **81**, 6299, 1976.

SMITH, B. E. and M. J. S. JOHNSTON, A tectonomagnetic effect observed before a magnitude 5.2 earthquake near Hollister, California, *J. Geophys. Res.*, **81**, 3556, 1976.

SMITH, B. E., M. J. S. JOHNSTON, and R. O. BURFORD, Local variations in magnetic fields, long-term changes in creep rate, and local earthquakes along the San Andreas fault in central California, *J. Geomag. Geoelectr.*, **30**, 539, 1978.

SOBOLEV, G., H. SPETZLER, and B. SALOV, Precursors to failure in rocks while undergoing anelastic deformations, *J. Geophys. Res.*, **83**, 1775, 1978.

SPENCE, W. and L. C. PAKISER, Conference report: Toward earthquake prediction on the global scale, *EOS, Trans. Am. Geophys. Union*, **59**, 36, 1978.

SPETZLER, H., G. SOBOLEV, B. SALOV, and A. KOLTSOV, Failure of a weak rock under triaxial loading, abstract, *EOS, Trans. Am. Geophys. Union*, **59**, 1206, 1978.

STEPPE, J. A., W. H. BAKUN, and C. G. BUFE, Temporal stability of P-velocity anisotropy before earthquakes in central California, *Bull. Seismol. Soc. Am.*, **67**, 1075, 1977.

STEWART, S. W., Real-time detection and location of local seismic events in central California, *Bull. Seismol. Soc. Am.*, **67**, 433, 1977.

STRANGE, W. E. and C. W. WESSELLS, Repeat gravity measurements in the area of the Palmdale Bulge, abstract, *EOS, Trans. Am. Geophys. Union*, **57**, 898, 1976.

SYKES, L. R., Aftershock zones of great earthquakes, seismicity gaps, and earthquake prediction for Alaska and the Aleutians, *J. Geophys. Res.*, **76**, 8021, 1971.

SYKES, L. R., Research on earthquake prediction and related areas at Columbia University, *J. Phys. Earth*, **25**, 513, 1977.

TALWANI, P., An empirical earthquake prediction model, *Phys. Earth Planet. Inter.*, **18**, 288, 1979.

THATCHER, W., Strain accumulation and release mechanism of the 1906 San Francisco earthquake, *J. Geophys. Res.*, **80**, 4862, 1975a.

THATCHER, W., Strain accumulation on the northern San Andreas fault zone since 1906, *J. Geophys. Res.*, **80**, 4873, 1975b.

THATCHER, W., Episodic strain accumulation in southern California, *Science*,

194, 691, 1976.

THATCHER, W., Systematic inversion of geodetic data in central California, *J. Geophys. Res.*, **84**, 2883, 1979.

THATCHER, W., J. HILEMAN, and T. C. HANKS, Seismic slip distribution along the San Jacinto fault zone, southern California and its implications, *Geol. Soc. Am. Bull.*, **86**, 1140, 1975.

TOKSÖZ, M. N., E. ARPAT, and F. SAROGLU, East Anatolian earthquake of 24 November 1976, *Nature*, **270**, 423, 1977.

TOKSÖZ, M. N., A. F. SHAKAL, and A. J. MICHAEL, Space-time migration of earthquakes along the north Anatolian fault zone and seismic gaps, in U.S. Geological Survey, open file report, 78-943, 829, 1978.

TURCOTT, D. L., Stress accumulation and release on the San Andreas fault, *Pure Appl. Geophys.*, **115**, 413, 1977.

U.S. GEOLOGICAL SURVEY, Earthquake prediction: opportunity to avert disaster, *U.S. Geol. Survey Circular*, **729**, 35, 1976.

VANWORMER, J. D., L. D. GEDNEY, J. N. DAVIES, and N. CONDAL, V_p/V_s and b-values: A test of the dilatancy model for earthquake precursors, *Geophys. Res. Lett.*, **2**, 514, 1975.

WALLACE, R. E., Patterns of faulting and seismic gaps in the great basin province, in U.S. Geological Survey, open file report, 78-943, 857, 1978.

WANG, C. Y., W. LIN, and F. T. WU, Constitution of the San Andreas fault at depth, *Geophys. Res. Lett.*, **5**, 741, 1978.

WANG, C. Y., R. E. GOODMAN, P. N. SUNDARAM, and H. F. MORRISON, Electrical resistivity of granite in frictional sliding-application to earthquake prediction, *Geophys. Res. Lett.*, **2**, 525, 1975.

WARBURTON, R. J. and J. M. GOODKIND, Detailed gravity tide spectrum between 1 and 4 cycles per day, *Geophys. J. R. Astr. Soc.*, **52**, 117, 1978.

WARD, P. L., Earthquake prediction, in *Studies in Geophysics, Geophysical Predictions*, p. 37, National Academy of Sciences, Washington, 1978a.

WARD, P. L., Volunteers in the earthquake hazard reduction program, *Earthq. Info. Bull.*, **10**, 139, 1978b.

WARD, P. L., editor, *Proceedings of Conference IV: The use of volunteers in the earthquake hazards reduction program*, U.S. Geological Survey, open file report, 78-336, 1978c.

WARE, R. H., High-accuracy magnetic field difference, *J. Geophys. Res.*, **11**, 6291, 1979.

WARE, R. H. and P. L. BENDER, Noise reduction techniques for use in determining local geomagnetic field changes, *J. Geomag. Geoelectr.*, **30**, 533–537, 1978.

WESSON, R. L., R. ROBINSON, C. G. BUFE, W. L. ELLSWORTH, J. H. PFLUKE, J. A. STEPPE, and L. C. SEEKINS, Search for seismic forerunners to earthquakes in central California, *Tectonophysics*, **42**, 111, 1977.

126 M. WYSS

WHITCOMB, J. H., New vertical geodesy, *J. Geophys. Res.*, **81**, 4937, 1976.

WHITCOMB, J. H., Earthquake prediction related research at the Seismological Laboratory, California Institute of Technology 1974–1976, *J. Phys. Earth*, **25**, 51, 1977.

WHITE, J. E., Elastic dilatancy, fluid saturation, and earthquake dynamics, *Geophys. Res. Lett.*, **3**, 747, 1976.

WINKLER, K. W. and A. NUR, Depth constraints on dilatancy induced velocity anomalies, *J. Phys. Earth*, **25**, S231, 1977.

WOOD, M. D. and N. E. KING, Relation between earthquakes, weather, and soil tilt, *Science*, **197**, 154, 1977.

WU, C. and R. S. CROSSON, Numerical investigation of dilatancy biasing of hypocenter locations, *Geophys. Res. Lett.*, **2**, 205, 1975.

WU, F. T., Gas pressure fluctuations and earthquakes, *Nature*, **257**, 661, 1975.

WU, F. T., L. BLATTER, and H. ROBERTSON, Clay gouges in the San Andreas fault system and their possible implications, *Pure Appl. Geophys.*, **113**, 87, 1975.

WYATT, F. and J. BERGER, Strain and gravity observations of a local earthquake of magnitude 4.9 on the San Jacinto fault, abstract, *EOS, Trans. Am. Geophys. Union*, **56**, 1019, 1975.

WYATT, F. and J. BERGER, Correlations of short based tilt measurements and long base strain measurements, abstract, *EOS, Am. Geophys. Union*, **59**, 331, 1978.

WYSS, M., Precursors to the Garm earthquake of March, 1969, *J. Geophys. Res.*, **80**, 2926, 1975a.

WYSS, M., Mean sea level before and after some great strike-slip earthquakes, *Pure Appl. Geophys.*, **113**, 107, 1975b.

WYSS, M., A search for precursors to the Sitka, 1972, earthquake: sea level, magnetic field and P-residuals, *Pure Appl. Geophys.*, **113**, 297, 1975c.

WYSS, M., Local changes of sea level before large earthquakes in South America, *Bull. Seismol. Soc. Am.*, **66**, 903, 1976a.

WYSS, M., Local sea level changes before and after the Hyuganada, Japan, earthquakes of 1961 and 1968, *J. Geophys. Res.*, **81**, 5315, 1976b.

WYSS, M., Interpretation of the southern California uplift in terms of the dilatancy hypothesis, *Nature*, **266**, 805, 1977a.

WYSS, M., The appearance rate of premonitory uplift, *Bull. Seismol. Soc. Am.*, **67**, 1091, 1977b.

WYSS, M., Sea level changes before large earthquakes, *Earthq. Info. Bull.*, **10**, 165, 1978.

WYSS, M. and J. N. BRUNE, Regional variations of source properties in southern California estimated from the ratio of short- to long-period amplitudes, *Bull. Seismol. Soc. Am.*, **61**, 1153, 1971.

WYSS, M. and R. E. HABERMANN, Seismic quiescence precursory to a past and a future Kurile island earthquake, *Pure Appl. Geophys.*, **117**, 1195, 1979.

WYSS, M. and D. J. HOLCOMB, Earthquake prediction based on station residuals, *Nature*, **245**, 139, 1973.

WYSS, M., R. E. HABERMANN, and A. C. JOHNSTON, Long term precursory seismicity fluctuations, in U.S. Geological Survey, open file report, 78-943, 869, 1978.

ZOBACK, M. C., Preliminary stress measurements in the western Mojave desert near Palmdale, abstract, *EOS, Trans. Am. Geophys. Union*, **59**, 328, 1978.

ZOBACK, M. C., J. H. HEALY, and J. C. ROLLER, Preliminary stress measurements in central California using the hydraulic fracturing technique, *Pure Appl. Geophys.*, **115**, 135, 1977.

ZOBACK, M. D., J. H. HEALY, J. C. ROLLER, G. S. GOHN, and B. B. HIGGINS, Normal faulting and in-situ stress in the South Carolina coastal plain near Charleston, *Geology*, **6**, 147, 1978.

EARTHQUAKE PREDICTION AND PUBLIC REACTION†

Janice R. Hutton,* John H. Sorensen,** and Dennis S. Mileti***

*Woodward-Clyde Consultants, San Francisco, California, U.S.A., **University of Hawaii, Honolulu, Hawaii, U.S.A., and ***Colorado State University, Fort Collins, Colorado, U.S.A.

Abstract Using social data from a variety of different settings and some "prediction" experiences, this chapter attempts to review the range of issues that center on human response to earthquake predictions. It is suggested that an earthquake prediction is information about risk which will cause people to re-evaluate what is and is not appropriate to do about the earthquake hazard. There are three general categories of prediction response: (1) relocation, (2) reallocation, and (3) reduction, as well as two reasons or goals for response: (1) increase emergency preparedness and (2) decrease vulnerability. Six factors account for why some people will and others will not take these prediction responses. These are: (1) image of damage, (2) insurance, (3) exposure to risk, (4) access to information, (5) commitment to target area, and (6) resources. On the basis of this research issue for public policy are presented including information management, earthquake insurance, and resources. Several alternatives along these lines are proposed to enhance human response.

†This chapter is a condensed and revised version of several other published works by the same authors. The major work on which this is based, *Earthquake Prediction Response and Options for Public Policy*, by Dennis Mileti, Janice Hutton, and John Sorensen will soon be published in the U.S.A. This chapter is also based on three papers presented at the International Symposium on Earthquake Prediction, UNESCO, Paris: April, 1979 by the same authors. The data on which these works are based was collected as part of National Science Foundation-RANN Grant No. AEN–7424079 which is gratefully acknowledged.

1. Introduction

1. 1   The social utility of prediction

The research effort in several nations continues to work toward developing a technical capacity to predict earthquakes. The development and use of the technology are defined by many as worthwhile. The desirability of this technology lies in the promise it holds to facilitate and propel two already ongoing goals. These goals are to reduce earthquake vulnerability, and increase earthquake emergency preparedness. Both goals exist independent of earthquake prediction technology and are pursued by other earthquake hazard mitigation adjustments, such as building codes and planning.

Earthquake prediction technology promises more specific information about the time, location, and magnitude of an earthquake occurrence than exists without a prediction. It provides the added specificity of hazard information on which can be mounted more intense efforts directed toward the reduction of earthquake vulnerability and increases in emergency preparedness. Herein lies the social utility of earthquake prediction technology. Based on the more specific risk information provided by a prediction, the technology can facilitate new decisions to further reduce earthquake hazard vulnerability and increase emergency preparedness.

The decision to initiate or intensify efforts to reduce vulnerability or increase emergency preparedness, because of the more specific risk information provided by a prediction, are obvious to some people and decision makers, partially obvious to others, and not obvious to yet others. The range of people that will be involved in making such decisions after a prediction have divergent levels of awareness about the decisions and their appropriateness to a prediction setting. As well, a prediction situation will contain numerous flows of information. Some of this information will be accurate and inaccurate, complete and incomplete, consistent and inconsistent. These mixes will create alternative expectations and mixed perceptions of level of risk by people faced with earthquake prediction decisions. The information available in a prediction setting—on which vulnerability reduction and emergency preparedness decisions will be made—can be upgraded by policy. Such policy could facilitate more appropriate decisions, enhance the adaptive actions that can occur because of a prediction, and help achieve the social goals for which the technology is being developed.

Some decisions to reduce vulnerability or increase emergency preparedness, even sound and appropriate decisions, will be constrained from becoming action because of current prevalent constraints in the social system. Policy to mitigate constraints and enhance the fruits promised by the technology will do so by enhancing viable and desirable choices by people and public and private decision makers.

By reducing earthquake vulnerability, we refer to decisions which speed up, intensify or initiate activities which reduce the potential for loss of property, life or dollars from some future earthquake. These decisions and activities cover a range of topics from taking pictures off the wall and packing away glassware to structurally modifying a building to better withstand an earthquake, or in the extreme, simply moving away. All such activities have two things in common. First, such decisions and activities reduce the potential for loss from earthquake impact. Second, an earthquake prediction need not exist in order for such decisions to be made and activities to reduce earthquake vulnerability begun. An earthquake prediction allows special decisions along these lines to be made that might not ever be decided in light of generic earthquake risk, such as temporarily lowering the water level in reservoirs, closing hazardous bridges, and temporarily evacuating particularly dangerous buildings.

Increased capacity for emergency response, or emergency preparedness, refers to decisions and activities to reduce or redistribute the cost and anguish of immediate and long range recovery after an earthquake. These decisions and activities also cover a wide range of specific topics. Some examples include stockpiling emergency supplies, buying insurance, and developing disaster plans. Other such decisions and activities along this same line can be made on the basis of a prediction that would likely never be decided in light of generic earthquake risk. These include, for example, moving emergency response equipment to the periphery of the target area, and having large scale emergency response organizations locate and begin operations on the target area fringe prior to the earthquake. All these actions have in common that they facilitate the ability to respond to earthquake loss once it has occurred, and that an earthquake prediction need not be made in order for decisions and activities along this line begun.

1. 2   The potential social costs of prediction

Earthquake prediction has been seen by some as problematic in terms of the potential it holds to impose negative economic, social, and political impacts on a local community. These concerns are not unfounded in that longterm information that a disastrous earthquake will occur in a community could elicit behaviors or responses by some people or groups that could create negative consequences. Early concerns about negative impacts have raised a variety of pointed questions. These include: (1) What will happen to the availability of earthquake insurance? (2) Will property values decline? (3) Will credit tighten for durable goods? (4) Will the sale of durable goods decrease? (5) Will mortgage money availability decline? (6) Will local property tax revenues decline? (7) Will construction fall off? (8) Will there be a decline in sales tax revenue? (9) Will unemployment increase? (10) Would there be a forced deline in public services? (11) Would some retail firms be forced out of business? (12) Will there be a net decrease in local population? Potential affirmative answers to just a few of these questions conjure up images of a severely disrupted local economy. It is easy to fall prey to speculative forecasts on these issues and join the ranks of those wishing prediction were only possible in the short-term (1 or 2 day announcements) without enough time to negatively effect an economy.

Other questions in the legal and political arenas are raised by earthquake prediction technology. For example, is an official liable for not taking some actions in the public interest because of a prediction? Is an official liable for taking certain actions in the public interest because of a prediction? How will a prediction affect the political process? What legislation is needed to put prediction to more effective use?

1. 3   Areas of social concern

Three arenas exist for concern over public policy related to earthquake prediction. Two of these center on maximizing the utility of knowing ahead-of-time that an earthquake may occur. Such beforehand knowledge can enable a variety of actions to be taken to increase the ability of the social system to respond to the earthquake disaster if it does occur. Efforts can be initiated or accelerated to increase emergency preparedness by people, families, business, and government agencies. Such efforts could decrease the disruption and anguish that would typically characterize the disaster. At the same time, a prediction could also provide the opportunity to take a variety of actions to reduce earthquake vulnerability, thereby, actually reducing the extent and degree of that which is damaged or lost when the earthquake strikes. Policy questions, therefore, fall into three categories: (1) how can the information contained in a prediction be used to increase emergency preparedness?, (2) how can a prediction be used to decrease earthquake vulnerability?, and (3) how can policy reduce the potential for negative secondary effects associated with the knowledge that an earthquake will occur?   Put simply, the challenge presented by the potential use and development of earthquake prediction technology is the question of how its benefits can be maximized while minimizing its potential negative impacts or costs.

2. Study Method

2. 1   The problems

The scientific study of human response to earthquake prediction is riddled with uncertainties and problems.   The most obvious is that it is difficult and perhaps impossible to study human response to an event which has not occurred.   On closer examination, several major obstacles work against scientifically accurate research, all of which are related to the lack of actual prediction situations to study.   These deal with the nature of the prediction characteristics, the " state-of-the " social system, the interrelatedness of social response, experience with prediction, locational and cultural differences, the effect of time and attituded-behavior inconsistencies.

One of the main obstacles in the study of before-the-fact earthquake prediction response is that we cannot be sure of the future character and nature of an actual prediction.   An earthquake prediction is comprised of a variety of different factors or variables important to consider from a human response point of view.   Some of these factors, which are discussed later, are the amount of lead time between the prediction and the anticipated earthquake, the reputation of the person or organization making the prediction, how certain the predictor is about if the earthquake will occur and where it will occur, how much consensus there is among seismologists about the characteristics of the prediction, and other factors.   These characteristics of the prediction comprise the set of factors to which human response is given.   If they differ, human response will differ.   If they change, human response will change.

Because it is almost impossible to know the actual characteristics of the first scientifically credible earthquake prediction now, it makes the study of human response problematic.　The ideal study of human response to earthquake prediction would be to observe an actual prediction and its characteristics and then to observe actual human response. It is difficult to speculate on future human response even if the characteristics of the prediction were known.　It is compoundly difficult to speculate on human response without knowing the character of the prediction.　Current uncertainty about the character and nature of future predictions makes before-the-fact study of human response to those predictions troublesome.

Another obstacle which introduces uncertainty into the current study of human response to earthquake prediction is that there is some chance that the nature of the state of causal system factors will change over time.　For example, holding the characteristics of an earthquake prediction constant suppose we were able to project that 10% of a citizenry would stockpile emergency supplies on the basis of a current study.　But it is very likely that stockpiling emergency supplies is conditioned by surplus capital.　It is likely that the extent of surplus capital in a citizenry changes over time.　It is troublesome in that there is no way to know how much surplus capital will exist in a citizenry if and when an actual prediction emerges.　This factor also makes human response projections difficult; although to now establish the relationship between, for example, surplus capital (independent variable) and stockpiling emergency supplies (human response dependent variable) is quite possible.

Social system changes of a different magnitude may also be important.　For example, normalization of relations between the U.S. and China may encourage the adoption of Chinese ideas about earthquakes among the U.S. citizenry and affect response in a subtle manner.

A third major source of uncertainty in the study of response to earthquake prediction is the interrelatedness of response.　Many actions are linked.　One response may be predicated on someone else's action. For example, the government may need to authorize the use of funds before local officials could take extensive precautionary action.　Other responses may limit subsequent action.　The person who purchases earthquake insurance may not do anything to secure household possessions.　Thus to predict response, it must be realized that events that take place will be influenced by the preceding response of others, and that some response will impact the future behavior of others.

To extend this notion further, behavior in light of a prediction is also interrelated with social values, technological development, and the pattern of earthquake adjustments adopted by society.

To some extent behavior will reflect prevailing social values, which are subject to change. Witness the environmental movement of the 1960's and 70's, in the United States, which greatly influenced public policy. A shift in social values, such as a distrust in science and technology, could have a large impact on prediction response.

Technological advances interact with possible social responses to predictions. For example, new earthquake resistant building techniques, improved building materials, better mass transit, or personalized mass communication may encourage behavior which is different from what we would predict at this time.

A general concept to recognize is that as the ability to reduce earthquake risk changes, different responses to a prediction could occur. For example, a low cost emergency sprinkler system for buildings, if developed, could reduce the need for public fire protection preparedness. In the extreme, earthquake hazard reduction techniques could eliminate the need for many preparedness and protective responses. Thus it is best to view prediction in terms of the current earthquake adjustment technology and availability. The total mix of adjustment practiced will influence public and private action. These mixes will likely vary from place to place.

Prediction experience injects another type of uncertainty for social research. Experience may occur directly by being in the area for which a prediction is offered, or more indirectly through media exposure. For example, the " Whitcomb " hypothesis or the " Palmdale Bulge " could affect how persons in Southern California will respond to a future prediction. Likewise, information dispersed about prediction success and failure on small quakes or the experiences in other countries such as China, will also have a bearing on future response. Such types of uncertainty are difficult to handle because it is unclear what the nature of the track record will be when the first significant prediction is issued.

It is reasonable to expect that response will be influenced by cultural and locational factors although the extent is unclear. Response to a prediction issued in the U.S. but outside of California would probably be different than one issued for the San Francisco area. Likewise, response could differ between Southern and Northern California, which have their own cultural settings.   At a larger scale, it would be impossible to infer what happens in China would apply to Japan, and so forth. On the other hand, there would be useful lessons that transcend cultural differences and can be universally utilized.   The Chinese experience with amateur seismologists provides a good example.

Time is another variable which is important in the calculus of measuring human response to a prediction.   It is easier and probably more accurate to estimate response to a prediction in the immediate future than one 5 to 10 years in the future.   Efforts to examine social consequences are based on assumptions that reflect existing social and technological circumstances.   These circumstances may be slightly or dramatically different in the future.   The further away in time, the more difficult it becomes to study the response of a complex social system to an earthquake prediction.

Finally, what people say about their behavior and their actual behavior are often different.   Uncertainties are almost certainly introduced into any study about human behavior in which behavior is asked about rather than observed.   A study which asked people to estimate probable but hypothetical earthquake prediction response would be on weak ground from which to project impacts of actual future human response to earthquake prediction.

2. 2   Minimizing the problems

These various problems and uncertainties should not prohibit research into human response to earthquake predictions.   They make it difficult to state any conclusive findings about the exact course of events that will follow any given prediction.   There are still meaningful questions current research can address, and in the course of investigation, techniques that can lend more confidence to the findings.

One strategy which can be utilized to study human response to earthquake prediction before any actual predictions emerge is to study human behavior in an analogous situation, such as a drought forecast, a flood warning, or response to projected estimates of shortages in imported oil.   There is great merit in conducting studies which seek to transfer knowledge of a more predictable situation in which behavior can be observed to an uncertain situation such as posed by earthquake prediction.   For example, a great amount of information about response to other hazard warning system already exists, and generalized findings may be applicable to the earthquake hazard.   However, uncertainty exists in the ability and confidence of such applications because of the situational and structural differences among different events, hazards, and warning systems.

A second means to overcome the aforementioned problems is to have a research design which controls for key factors which may affect response. For example, to vary one prediction characteristic while holding other variables constant. The efficacy of this approach is, however, severely hampered by a lack of knowing what factors to control for and because of the huge effort required.

A third means is to employ a variety of data types, from different sources, under different circumstances, and compare them for consistency and differences in findings and interpretations.

Problems exist in an attempt to study human behavior in response to an event which has as yet not occurred. However, such studies may still be worthwhile even if projections of future behavior are problematic. Studies could still be quite successful, not in predicting what people will do, but rather in the discovery of what factors will influence their future decisions to behave. Caution is still warranted in interpretations made on data of this type.

2. 3   Data sources

The analyses, interpretations, and findings in this report are based on a variety of sources of social data which differ with respect to situation, societal level, and method of collection.

Four general situations provide a context for response. First, the ideal situation is to observe actual responses or behavior in a credible prediction situation. In general, circumstances did not permit this except for limited secondary information from the experience in China with the 1976 Haicheng earthquake.

Second, it was possible to use data on actual responses to predictions which did not possess scientific credibility. The three situations examined were Whitcomb's " hypothesis " for Los Angeles, a prediction for Wilmington, North Carolina, by a " psychic," Clarissa Bernhardt, and a prediction in Tokyo and Kawasaki which failed to receive scientific endorsement.

The third situational type involves a hypothetical response to a real situation. In this instance, we asked various people and organizations in San Fernando what they would have done if they had known the 1971 earthquake were coming in advance of its impact.

The final, and least realistic, situation is characterized by hypothetical response to hypothetical predictions.   The majority of the available data on prediction response is of this type.   Responses to several different predictions were surveyed in two local areas in the U.S. (San Bernadino-Riverside, and Santa Clara, County, California) and for the State of California in general.   To a limited extent, additional data of this nature was collected in Tokyo and Kawasaki, Japan.

Data was collected for three general social system levels: families and individuals; private organizations and businesses; and public officials and agencies.   These can be further refined into specific groups.   Private organizations included businesses localized to one community and ones which operated on a regional, or larger, basis.   A diversity of organizations were chosen in a non-random fashion to give a diversity of function and purpose.   The range included local newspapers as well as large television stations; and corner pharmacies and multi-national conglomerates.

Public groups and officials included those at the Local, County, State and Federal governmental levels, and various non-governmental organizations.

In addition, several different types of data-collection techniques were utilized.   In some instances secondary data and unsystematic field observation was all that resources permitted.   Informal interviews and discussions were conducted with individuals and spokespersons of organizations.   Formal questionnaires were also administered.   In some cases several techniques were utilized with the same sample.   In addition, data from informal interviews was utilized to construct questionnaires and scenarios of alternative prediction possibilities.   These scenarios were utilized to provide a context for the respondents to answer the structured questions on earthquake prediction response.   The formal questionnaires include questions on response and a wide variety of background data on the organizations or families in the samples.   In a later section some of the key relationships which help explain response are discussed.

2. 4   Techniques of analysis

Given the wide variety of data types, more than one single method of analysis was utilized.   Those employed are best discussed in relation to the different types of data collected.

Little in the way of systematic analysis was performed on the secondary data collected.   It was chiefly used in the study as background information and to provide a descriptive picture of events and situations. To a minor extent content analysis of newspaper accounts was conducted as in the case of the Los Angeles Time's reporting on the Whitcomb " hypothesis."

Informal interviews provided a mass of information on response to earthquake predictions.   One way the data was examined was by the use of non-quantitative classification techniques.   For example, responses from organizations in San Fernando were placed on cards which were examined for common dimensions and concepts.   From this analysis a response typology was generated which provided the basis for developing the more formalized response questionnaires in the study.

Second, data was arranged and worked into speculative scenarios of possible prediction situations.   The scenarios were utilized as research tools to help respondents conceptualize a hypothetical prediction situation.   In addition, rough forms of content analysis were selectively applied.   For example, the interviews with public official were analyzed to determine an agency's perceived responsibilities for conducting vulnerability assessments of buildings and other structures.

Quantitative statistical analyses were made with the data collected through structured questionnaires.   Similar procedures were used for the data from private organization and public agencies, Santa Clara families, and Wilmington households.

The organization data consisted of a set of dependent variables measuring intended responses to a hypothetical prediction and independent variables measuring organizational characteristics.   Descriptive aggregate statistics of all dependent and independent variables were compiled.   Next, zero-order correlation coefficients were produced for most pairs of independent and dependent variables.   This was done for all oraganizations taken together, and, then, for a number of subgroups such as local government agencies and officials or non-local businesses. This pointed out which independent variables were associated with each category of hypothetical response.

The correlation analyses provide the grounds for then determining the relative importance of the associations between independent and dependent variable.   Multiple regression techniques were utilized to assess the strength of the relationship between each dependent variable and all independent variables which were significantly correlated to it. Further types of multivariate analysis were conducted, but do not provide the analytical basis for this chapter.

The family data also consisted of a set of dependent variables on hypothetical response and independent variables describing family characteristics. The procedure for analysis was almost identical to that for the organizations.

2. 5 Limitations and utility

The conclusions drawn from this data must be viewed in light of several cautions. First, hypothetical responses to a hypothetical prediction is not the same as what will be actual response to an actual prediction. What people think they will do in a situation and what they actually will do are often different. Second, response to a non-scientific versus a scientific prediction are not the same. Different dimensions of credibility in prediction types alter response to them by people. Even with these shortcomings, however, the data assembled do have utility. First, they identify the range of responses with which people and decision makers will be faced in an actual prediction. Most of the possible behaviors from which to choose in an actual prediction have been identified. As well, these data identify the key causes of deciding to do or not do these behaviors in an earthquake prediction. The range of possible response and the factors which cause them, therefore, have been identified. One limitation remains, we can not predict the extent of any one kind of response in an actual future prediction. For example, it is possible to know that some will do a certain behavior, and even why they will do it, but it is impossible to predict how many.

Overall, we have attempted to make the greatest use out of the assembled data within the bounds of these limitations and problems. The chapter reflects this in that it varies in what can be said about any given subject related to social response to prediction. We are, however, confident that what is said has a reasonable data base for support.

3. Prediction and Social Response

3. 1 Earthquake prediction warning systems

Limited experience with earthquake predictions leaves public warning officials with scant evidence on which to base the design and execution of an earthquake warning system. However, findings from the limited investigations of public response to earthquake warnings indicate that despite the absence of many specific earthquake warning experiences, officials can be guided by experience and studies of warnings of other natural hazards such as floods, tornadoes, and tropical cyclones.

The overriding aim of hazard warning systems is to alert as many people at risk as possible to the likelihood of an impending disaster, and to tell them what protective actions to perform. Warning system effectiveness, therefore, is measured by the extent of appropriate actions taken which result in reduced casualties and losses. The most elaborate prediction and forecasting procedures are irrelevant if the system does not adequately warn members of an endangered public and those warnings elicit appropriate response.

An intergrated warning system performs three basic functions: (1) evaluation of the physical information about the threat, (2) dissemination of the threat information through predictions or warnings, and (3) human response to the threat information (MILETI, 1975: 11).

Evaluation is the projection of the extent of threat from a hazard to people in an area-at-risk. The processes involved in evaluation are: detection, measurement, collation, and interpretation of available physical information about the threat posed by the impending hazard.

Dissemination of a warning message to people in danger involves a decision about whether or not the evaluated threat warrants alerting the public to the possible danger, and telling them about it and what protective actions to take (NATIONAL ACADEMY OF SCIENCES, 1975).

Response, the third function of a warning system, is the taking of protective action by the people who receive warnings. People's actions, however, are preceded by their interpretations of warnings. These interpretations are directed by many social and psychological factors. To date, the response function has been paid least attention in designing and implementing hazard warnings of any type, even though the main purpose of any warning system is to elicit adaptive responses from people at risk.

These three system functions (evaluation, dissemination, and response) are as applicable for an earthquake warning system as they are for warning systems associated with other types of natural hazards: all three processes will occur and each subsequent process is linked to the prior process.

Warning system effectiveness is, to a large extent, based on performance of the operational linkages between the three basic system functions. Evaluations and subsequent predictions are typically conveyed by a formalized hazard-related organization to emergency groups in the area to be affected. They, in turn, are charged with alerting the endangered public and suggesting appropriate actions.

The link between the evaluation and response functions is normally performed by the dissemination function. The dissemination-response link is as important as any link in a warning system. Yet the dissemination of adequate warnings to endangered publics has also been shown to be a weak link in warning systems as they are currently structured.

3. 2 Disseminating earthquake prediction information

An important concept in explaining human response to hazard warnings is credibility. Is the prediction and or warning perceived by those persons potentially affected as being plausible? Is the prediction or warning believable by those who receive it? In order to identify factors which affect the perception of credibility for an earthquake prediction we talked with 35 officials of organizations selected in a purposive sample of a variety of organizational types.

These persons were asked to rank the relevance of various dimensions of a prediction in order of importance to their believing the prediction. Each respondent was given a list of dimensions to rank. The dimensions were: (1) reputation of the person or group making the prediction; (2) length of leadtime, the amount of time between the prediction being made and the possible earthquake; (3) the magnitude of the predicted earthquake; (4) how specific the prediction was about the date when the earthquake was to occur; (5) how certain the people making the prediction were about whether the earthquake would actually occur; (6) how specific the prediction was as to where the earthquake would happen; (7) how much agreement there was among scientists about the prediction; and (8) observable actions being taken by government organizations behaving as if they believed the prediction. Respondents were asked to rank these factors with other factors they thought would be important to them.

Our interviews supported previous warning response literature in identifying three key elements which influence public perceptions of credibility; for an earthquake prediction factors of major importance will be: (1) the reputation of the prediction; (2) confirmation of the information given in the prediction; and (3) certainty of the threat or how sure the predictor is that the earthquake will occur. To a smaller degree, the size of the predicted earthquake will be an influence in determining credibility (MILETI et al., 1980).

Several factors establish the reputation of the prediction: the professional reputation of the person(s) or organization(s) making the prediction; the validation of the prediction by scientific peers; and the historical track record of earthquake predictions.

By and large the public recognizes the difference between scientific predictions and those put forth by seers or prophets. The public gives much more credibility and has a greater propensity to take adaptive action in response to predictions based in scientific evidence. Furthermore, the greater the scientific reputation of the persons issuing and validating the prediction, the more credibility will be established with the public.

It is not surprising that the track record of earthquake prediction successes and failures will be an important part of the public evaluation of credibility. Since our collective experience with earthquake predictions is scant and will mount slowly, the absence of a track record will impede the establishment of public credibility for the first several predictions.

Worldwide experience will undoubtedly be taken into the public's account of reliability.

The idea behind the importance of confirmation is that there is a general human tendency to seek evidence in support of maintaining the status-quo. People will not want to decide to make changes in their lives in response to an earthquake prediction, so they will use all available information to find reasons to not take actions. For instance, if, in a flash flood warning, some people are urged to leave their homes and the people being warned see no rain, it will be difficult for them to decide to evacuate.

Information that is vague or conflicting in either news of the prediction or appropriate actions to take in response to the prediction will impede public perceptions of credibility. In turn, adaptive public responses will be minimized. Confirmation presents a special problem for early earthquake predictions because the parameters of an earthquake prediction, such as timing of the event, are very likely to change over the duration of the prediction. The more changes in prediction parameters and the more debate about what the parameters really are, the less adaptive responses can be expected.

Certainty of threat refers to how certain the predictor(s) is that the earthquake will really happen and how specific the prediction is about when and where the event will happen. The higher the probability of occurrence the more credible the prediction will be viewed by the public.

The nature of the relationship between intensity of the threat (how large or potentially damaging the earthquake) and prediction credibility is currently too difficult to define, although there appears to be a relationship. The evidence assembled thus far is mixed so the direction of the relationship cannot be specified. Furthermore, for the first several predictions the relationship of precursor data to earthquake magnitude will be imprecise and may vary by as much as ± 1 Richter magnitude during the course of the warning period.

In addition to its influence on adaptive response, the degree of perceived prediction credibility will have a major impact on decisions about whether or not a warning will be issued. It is inconceivable that an earthquake warning would be issued by government officials in the absence of high levels of credibility perception among scientists, at least.

In the United States, current policies indicate that official predictions will be issued mainly by scientists (who may also be Federal officials) and that warnings will be issued by officials of state and/or local governments. When an earthquake warning is issued it will include normative statements concerning what citizens should do to minimize adverse impacts of the potential earthquake. Whether persons take such actions will, in part, be determined by their perceptions of those giving the advice. If individuals perceive that the warning originates from a reputable source they are more likely to perceive the warning as credible and are, in turn, more likely to take adaptive responses. An important question for warning system design is, therefore, will perceptions of reputation of warning source vary systematically for different kinds of citizens who are to receive the warning?

To provide an answer, 243 families in a larger study of earthquake prediction response were questioned about their perceptions of credible information sources. The data were analyzed to determine if preferred sources differed along measures of socio-economic status. Results were compared to the existing structure of earthquake information dissemination to determine if de-facto inequities exist (HUTTON and SORENSEN, 1979).

Family decision makers were asked who would be their preferred source of information about what to do because of an earthquake prediction. The distribution of their responses appears in Table 1. Local officials, including police and fire personnel, were named by over a third

TABLE 1. Preferred warning source.

Source	Response
Local government (including police and fire officials)	36%
Private volunteer organizations (Red Cross)	24%
State government	11%
Federal government	10%
Self	9%
Church	6%
Scientists	2%
Business / chamber of commerce	2%

TABLE 2. Relationships between socio-economic variables and information credibility.

1) Persons with high occupational prestige are likely to view the government as a credible source of advice (sig. at .005).
2) The young and the old are likely to view private volunteer organizations as credible sources of advice (sig. at .05).
3) Persons of middle age are likely to view federal agencies and scientists as credible sources of advice (sig. at .005).
4) Persons with high incomes are likely to view the government as a credible source of advice (sig. at .001).
5) Persons with low incomes are likely to view private volunteer organizations as credible sources of advice (sig. at .01).
6) Persons with high incomes are likely to view federal agencies and scientists as credible sources of advice (sig. at .05).
7) Minorities are likely to view private volunteer organizations as credible sources of advice (sig. at .01).

of the respondents and organizations like the Red Cross were named by the second largest group of respondents.

These data were then analyzed with respect to socio-economic status of the respondent and preferred warning source. Table 2 summarizes the significant findings. Generally, persons of high and low socio-economic status differ in their perceptions of preferred warning sources. Higher status citizens prefer government sources while persons of lower socio-economic status prefer information from the Red Cross.

When these results are compared to the existing U.S. structure for the dissemination of earthquake warnings it appears that there is de-facto inequity inherent in the current warning system. We infer from the analysis that high status groups will be more apt to respond adaptively to earthquake warnings than persons of lower social status.

Issues for concern. The main conclusion from our studies of the dissemination-response link in an earthquake warning system is that earthquake warning systems will have the same basic problems of other hazard warning systems. The lessons from past warning experience will apply to earthquake warnings. The current policies do not reflect enough knowledge of past experiences. For instance, no one agency has responsibility for the ultimate effectiveness of the earthquake warning system. In the absence of ultimate responsibility there will be breakdowns in operation of and between the three functions resulting in more disruption, damage and casualties than should occur. In the United States, the Geological Survey still has the opportunity to extend its responsibility and to apply more of what is known to perfect its prediction and warning policy.

The major problem we anticipate in achieving warning system effectiveness is how to get endangered persons to respond adaptively to the potential threat. As was just reviewed, credibility and equity in the warning process are important factors which promote adaptive response. At the same time, studies have shown that a major constraint to effectiveness is lack of practice. Hazard warning systems deal with very infrequent events. As a consequence, agencies and jurisdictions give a low priority to the adoption and maintenance of warning systems. This constraint also serves to weaken the links between whatever components of a warning system do exist, thereby reducing the effectiveness of the warning system when put into operation. Poor communication exists between those who evaluate threat and those who carry that information to the public. Public warning disseminators usually proceed without sufficient knowledge or training in what information should be contained in public warnings, or the best means for warning delivery. The result is an inadequately warned public and needless deaths, injuries, and disruption.

We raise the issue of the need for practice in warning system operation because of current talk among earthquake scientists that predictions may cause more costs and disruption than they prevent. We realize this talk is at the level of professional gossip rather than policy. But from the data we have there is little indication that people will over-react to earthquake warnings. The problem with earthquake warning systems will be getting people to respond at all. This is the same problem dealt with by warning systems for other hazards.

To illustrate this point we draw again on findings from interviews

TABLE 3. Family response to a perfect earthquake prediction.

Response	Percent distribution
Stockpile supplies	76
Take seriously	75
Stay home more than usual in last few weeks	54
Delay large purchases	52
Save more money	33
Safeguard possessions	31
Buy real estate if prices are down	29
Ask for time off from work in last few weeks	29
Leave area temporarily—mostly vacations	21
Sell home	2
Permanently leave area	0

with citizens from a large urban area of the U.S. The parameters of the prediction to which citizens were responding represent an extremely concise warning compared with what will likely be the case in the first earthquake warning in the U.S. The scenario presented respondents was for a 6.3 R magnitude earthquake with five months leadtime, a narrow, 10 days, time window and 50% level of scientific confidence that the earthquake would occur as predicted. The scenario featured near perfect prediction and warning circumstances to produce widespread perceptions of credibility.

As Table 3 reveals, likely citizen responses are minimal compared with the near perfect parameters of the hypothetical prediction. The main activity is stockpiling of emergency supplies and making family disaster plans. Not one respondent thought that they would move from the area in response to the prediction. The main reasons for illustrating the problem of getting people to respond adaptively is to encourage stepped up action by warning officials to inform the public about earthquakes, earthquake damages, and earthquake prediction.

Public understanding and the development of the new science application will fare better if all information is shared with the public. In the long run, an open and candid approach to publicly issuing even the first few predictions and warnings, despite low levels of statistical confidence, will alleviate some of the natural reluctance of warning officials and make their judgements and responsibilities easier to live with. In turn, a candid approach to the release of prediction information allows the public to learn with the scientists. This open approach could lessen the disruption created by a so-called " cry-wolf " attitude toward predictions that may impede adaptive response.

Current policies which minimize opportunities for official scientific validation of predictions, even those for earthquakes of less than damaging magnitude, may impede public understanding of the developing technology. Most U.S. earthquakes are of non-damaging magnitudes so early predictions may be made more frequently for non-damaging quakes. Opportunities for gaining public experience with which to increase warning system effectiveness should not be missed since these opportunities will be so rare.

3. 3   Interpreting earthquake prediction information

Earthquake risk is a multi-faceted as well as ambiguous concept. Rarely is the term used in a concise and consistent manner. In addition, there are many ways to measure and display earthquake risk.

To many people risk is an event which poses a threat to life, property and well-being. It exists as a cognition or mental image and has no well-defined bounds. To most people in the world earthquake risk assumes this form and can only be defined in view of a person's knowledge of and experience with earthquakes. To others earthquake risk is a set of consequences: the ground shaking, houses swaying, buildings collapsing, and so forth. The risk has a more precise and objective definition and is measured by human factors. To others earthquake risk is defined by a physical system. It is the probability a given event will occur over a specified time period. Within this view risk is a rigorous and quantifiable concept.

Given these different notions concerning earthquake risk we can see that there exists many ways of measuring and displaying risk information. Some of these will come to light in this paper.

It is also useful to view earthquake risk in a space/time perspective. Risks vary with respect to geographical location and change through time. For example persons living in San Francisco have a different image of earthquake risk than residents of Peking. Likewise, the probability of a large magnitude event changes immediately after one occurs. This earthquake risk is dynamic and subject to change via geographical setting.

Earthquakes will occur regardless of whether they are or are not predicted. Objectively a prediction does not change the risk. A prediction does, however, provide more information about the nature of the risk. In other words, a prediction reduces the uncertainty of the event.

Any system which is associated with risk also contains uncertainty. For example we may know that an 7.0 R magnitude earthquake will occur next year but can't specify when and where. Conceptually uncertainty is different from risk.

It is important in understanding the link between risk information and decisions to distinguish between two types of information; that which relates to people's images of the consequences of an earthquake (image of loss); and that related to the levels of uncertainty (image of uncertainty) associated with the prediction.

Subsequent to any prediction a number of activities will ensue which include collecting information, measuring impacts and making decisions to alter the normal course of daily life.   We feel this process is best described and analyzed in a framework proposed by OTWAY (1973) and refined by KATES (1978).

Risk assessment involves a four stage process of identifying, measuring, evaluating, and managing risks.   In the context of a prediction identification is the canvassing and revealing of the possible consequences of an earthquake.   This includes damage from fault-slippage and ground shaking and secondary hazards such as landslides, tsunami, liquefaction, fire, and so forth.   Measurement is the way people attack meaning and likelihood to the identified risks.   This may be in quantified problablistic terms; or in more qualitative terms, such as " likely " or " high."   Evaluation is how individuals and organizations utilize risk information in arriving at a decision about responding to the prediction. This process is very difficult to understand or measure.   Finally, management encompasses the methods to reduce earthquake risks to an acceptable level.

Figure 1 provides a model which links a prediction with risk concepts and risk assessment process.   An earthquake warning or prediction will generate an increase in risk identification and measurement activities.   These may be relatively simple, as in the case of the homeowner surveying the extent of unstable furniture or decorative objects; or complex, when computer simulation models are utilized to evaluate the earthquake resistivity of a dam.   Furthermore the activities may be conducted unwittingly or without conscious design.   These two activities produce a wide range of risk information which will vary in scope and sophistication.   Such information functions to shape people's cognitions of risk, which, in conjunction with other factors, guides risk management

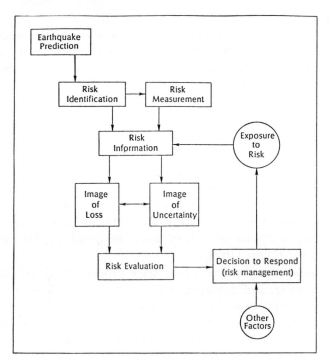

FIG. 1.   Risk information and decisions.

efforts.   The system is dynamic.   New risk information is generated, or changes in the physical system produces new prediction parameters. Likewise actions may reduce exposure to risk which can further alter risk cognitions.   Utilizing this conceptual framework we can now proceed to investigate the process in greater detail, utilizing, when possible, empirical findings from our previous research efforts.

As defined in our model, two broad categories of risk information exist: that affecting images of uncertainty, and that influencing images of loss.   Examples of these two types are provided in Table 4.

Due to a lack of experience with, lack of knowledge on, and the probabilistic nature of predictions, large degrees of uncertainty will accompany a prediction.   Public officials, families, and corporate groups will be bewildered or confused regardless of the scientific parameters that are given.   Several factors will, however, influence levels of uncertainty.   These include lead time, time window, location, and probability.   The lead time of an earthquake prediction may range from a simple

TABLE 4.   Risk information types and communication.

Risk information		Channels of communication	
Image of uncertainty	Image of loss	Source	Mode
Lead time	Physical location:	Scientists	Radio
Time window	fault lines	(predictors other)	TV
Location	Environmental factors	Federal agencies	Newspaper
Probability	Construction	State agencies	Periodicals
	Land use	Local agencies	Interpersonal
	Hazard adjustment	Private firms	Special publications
		Soothsayers psychics	

day or less to perhaps a decade.   As the lead time increases the amount of perceived uncertainty will rise, but, paradoxically, so does the ability to make adaptive responses.   Shorter lead times may, however, raise the propensity to take actions.   Evidence suggests residents of San Francisco would prefer short lead times (JACKSON, 1977).   In addition, in our research business professionals in San Fernando expressed a preference for short lead times while public officials were split over short and longer times.   Time window, the bracket of time in which the earth-quake is expected to occur, may range from a day, to a year, at the extremes.   As time windows grow larger, the amount of uncertainty increases.   In addition, there is a rough correlation between lead times and time windows.   The Chinese experience implies short time windows may accompany short lead times.   Scientific speculations hold that with large lead times, time windows will be greater.   Uncertainty will be generated by vague delineation of the area in which the predicted earthquake could occur.   The more precise the prediction of the epi-center location, the less the uncertainty.   Scientific consensus does not allow speculation on how accurately the target area can be defined.   Whitcombs " hypothesis " covered an irregular area with an 87 mile diameter.   Future predictions may be for certain stretches of a fault line or, for a general metropolitan or county-wide area, for a specific community or for an area overlapping political boundaries.   It is com-monly felt that predictions will contain a confidence level or probability.   One difficulty will be in the interpretation of the probability—what does it mean to the decision-maker?   A large body of psychological literature says that people are inconsistent in their use and interpretation of probabilistic and quantitative information.   In light of this it may be desirable that the warning is very specific about the definition of probability and how to utilize probabilistic concepts in making decisions.

Images of loss are those cognitions which concern the consequences of the earthquake event. Foremost are the personal risk of death and damage to possessions and property. Theoretically images of loss can be viewed as subjective estimates of exposure to risk. Important types of information on exposure to risk likely to be available or disseminated following a prediction include magnitude and intensity data, damage maps, and vulnerability assessments.

In scientific terms there is a rough correlation between earthquake magnitude and damage levels. It is less clear how images of loss are influenced by increasing magnitudes. It is more likely that people perceive different thresholds at which there are marked changes in images of loss than perceive a continuous linear relationship between these variables. Confusion may arise over the distinction between magnitude and intensity scales. Intensity scales may be a more viable means of portraying the potential damages from the predicted event because they are cast in human terms and not in abstract unity of energy.

Following a prediction, maps of potential damage will be published. Such maps could vary from simple ones depicting two or more classes of potential damages to highly sophisticated maps of isoseismic intensities or strong ground motion. Maps released as public information can vary on the basis of underlying assumptions and definitions which may not be well understood by their users. For example, classes of damage potential may be based on computer-generated estimates of strong ground motion, while another is based on intensity data from historic earthquakes. Our research shows damage maps are an important factor in shaping people's image of loss. In a hypothetical situation people were informed that they were living in either a potentially heavy, moderate or low damage area. People in the higher potential damage areas had a stronger belief that the earthquake would occur as predicted. In addition they expressed a greater desire to search for additional information on potential damages. Some people or organizations will seek more detailed or personalized information on damage potential. One of the more common requests may be for engineering assessments of the vulnerability of houses, and public, commercial or industrial buildings. Engineering studies of vulnerability will be easy to obtain if one has the monetary resources and the prediction lead time is sufficiently long. Based on our interviews with engineering firms an assessment of a typical single family house would be highly variable in its cost depending on the detail of the work.   For large structures such as high rises or dams the expense and effort will be considerable.   Because prediction parameters may be imprecise and factors determining damage extent are numerous and complex, such assessments may not provide as sound of basis for estimating damage as the decision-maker might desire.

In this framework we are treating information generally, or as a message.   This information originates from different sources and can be transmitted through several alternative channels.   Examples of sources and channels are provided in Table 1.   While undoubtedly source and channel have a profound effect on utilization and interpretation of risk information space limitations do not allow a discussion of this topic.

While to some extent everyone will be a " risk assessor " following a prediction, some individuals and groups will have greater responsibilities for providing information or take a more active role in identifying or assessing risks.   On the other hand some individuals and groups will have greater level of demand for risk information and greater capacities for understanding and utilizing it.   The following discussion represents an attempt to address these observations in greater detail based upon our field work in California.

Who will make risk assessments and provide information?   At a general level we can distinguish between the assessors in the public and private sectors.   Within the public sector risk information will come from individuals and agencies in different levels of government—Federal, State, and Local.   Within the private sector information will originate from professional sources and amateures including the soothsayer or the local barber shop.

To gain a better picture of what may ensue following a prediction let us examine the suppliers or risk assessors in Federal and State governmental levels.   Figure 2 depicts the perceived agency responsibilities for assessing the physical vulnerability of various segments of the built environment.   These agencies will of course, vary in their abilities and the intensity at which they can provide risk information.

From viewing Fig. 2, it is apparent that there is a differential degree of interest in the various segments of the built environment.   For example four agencies expressed that they would assess the vulnerabilities of dams and related engineering works.   Other segments such as schools and hospitals are of singular agency interest.   When capability to evaluate risk is added to the picture an even greater difference will

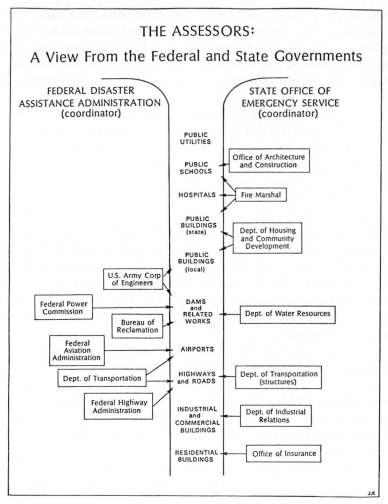

FIG. 2

emerge.   Because of the existing structure of government agencies there will be uneven efforts at assessing risks from a predicted earthquake. A second problem area will be one of coordination.   Our interviews reveal this will be a function of perceived organization role, competition, and limited resources.   A situation that emerged in our interviews was that Agency A said they are not responsible for determining the safety of hospitals but Agency B was.   Yet Agency B did not perceive that duty as part of their responsibility.   This problem is exacerbated when we discovered that hospital officials did not know who was responsible or who could provide information concerning their vulnerability. Such problems can be avoided through better emergency preparedness planning.

Another problem may arise when agencies have overlapping or competing interests.   Coordination is required so that multiple assessments are not made of one dam and other structures are ignored.

A third problem of coordination relates to limited resources and time.   The shorter the lead time of a prediction, the less the amount of risk assessment activities that can occur.   In the short run evaluating vulnerability to the predicted earthquake may be hampered due to competition with agencies' existing mission and duties.   The lack of adequate monetary resources may constrain risk assessment activities as well.

The view from local government presents a more confusing picture. Within local government there are great uncertainties over which individuals or agencies are responsible for identifying and measuring risk. It appears that many departments will make fairly superficial investigations of damage potentials.   A dominant attitude is that " someone else will probably tell me what is safe."   In city and county governments coordinating of activity may be more difficult than at the Federal and State levels.   Politicians, in attempts to minimize public concern, may act to hinder risk assessment activities, and information flows.

Overriding the question of agency responsibilities and functions are the issues of legal responsibility for assessing risks and providing risk information and financing.   To what degree are government bodies legally responsible for identifying and measuring risks?   How active must they be in disseminating information on various risks?   How accurate must the information be?   These and other questions are not likely to be readily answered.   The question of financing is more straightforward.   Given decreasing public revenues, how will risk assessments be financed?   To a very limited degree internal reallocation of funds can be made.   This, however, will not be sufficient.   Added revenues must come from higher levels or different branches of government.   An alternative would be a pricing scheme to bill the recipients of risk information.

The most important involvement within the private sector will come from engineering and construction companies who will be capable of identifying and measuring risks, and from the media, who will disseminate risk information. Some organizations in the private sector will be capable of internally generating risk information. For example, many large utility companies have sufficient expertise to act in this capacity. Most private organizations, however, will not formally act as suppliers. More difficult to identify and trace are the potential activities of soothsayers or others without engineering expertise. It is likely that persons will offer "psychic" services of predicting damages or human fates in non-scientific manner. The extent to which this should be regulated is unclear.

Everyone, with the exception of a hermit or total recluse, will receive risk information following a prediction. The level and type received will be a function of many factors. Some people and organizations will, in addition, exhibit a demand for greater levels of risk information. From our research we can begin to identify some factors which correlate with different levels of demand for such information.

Within the public sector, many of the organizations interviewed expressed a desire for risk information beyond what would be received through mass channels. Those who will be information adversive are those who perceive that their locations and buildings would be safe from earthquake damage, which may be an erroneous assumption. In addition, some will be capable of internally generating the information. Many agencies or public organizations would not know where to go in seeking risk assessments and information. This seems particularly true for organizations removed from the main stream of political activity such as schools and hospitals, but less true for State and Federal agencies. Demands from the private sector will chiefly eminate from commercial and industrial businesses and from homeowners. It is impossible to estimate the level of demand or any price-quantity relationships. That will depend on the prediction circumstances and location. Nevertheless, we can identify business organizations and individuals who are more or less likely to take certain actions to gain added information.

Table 5 summarizes factors which are strongly correlated to business organizations' propensity to: (1) assess their physical vulnerability, (2) assess their financial vulnerability, and (3) assess the consequences of a prediction on their business activities. Several key observations emerge. Organizations with the highest demand for information will be those with large amounts of property which are owned and difficult to move. Those which traditionally acquire large amounts of information before making decisions will have a higher demand. In addition, "hazard conscious" organizations, those with "earthquake-resistant" structures or disaster preparedness plans are more likely to seek risk information.

TABLE 5. Factors related to business organizations' propensity to seek risk information.

Assess vulnerability of physical property orgs. with (which)		Assess financial vulnerability orgs. with (which)		Assess consequences of EQP on business orgs. with (which)	
Rapidly increasing assets	−.39	Larger assets More likely	+.34	Policies are affected by "environmental" information More likely	+.63
Less likely					
Earthquake insurance	−.58	Earthquake insurance	−.50		
Less likely		Less likely			
Property which is difficult to move	+.45	Fire insurance Less likely	−.56	Complex or intricate operations	+.34
More likely				More likely	
Earthquake resistant constructed buildings	+.32	Disaster preparedness plans	+.40		
More likely					
		Use more "environmental" information in making decisions More likely	+.57		

Conversely Fire and Earthquake Insurance are negatively related to demand suggesting that a confidence in recouping loss may discourage such efforts.   Other factors, such as organization size, experience, age, or functions are not linked to demand for risk information.

Our research results do not yield any factors which are strongly correlated with the propensity of individual homeowners to seek added risk information.   We can say, however, it is not related to socio-economic status, dwelling type, age, or belief in scientific ability to predict earthquakes.

The numbers of homeowners who would pursue an assessment will likely keep professional engineers busy.   Our data suggests at a minimum, three percent of the homeowners of Santa Clara County would hire an engineer to evaluate the safety when actively solicited followed a credible prediction.   With more severe risks or greater levels of certainty that figure would likely climb.

It has been established that people and organizations will have varying levels of information because of supply and demand factors. It is also believed that there will be a variation in the ability to utilize risk information. Prior research has solidly demonstrated that people and organizations vary in intellectual skills and the ability to make decisions. Some are better at interpreting quantitative data, or comprehending concepts or synthesizing information. Our research suggests that persons will vary in the comprehension of earthquake prediction information as well. For example when a person claiming psychic powers issued a prediction for an earthquake to hit Wilmington, North Carolina and a university geologist lent credibility, we discovered not everyone had the same understanding of the prediction time, place, and magnitude, nor who predicted it under what circumstances. Furthermore the accuracy of prediction knowledge was related to both ethnicity and socio-economic status. As socio-economic status increased, the accuracy of knowledge did likewise. Closely related to this relationship was that non-whites had less accurate knowledge than whites.

Issues for concern. Government institutions can improve response to an earthquake prediction by regulating and managing risk information. Steps can be taken immediately as well as at the time of a prediction which can alleviate gaps in communication, insure better information is provided and assist people in assessing the level of potential risk.

The following problems have been identified in the course of our research. Where feasible changes in policy which will alleviate the problem are recommended. People will have different capabilities for identifying earthquake and prediction related risks. Few will identify the complete range of possible consequences. Governmental efforts should provide the public a mechanism for systematic identification of the consequences of an impending earthquake. People will not always understand the meaning of risk concepts such as earthquake probability, magnitude, or intensity. Policy should insure that adequate explanations and communication are made public. This requires preliminary research on the optimal manner to issue earthquake warnings. It is very important that the public has a clear grasp of the prediction situation in order to avoid panic or other undesirable responses.

We are quite certain damage maps will be widely publicized following a prediction. Their style and information contents will have a strong influence on public perceptions and response. Policy must insure that damage maps contain the best available information, are not conflicting, identify the assumptions on which they are based, and reflect a confidence level or the accuracy of the information.

The services of detailed assessments of vulnerability and potential damages will be available following a prediction. There is a danger that not all will be provided by a reputable source. Public policy should insure that such services meet a professional standard and reflect ethical business practices. In addition, the assessors should be required to state the confidence level of their work. Because of education, employment, or other situation factors some people and organizations will have easier access to information. Government efforts should provide a means for enhancing the access to information to the disadvantaged.

These recommendations have been derived from field work and analogous warning situations such as hurricanes or flooding. They are general and need detailed refinement. Less important issues are omitted to conserve space. Public policy as measured by governmental actions should be made now in order to be adequately prepared for prediction situation. Furthermore, mechanisms for carefully monitoring risk assessment activities following a prediction should be formulated to grasp a better understanding of how risk information influences perception and behavior.

3. 4 Responding to earthquake prediction information

Public (local, state, federal) and private (corporate and family) decision makers will be faced with decisions that have the potential to reduce earthquake vulnerability and increase emergency preparedness on the basis of an earthquake prediction. Three response strategies are suggested by our work as being available to decision makers to achieve these goals; the strategies can be employed separately or as a mixed set. These strategies are the relocation, reduction, and reallocation of activities, people, and resources. Figure 3 illustrates the potential use of

| | GOAL STRATEGIES | | |
	RELOCATION	REDUCTION	REALLOCATION
REDUCE VULNERABILITY	· activities · people · resources	· activities · people · resources	·activities ·people ·resources
INCREASE EMERGENCY PREPAREDNESS	·activities ·people ·resources	?	·activities ·people ·resources

GOAL

FIG. 3. The strategies and goals of earthquake prediction technology.

these three strategies to achieve the goals of earthquake prediction technology (MILETI *et al.*, 1980).

The act of relocating or moving activities, people, and resources out of the area-at-risk, or to a safer part of the house, or from the nation to the periphery of the area-at-risk, can serve to respectively, reduce vulnerability or increase emergency preparedness. This strategy refers to any decisions or actions which serve to relocate things, people or the things people do in order to reduce risk or increase the ability to respond to earthquake loss.

A second strategy exists which can reduce earthquake vulnerability. Reduction in activities, people, or resources refers to decisions such as temporarily discontinuing the manufacture of vulnerable goods, reducing the labor force, reducing consumption of natural gas, and other activities which would otherwise increase loss potential. We have no evidence to suggest that reduction is a strategy applicable to emergency preparedness.

The reallocation of priorities is a third strategy which can reduce vulnerability and increase preparedness. Organizational, familial, corporate, or government resources can be reallocated or reprioritized to reduce vulnerability, for example, having building inspectors inspect buildings for earthquake risk rather than performing routine inspections, spending money to have a structure appraised for seismic resistance rather than buying a new block wall, or other such decisions. Reallocation can also enhance emergency preparedness. For example, by establishing or realigning mutual aid agreements between organizations, saving money, buying insurance, and so on.

Six key factors are pointed out by our data which explain why some decision makers will make some choices on the basis of a prediction and others will not. These factors will determine what will happen and not happen to reduce vulnerability and increase preparedness after a prediction. Since some actions to reduce vulnerability also carry the potential of negative secondary inpacts, these same factors will play a key role in maximizing the goals of earthquake prediction while minimizing negative consequences. These factors are: (1) image of damage, (2) exposure to risk (insurance), (3) exposure to risk (other factors), (4) access to information, (5) commitment to target area, and (6) resources. Figure 4 summarizes the direction of the causal effects that these factors have on earthquake prediction response variables (MILETI *et al.*, 1980).

What decision makers think will happen if the predicted earthquake occurs will be a major factor in deciding what they think should be

	Image of Damage	Insurance	Exposure to Risk	Access to Information	Commitment to Target Area	Resources
RELOCATION TO REDUCE VULNERABILITY	+	-	+	+	-	+
REDUCTION TO REDUCE VULNERABILITY	+	-	+	+	+	+
REALLOCATION TO REDUCE VULNERABILITY	+	-	+	+	+	+
RELOCATION TO INCREASE EMERGENCY PREPAREDNESS	+	+	+	+	+	+
REALLOCATION TO INCREASE EMERGENCY PREPAREDNESS	+	+	+	+	+	+

FIG. 4. Summary of relationship of causes to responses.

done because of a prediction.   Unfortunately, evidence suggests that images will be inaccurate and inconsistent.   Our data lead us to conclude (see Fig. 4) that image of damage is directly and positively related to all prediction responses.   That is, the more damage that is anticipated, the more likely decision makers will decide to reduce vulnerability and increase emergency preparedness.   Evidence is strong that some will over-estimate damage, many will hold accurate images, and still more will under-estimate damage on the basis of a prediction.   Some decision makers will over-react and more will under-react because of inaccurate images of damage.   Over-reaction to a prediction has costs.   Taking greater actions than are appropriate to the risk could increase the negative secondary effects.   Under-reaction is also costly; it will constrain the goals of prediction, e.g., vulnerability reduction and increased emergency preparedness.   Appropriate response to a prediction will only proceed on the basis of *accurate* images of damage.   Only accurate images will allow the benefits of the technology to be maximized while imposing costs due to sccondary negative impacts well worth the benefits gained.

Surprisingly, evidence suggests that holding earthquake insurance, for some decision-makers, may constrain decisions to reduce vulnerability because of a prediction (see Fig. 4). Corporations and businesses which are insured against earthquake loss are not likely to take actions to further reduce vulnerability. On the other hand it also appears that possessing earthquake insurance has no effect on directing vulnerability reduction or preparedness decisions for citizens (families) or state and federal agencies. Interestingly, having insurance serves to enhance decisions to take emergency preparedness measures for large corporations, local businesses and local government agencies. Holding earthquake insurance is likely the result of hazard awareness and this same awareness, in a prediction setting, enhances decisions to upgrade preparedness by most decision-makers. However, earthquake insurance held by businesses appears to inhibit decision makers from making decisions to further reduce the vulnerability of their physical properties. The strategy operating may be that there is not a practical cost-benefit ratio to inhibit damage if insurance exists to cover the cost of potential damage.

Other factors interplay beyond insurance to determine the risk held by a decision maker to the earthquake hazard. Among these are other adjustments to the hazard, such as building design, value of the structure, proximity to the fault and to secondary earthquake hazards, and other factors.

The effects of exposure to risk on vulnerability reduction and emergency preparedness decisions because of an earthquake prediction are presented in Fig. 4. Vulnerability reduction and emergency preparedness decisions seem more likely to be made by all decision-makers as exposure to risk increases. The greater the extent of already in-place hazard adjustments, however, the lesser the risk, but the more likely to engage in additional activities to reduce risk further. Once again, having installed other earthquake adjustments while lessening risk is likely the result of hazard awareness, and this awareness, in the prediction setting, enhances decisions to further reduce vulnerability and increase emergency preparedness. At the same time, the larger the risk held by decision-makers, the more they have to lose, and the more likely are decisions to decrease vulnerability and increase emergency preparedness.

Information will play a key role in determining who does and does not take strategies to reduce vulnerability and increase emergency preparedness in a prediction. First, different kinds of social units by their nature have more access to information during the routine of life than others. Some organizations, for example, have resources to employ staff whose job it is to get and process information of interest to the organization. Others lack the need for such employees or the resources to hire them. Decision-makers fortunate enough to have good information access as a part of normal life will likely have it in an earthquake prediction. These decision-makers are more likely to reduce vulnerability and increase preparedness. Such is the case for all types of decision-makers for increasing emergency preparedness (see Fig. 4). Such is the case only for business organizations for vulnerability reduction.

A family, a large corporation, a local business, a state or federal agency, or a branch of local government could be tied to a locality for many reasons. A business may be reliant on the local area for all of its income. A corporation may not gain any income from its geographical location; its market could well be kilometers away. Branches of state or federal organizations could be tied to a local area or just happen to be located in that community. How commitment to the local community at risk in an earthquake prediction effects decisions to reduce vulnerability and increase preparedness is presented in Fig. 2. Commitment to the area at risk has no effect on the decisions of public decision makers with one exception. State and federal agencies, local businesses, corporations, and families are more likely to relocate to an area of lesser risk on the basis of a prediction if they have few ties to the local community. Commitment enhances all other adaptive response strategies. These findings suggest that decision-makers with little local commitment are less likely to decide to increase preparedness or decrease vulnerability by reduction and reallocation. They are more likely to reduce vulnerability by relocation to another area. We have no way to speculate how much relocation will go on. We know that it is a possibility, and that too much relocation could elicit negative secondary economic effects.

The findings presented in Fig. 4 are quite consistent. Applicable to all decision-makers, some decisions to reduce vulnerability and increase preparedness, no matter which strategy is used, require resources. The most fluid resource for families, businesses, and government agencies are money. The more resources that are available, the more likely earthquake prediction will result in the goals of increased preparedness and decreased vulnerability. Without a program to make such resources available, the goals of prediction will befall the affluent and the costs will be incurred by the poor.

Issues for concern. Earthquake prediction holds promise to assist in accelerating the already ongoing processes of earthquake hazard vulnerability reduction and increased emergency preparedness. The range of responses to a prediction will help achieve these goals. Some of these same behaviors, however, have the potential to induce negative secondary consequences. The time is ripe to develop effective public policy about what are likely prediction responses and their causes. Our data suggest three arenas in which public policy needs exist.

Information can and must be effectively managed and disseminated in a prediction setting in order to create accurate images of likely damage and loss, and to correct for inequities in existing access to information. Inaccurate images of damage will result from several causes. One is that people deny or diminish threat. Denial will cause some to under-estimate damage (OFFICE OF SCIENCE AND TECHNOLOGY POLICY, 1978: 10). People also tend to assume that their last hazard experience will be what subsequent ones are like. This will cause some to over-estimate and others to under-estimate damage. A prediction will also cause a surge of media attention to the subject of earthquakes. Pictures and stories of damage in other earthquakes in other countries will be shown. Such histories will alter damage perceptions for the predicted earth-quake even though building designs, magnitudes, and intensities of historical quakes may be different from the one predicted. Conflicting damage maps will occur. Images of damage could be changed every time another map is seen or an existing map changed. Decision-makers will be able to choose the map which describes the extent of damage to themselves they would most prefer (none, some, total). In sum, a prediction will provide confused images of damage. No one policy related to earthquake prediction is more important than policy about public information to enable the accurate portrayal of damage to decision-makers.

Our findings also suggest that equity in access to information will be a troublesome concern for policy in a prediction setting. Put simply, because decision makers have different levels of access to information, and because access to information will be influential in deciding to increase emergency preparedness especially, public information campaigns and packages which are presented on the basis of a prediction must be delivered and designed in such a way to maximize equity in information access. This suggests a hearty chore for an earthquake prediction warn-ing system.

Information which is prepared to create accurate images of damage and convey notions of what people can do to increase emergency preparedness and decrease vulnerability must be: (1) prepared in different ways such that different kinds of people all *perceive* the same thing when they are exposed to the information and (2) delivered through different channels such that decision-makers who have different levels of access to information have an equitable chance of receiving it.

Already ongoing efforts to enhance emergency preparedness and decrease earthquake vulnerability hold promise to enhance accurate images of loss and equity in earthquake hazard information in a prediction. Prediction-related policy could achieve desired ends by capitalizing on and enhancing ongoing public information campaigns. An example would be to encourage emerging earthquake hazard microzonation activities (HUTTON and MILETI, 1978).

Our data suggest that the more earthquake and earthquake-related hazard insurance that is in effect in a community for which a prediction is issued (excluding citizen and government policy holders), the fewer decisions will be made to take steps to reduce losses because of a prediction. Since steps to reduce losses on the basis of a prediction will reduce vulnerability, less destruction and disruption will befall a community if less insurance is held by the businesses who reside there. Any program of nationally sponsored earthquake insurance would do well to: (1) require some set of minimal and inexpensive vulnerability reduction strategies in order to be eligible and (2) sanction holders, perhaps by a larger deductible clause, who do not proceed with more vigorous vulnerability reduction efforts in light of a scientifically sanctioned earthquake prediction. Such policy, however, must be well reviewed and appraised such that requirements and sanctions have a maximizing effect on other vulnerability reduction strategies, but a minimal effect on constraining the purchase of insurance in the first place. Prudent review of these issues by national agencies such as the Federal Insurance Administration in the United States is warranted at this time.

The findings of our study also strongly suggest that actions to reap the benefits of earthquake prediction (increased emergency preparedness and vulnerability reduction) will be dramatically constrained by the availability of resources and money. Policy needs exist to address the most practicable means by which to provide predisaster national financial assistance to people, families, government agencies, and businesses to enhance actions in a prediction setting to reduce vulnerability and increase emergency preparedness.

REFERENCES

HUTTON, J. R. and D. S. MILETI, Social aspects of earthquakes, *Proceedings of the Second International Conference on Microzonation*, San Francisco, California, 1978.

HUTTON, J. R. and J. H. SORENSEN, Equity and earthquake warnings, San Francisco, California, Woodward-Clyde Consultants, 1979.

JACKSON, E., Public response to earthquake hazard, *California Geology*, December, 278–280, 1977.

KATES, R., *Risk Assessment of Environmental Hazard*, John Wiley and Sons, New York, 1978.

MILETI, D. S., *Natural Hazard Warning Systems in the United States*, Institute of Behavioral Science, University of Colorado, Boulder, Colorado, 1975.

MILETI, D. S., J. R. HUTTON, and J. H. SORENSEN, *Earthquake Prediction Response and Options for Public Policy*, Institute of Behavioral Science, University of Colorado, Boulder, Colorado, 1980 (forthcoming).

NATIONAL ACADEMY OF SCIENCES, *Earthquake Prediction and Public Policy*, A report by the Panel on the Public Policy Implications of Earthquake Prediction, Advisory Committee on Emergency Planning, Commission of Socio-technical Systems, National Research Council, National Academy of Sciences, Washington, D.C., 1975.

OFFICE OF SCIENCE AND TECHNOLOGY POLICY, Earthquake Hazards Reduction, *Issues for an Implementation Plan*, A report by the Working Group on Earthquake Hazards Reduction, Executive Office of the President, Washington, D.C., 1978.

OTWAY, H., Risk estimations and evaluation, IIASA Conference on Energy Systems, International Institute for Applied Systems Analysis, Schloss Laxemburg, Austria, 1973.

GEODESY AND EARTHQUAKE PREDICTION

Takeshi DAMBARA

Institute of Geosciences, Faculty of Science, Shizuoka University, Ooya, Shizuoka, Japan

Abstract A number of upheavals or subsidences of the ground associated with a great earthquake have been reported in Japan's history. In addition to these historical events, the geodetic surveys have brought out vertical and horizontal deformations of the earth's crust accompanied by many great earthquakes since 1891. In the case of the Niigata earthquake of 1964, for example, it is found that the rate of movement of slow and gradual changes in height of bench marks over a long period of time, at least since 1900, changed around 1956, that is, about 8 years before the earthquake. The subsequent very large subsidence in the mountainous area accompanied by the shock seems to compensate the slow upheaval movements lasting over many years.

In the light of the above data, it is reasonable to consider that the rate of crustal movement which is due to the regional tectonic force is normally small and almost constant, and that after accumulation of a certain amount of strain in the crust, anomalous crustal movements can be expected.

The geodetic approach to earthquake prediction is superior to other means of continuous observation of crustal movement in the fact that observation points cover a vast area of land. As a result of geodetic surveys, an area, over which an anomalous movement is taking place, can be identified. Such information enables us to estimate the possible focal area of a coming earthquake and its magnitude. A relation between earthquake magnitude and the extent of crustal deformation associated with the earthquake is given in Fig. 3 and Eq. (1). It is reasonable to consider that the same relation holds good for precursory movements.

It is certainly a matter of the deepest social concern to know the occurrence time of a coming earthquake.   As for long-range prediction, somewhat deterministic approaches have been suggested by many investigators.   There is no large difference among their results for $M<6$. However, the difference becomes fairly large for $M>7$, and all their relations result in very long precursor times amounting to several decades or more for $M>8$.   Although no conclusive evidence can yet be found, the possibility that the precursor time of an earthquake of $M\geqq$ 8 takes on a value around 10 years cannot be ruled out.   As short-range prediction can be achieved only by continuous observation, no discussion about precursors of this class is attempted here because the geodetic method aims at prediction of moderately long-range.

After some comments on the accuracy of levelling surveys, vertical movements by this survey associated with real earthquakes are discussed in Subsections 5. 1–5. 4.   Vertical movement by tide observation is discussed in Section 6.   As a practical application of this movement to earthquake prediction, method and accuracy of taking differences in the yearly or monthly mean sea levels between two stations are stated in Subsections 6. 1–6. 4.

Horizontal movements and strain obtained by triangulation or trilateration, and horizontal strain by observation of baselines are discussed in Sections 7 and 8, respectively.

1. General Remarks

That abrupt and large changes in the position of triangulation stations and the height of levelling bench marks are often associated with a great earthquake has been well known.   Such changes are certainly caused by a sudden release of crustal strain accumulated to the breaking limit by some tectonic forces working over a long period of time. The above facts lead geodesists to an idea that some anomalous change in the geodetic elements may possibly appear before an earthquake while the strain is approaching to its ultimate value.   Even if such precursory changes may not necessarily be quite large, geodesists are making every effort toward detecting them by means of whatever geodetic methods now available.

A number of upheavals or subsidences of the ground associated with a great earthquake have been reported in Japan's history. Such land deformations were found by watching changes in the height of seashore relative to the sea level, unnavigable ships at a river port, a newly-made waterfall in the river and the like. In addition to these historical events, the geodetic surveys by the Military Land Survey (the predecessor of the present Geographical Survey Institute) brought out vertical and horizontal deformations of the earth's crust accompanied by such great earthquakes as Nobi (1891), Kanto (1923), Tango (1927), North-Izu (1930), Tottori (1943), Nankai (1946), Fukui (1948), Niigata (1964), and so on.

As one of the typical examples, the vertical movements before, at and after the Niigata earthquake of 1964 ($M=7.5$) will be outlined below. In the early stage of the geodetic work in Japan, levelling surveys were carried out to measure the elevations of bench marks only for application to constructing topographic maps, so that resurveys were seldom conducted. Over the Niigata area, however, levelling surveys had been repeated many times since 1954 in order to check ground subsidences probably caused by over-pumping of the underground water. The speed of subsidence was at one time so great that hardly any data near Niigata City could be applied to investigating crustal movements of geophysical significance. At the most violent stage of sinking, the speed amounted to about 1 mm per day. It has become noticed, however, that bench marks along the levelling route running through the mountainous area around Niigata City tended to uplift, starting in 1956. After the earthquake, TSUBOKAWA et al. (1964) claimed that the abnormal change of upheaval rate was a kind of precursor. It appears to the present author, however, that the way in which the data was reduced to a fixed station at Kashiwazaki was incomplete in their study, and that the starting year of the abnormal change was not shown very clearly. DAMBARA (1973a) recalculated all the results of levelling surveys concerned, and added new data obtained after the earthquake. The result is shown in Figs. 1a and 1b.

It is seen from Fig. 1b that the slow and gradual changes in height for both the cases of upheaval and subsidence depending upon bench marks had been taking place over a long period of time at least since 1900, while the rate of movement changed around 1956, that is, about 8 years before the earthquake. The very large subsidence in the mountainous area accompanied by the earthquake seems to compensate the upheaval movements lasting over many years. In the light of the above

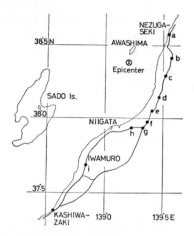

Fig. 1a.   The levelling lines in the Niigata area, and the location of the epicenter of the Niigata earthquake of 1964.   Changes in height at bench marks
with alphabet are shown in Fig. 1b.

data, the following story about the occurrence of the Niigata earthquake may not be very far from the truth.

(1)   The rate of crustal movement which is due to the regional
tectonic force is small and almost constant with small fluctuations.   Such
a movement probably started soon after a large earthquake in the same
area in the past.   In the case of the Niigata area, an earthquake of $M=$
7.4 occurred in 1833 at about 30 km north of the epicenter of the 1964
quake.   It may therefore be assumed that the slow accumulation of
strain seems to have started about 130 years ago.   The movement in
this stage is, so to speak, a normal movement.   It is undoubtedly important to follow up the normal movement by periodic surveys, because the abnormality is to be detected only in comparison with the
normal movement.

(2)   After accumulation of a certain amount of strain in the crust,
it is not unreasonable to expect anomalous crustal movements or other
geophysical phenomena although the physical mechanism of such anomalies is not as yet fully understood.   Roughly speaking, abnormality can
be discriminated by a change in the speed of uplift or subsidence from
a normal value of 2–4 mm/year to an anomalous one of 7–8 mm/year
or more in the case of vertical movement.   In the case of horizontal
movement, a speed larger than the vertical one by 3–4 times may well
be expected by virtue of the Poisson ratio when the direction of tectonic

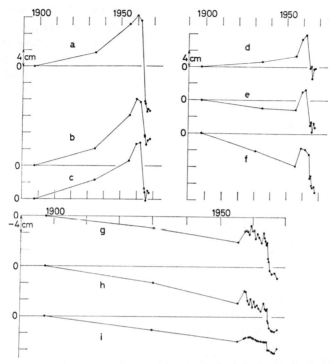

FIG. 1b.   Changes in height at several bench marks associated with the Nii-gata earthquake of 1964.   At bench marks g, h and i, all surveys for investigation of the ground subsidence are used.

force is horizontal.   The Poisson ratio is about 1/4 in the earth's crust.

(3)   As the movement in a normal stage is small, frequent re-surveys of short time-interval do not usually lead to significant results. Taking into account the accuracy of survey, a moderately-long period of time-interval must, therefore, be chosen.   Such a time-interval can naturally be shortened in an abnormal stage of crustal movement.   It is clear, however, that the exact time of the beginning of abnormality can only be more accurately determined by means of continuous observation of crustal movement, such as sea level, extension and tilting of the earth's crust, and so on.

(4)   The geodetic approach to earthquake prediction is superior to any other means of continuous observation of crustal movement in the fact that observation points, i.e. levelling bench marks and triangulation stations, cover a vast area of land.   Meanwhile, tiltmeters, strainmeters and the like can represent changes only at a single point. It is therefore possible to obtain smoothed, or general, so to speak, movement of the earth's crust over a wide area eliminating accidental anomalies. As a result of geodetic surveys, an area, over which an anomalous movement is taking place, can readily be identified, such information leading us to a guess of the possible focal area of a coming earthquake and its magnitude as will be mentioned in later chapters.

(5) Among the various aspects of geodesy, the following means are now believed to be the most useful for earthquake prediction.

(a) Trilateration with an optical distance measuring device (hereafter this device is abbreviated as ODMD).

(b) Rhombic, quadrilateral, or radiant base-line net with ODMD.

(c) Triangulation with a precise theodolite.

(d) Combined method of trilateration and triangulation.

(e) Levelling survey with a precise level.

(f) Observation of mean sea level with a tide gauge.

As stated in (3) of this section, it should be emphasized that continuous observation of crustal movement must be used for obtaining data supplementary to geodetic work, especially for finding the beginning time of an anomalous movement.

2. Prediction of Rupture Zone and Seismicity Gap

It is needless to say that an earthquake prediction should provide the following three elements; rupture zone, magnitude and time of occurrence. The hypocenter as determined by seismic observations is the point from where the first seismic waves are generated. It is important, therefore, to identify the volume, in which the strain energy is stored, rather than the hypocenter for earthquake prediction. In practice, it is essential to specify the epicentral area or rupture zone, as called by the present author, for actual prediction. After the actual occurrence of an earthquake, the rupture zone can be identified by the area which includes epicenters of foreshocks, main shock and aftershocks, or by the area of the pre- and co-seismic crustal deformation.

In order to find an anomalous crustal deformation over a non-identified region, it is necessary to conduct frequent geodetic surveys on a nation-wide scale. This kind of routine work may be called fundamental observation. Although the interval between successive surveys cannot be made extremely short because of expense and labour, this fundamental observation is important in a country such as Japan, most areas of which are under menace of a destructive earthquake.

In addition to the fundamental observation, more frequent surveys must be carried out over special areas where the probability of an earthquake occurring in the near future is believed to be high judging from the seismicity in the past. The Coordinating Committee for Earthquake Prediction (CCEP) in Japan had nominated seven areas of special observation in 1970. These areas were revised in 1978.

Supposing that an anomalous crustal movement is detected in a certain area by geodetic surveys conducted as a part of either fundamental observation or special area observation, the area is nominated to an area for intensified observation. The CCEP nominated the South Kanto District in 1970, and the Tokai District in 1974 to the intensified observation area.

Most earthquakes occur along limited zones of the earth. When these zones are investigated in detail, it is seen that there are special zones where many earthquakes of small magnitude occur at moderately short time-intervals. In contrast to such zones, history indicates that there are areas in which large earthquakes occur repeatedly at almost the same area with more or less similar magnitude. The return period of these quakes is not necessarily constant throughout the areas concerned, but it takes on a value of about 50, 100, and 150 years, respectively, depending on regional characteristics.

The return period as stated above represents the time during which crustal strain is accumulated to the limit of rupture. After the release of accumulated strain by a crustal break, the ruptured zone becomes an area which is called the seismicity gap. The area is subsequently subjected to a new build-up of crustal strain probably because of the effect of plate motion, so that recurrence of an earthquake will only be expected after a period of time long enough for accumulating strain in the earth's crust to its ultimate value. In any case a seismicity gap seems likely to identify a target of a coming earthquake.

As a typical area of a seismicity gap, recent seismicity around the Tokai District as presented by the Japan Meteorological Agency is shown in Fig. 2 (JMA, 1977). It is clearly seen in the figure that there is a vast area of seismicity gap in the sea region off the Tokai District, that is, in the so-called Enshu-Nada including the southern part of the Suruga Bay. Extremely great earthquakes occurred in that region in 1096 (M=8.4), 1498 (M=8.6), 1605 (M=7.9), 1707 (M=8.4), and 1854

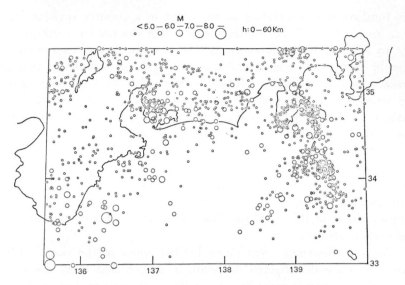

FIG. 2.   Seismic activity in the Tokai District and its outskirts area during the period from Jan., 1926 to June, 1976 (JMA, 1977).

($M=8.4$).   These historical events show that great earthquakes tend to recur in this region with a return period of approximately 120 years or so.   As 126 years have past since the last earthquake of 1854, it is feared that a great earthquake might recur there in the near future.

A number of seismicity gaps can be found in Japan.   A seismicity gap, however, does not always indicate a possible rupture zone because of the following ambiguities: (a) The boundary of a gap is not always clear.   It is possible to get at the feature of a gap which varies depending upon the adopted range of magnitude and period of observation. A seismicity gap should therefore be determined by taking into careful consideration the time-interval of recurrence and magnitude of the main earthquakes in the area concerned.   (b) There is no guarantee that the whole area of a seismicity gap is directly connected to the rupture zone of a single earthquake.   To fill the seismicity gap with a number of earthquakes having magnitudes smaller than that of the hypothetical shock, which releases the whole stored strain energy, may also be possible.   (c) When past historical records of earthquakes is scarce, it is hard to identify a seismicity gap with any certainty.

In Japan, a number of recurrence examples of earthquake of large magnitude are known.   It is said that the return period amounts to

1,000 years or thereabout for some class of earthquakes.   One example is the Kanto earthquake of 818 ($M=7.9$; epicenter 35°2N, 139°3E) and that of 1923 (the same magnitude and location of epicenter) although the exact location of epicenter for the former is not known. Another example is the North-Izu of 841 ($M=7.0$; epicenter 35°1N, 138°9E) and that of 1930 ($M=7.0$; epicenter 35°1N, 139°0E).   Rupture of the earth's crust is, of course, an irreversible phenomenon, so that no recurrence of earthquake in an exactly similar manner to the past one can be expected.   It appears to the author that a period of 1,000 years may have produced a state of crustal strain similar to that of the past one at least for a number of Japanese earthquakes.   It could be speculated that the earthquakes such as Off Izu Peninsula (1974, $M=6.9$) and Izu-Oshima Inshore (1978, $M=7.0$) might be the recurrent ones that occurred some 1,000 years ago.

Another key-point for foreseeing possible rupture zones is provided by active faults although it is usually hard to guess the time of the actual break.   In the case of the San Andreas fault, which is one of the best-known active faults in the world, it is known that some portions of the fault are subjected to a small lateral movement along with very low micro-seismicity.   It may be that much strain is accumulating over such an area, consequently the probability of an earthquake of fair magnitude occurring there must be high.   Although there are many active faults in Japan, no fault which is constantly creeping is found. Active faults in this island arc seem to move only at the time of an earthquake.   Geological and topographical evidences indicate that many of the first-class active faults in Japan seem likely to move with an interval of 1,000 years or so.   It is therefore surmised that an active fault zone where there is no historical record of occurrence of a large earthquake may be the possible target of a future earthquake.

In conclusion, areas related to a seismicity gap or an active fault play an important role when predicting the rupture zone of a future earthquake.   It is certainly more important, however, to detect the regions over which anomalous changes in geodetic and geophysical phenomena tend to occur, but since such regions are often adjacent to a seismicity gap or an active fault, the importance of the latter elements may be appreciated.

3. Prediction of Magnitude

As the strain that can be stored in a unit volume of the earth's crust is more or less the same everywhere, it is physically acceptable to assume that the larger the magnitude of an earthquake is, the wider is the area of crustal deformation associated with that earthquake. Such a relation between earthquake magnitude and the extent of crustal deformation was first obtained by DAMBARA (1966). DAMBARA (1978) recently revised the relation as follows;

$$M = 1.91 \log_{10} r + 4.43 \qquad (1)$$

or

$$\log_{10} r = 0.51 M - 2.26 \qquad (2)$$

where M is the magnitude of earthquake, and r is the mean radius of the area of crustal deformation in units of km. The relation between $\log_{10} r$ and M for 34 earthquakes is shown in Fig. 3, in all of which the crustal deformations were detected by means of geodetic surveys. The standard deviation for a single value of M from the relation as expressed by Eq. (1) is about ± 0.3.

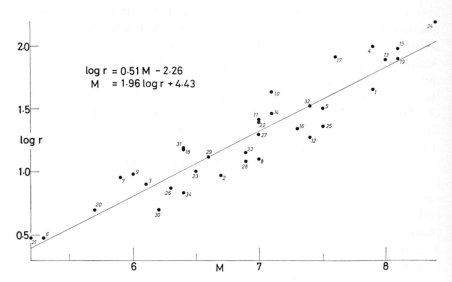

FIG. 3. The relation between $\log_{10} r$ and magnitude, M, where r is radius in units of km of a circle-approximated crustal deformation area accompanied by an earthquake.

Although equation (1) is obtained from data of crustal deformation accompanied by an earthquake, it is assumed that the same relation holds good for precursory movements. Such an assumption may be justified by the fact that anomalous crustal movements tend to occur over the epicentral area where the crustal strain is stored. In fact, at the second active stage of the Matsushiro swarm earthquakes, M was estimated as about 6.6 from Eq. (1) from the extent of the area of abnormal movement at that time. The cumulative magnitude of these swarm earthquakes is 6.3 in total according to the estimate of the Japan Meteorological Agency.

The mean radius of a seismicity gap may also be applied to Eq. (1) for estimating the magnitude of a coming earthquake. In this case, however, other evidence is needed in order to presume that the seismicity gap is related to a single shock.

Equation (2) may provide a measure for estimating the appropriate density of observation stations for prediction of earthquakes of different magnitudes. For example, the radii of anomalous crustal deformation for various magnitudes become as follows;

M	...	8	7	6
r	...	65 km	20 km	6 km

In order to make a distinction between an anomalous and normal area, it is certainly required that at least several points among many stations show abnormal movements. Accordingly, if we are going to predict an earthquake of magnitude 7, it is clear that a triangulation net with a side-length of 10 km or less is required. In the case of levelling survey, bench marks are usually set up along a route with a 2 km interval. But there is no guarantee that the route passes through the central part of the anomalous area, so that a fairly dense net of levelling routes is required for predicting the correct magnitude of a coming earthquake. If the expected magnitude is smaller than 6, the interval of distance between bench marks should be shortened, otherwise we would miss the anomaly altogether even if it does actually take place.

4. Prediction of Occurrence Time

It is certainly a matter of the deepest social concern to know with fair accuracy the time of occurrence of a coming earthquake even if both the rupture zone and the magnitude are only vaguely predicted. As an earthquake is caused by rock fracture, it is inevitable that the very moment of an earthquake occurrence is subjected to a law of probability. Short-range prediction of occurrence time is naturally affected by such a law.   As for long-range prediction, however, somewhat more deterministic approaches have been suggested by a number of investigators.

TSUBOKAWA (1969, 1973) first pointed out that there seems to exist a linear relation between the earthquake magnitude, M, of main shock and $\log_{10} T$, where T is the duration time of abnormal crustal movement.   SCHOLZ et al. (1973) also obtained a linear relation by using many data of plausible precursor time.   Although they did not give an equation, the relation can be empirically expressed as follows;

$$\log_{10} T = 0.65\,M - 1.2 \qquad (3)$$

where T is the precursor time measured in units of day.   WHITCOMB et al. (1973) obtained a relation similar to Eq. (3).   RIKITAKE (1979) who analysed all the available precursor data, the number of which is one order of magnitude larger than that of the above studies, obtained the following equation.

$$\log_{10} T - 0.60\,M - 1.01. \qquad (4)$$

According to Eqs. (3) and (4), the precursor times for various magnitudes are calculated as given in Table 1.   It is seen that there is little difference between the results based on the two equations for $M <$ 6.   However, the difference becomes fairly large for $M > 7$, and both equations result in very long precursor times for $M > 8$.   Such long precursor times amounting to 20 years or more for $M > 8$ do not seem practical for earthquake prediction.   Although the results by SCHOLZ et al. are approximately supported by the dilatancy theory in which the logarithmic precursor time is proportional to earthquake magnitude, it is not necessarily established that the dilatancy theory holds good for a great earthquake of interplate origin.   It is not always self-evident that the earlier the starting time of precursor phenomenon is, the larger is the magnitude of the earthquake.   Although no conclusive evidence can yet be found, the possibility that the precursor time of an earthquake of $M \geq 8$ takes on a value around 10 years cannot be ruled out.

TABLE 1.   Precursor time in units of year calculated from Eqs. (3) and (4).

M	5	6	7	8	8.5
Eq. (3)	0.3	1.4	6.1	27	58
Eq. (4)	0.3	1.1	4.2	17	34

According to RIKITAKE (1979), there is a group of A_1-type precursors, which cluster around $\log_{10} T = -1$ (i.e. T amounts to a few hours) for $M > 5$. It is clear that the precursor time does not depend on earthquake magnitude. Such precursors having a precursor time of a few hours are, of course, very useful for imminent warning of a coming earthquake. As such a short-range prediction can only be achieved by continuous observation, no discussion about precursors of this class is attempted here because the geodetic method aims at earthquake prediction of a moderately long-range.

As already stated, the meaningful time-interval between two successive geodetic surveys is determined by the accuracy of the survey and the speed of crustal movement. It is widely accepted by geodesists that an interval of 3–5 years is usually adequate in the normal stage of movement. When a crustal movement having an abnormally large speed continues over a period of several years, the interval can be shortened. In that case considerably detailed changes in vertical or horizontal movements of the earth's crust can be determined. Moreover, it is even possible to estimate the time when an anomalous crustal movement started by a series of successive surveys at short time-intervals.

In Sections 1–4 the author presented a number of general remarks on earthquake prediction work to be accomplished by geodetic methods. We are now in a position to look into the geodetic methods along with its accuracy, possibility and special notices for earthquake prediction on the basis of actual examples of levelling survey, mean sea level observation, triangulation, trilateration and base-line net measurement in the following sections. The data used are mostly those taken in Japan.

5. Accuracy of Levelling Survey

A levelling survey is usually carried out on the basis of the double check measurements. One of such checks is the restriction of discrepancy in the error of closure along a circuit of levelling route. The restrictions currently adopted in Japan are as follows, where S is distance of a route measured in units of km:

Discrepancy of forward and backward levellings $< \pm 2.5 \sqrt{S}$ mm

Error of closure $< \pm 1.5 \sqrt{S}$ mm

By these restrictions, elevations of bench marks and their changes obtained after an adequate net-adjustment are generally highly reliable. But some cautions are occasionally necessary in actual work.

In spite of the severe restrictions of measurement stated above, a one-sided inclination along a route is sometimes brought out. For example, an almost linear increase of elevations of bench marks along a route of J_1–J_2–J_3 is schematically illustrated in Fig. 4.

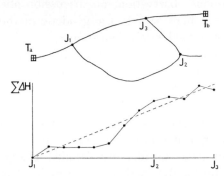

FIG. 4. An erroneous increase of elevations of bench marks along a signle route J_1–J_2–J_3. T_a and T_b are tidal stations where changes in height are given independently.

When the survey result for J_3–J_1 is added to that of J_1–J_2–J_3, the route is closed, and so the erroneous inclination can be extinguished by distributing the error of closure to the elevation of each bench mark according to its distance from the origin point. To use the elevations of J_1 and J_2 as determined by an independent survey along a route connecting two tidal stations T_a and T_b is another way for correcting the erroneous inclination. Even after such treatment, there is a case for which an error-like inclination still remains. Accordingly, an observation along a single route should be avoided. When an erroneous movement seems to overlap on the true movement, the latter can be obtained by setting two bench marks at adequate locations as fixed stations.

Reversed shape of movements between two successive periods in a certain levelling route is another problem. An example is schematically illustrated in Fig. 5, where movements ΔH_{2-1} and ΔH_{3-2} are obtained by subtraction of successive elevations of H_1, H_2 and H_3. When elevations of H_2 are systematically too high by some observational condition, such a reversed shape as shown by curves of ΔH_{2-1} and ΔH_{3-2} will be obtained. If such movements are unnatural from the movement over

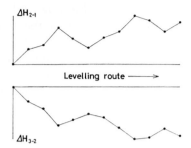

FIG. 5. A reversed relation of vertical movements between ΔH_{2-1} and ΔH_{3-2} obtained by successive elevations H_1, H_2, and H_3.

a long period, it is advisable to abondon H_2 altogether by calculating $\Delta H_{3-1} = \Delta H_{2-1} + \Delta H_{3-2} = H_3 - H_1$.

When a levelling route crosses a mountainous region, an apparent vertical movement, of which the profile along the route resembles very much the topography, is sometimes obtained (Fig. 6). When there are index errors of the two staffs, it is known that an apparent change as stated here may well occur. Whenever such a change is obtained by successive levelling surveys, a careful examination of the survey results is compulsory.

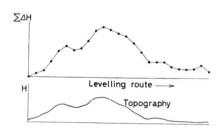

FIG. 6. Resemblance of shapes between vertical movement and topography.

5. 1 Anomalous vertical movement before an earthquake

Geodetic surveys with a comparatively short time-interval have recently been repeated over various portions of Japan under the national programme of earthquake prediction. The author fears that only a few earthquakes in which an anomalous crustal movement occurred before the earthquake are known. This is partly due to the fact that the observation network is incomplete. One of the best examples of precursory crustal movement is certainly the one that foreran the 1964

Niigata earthquake as illustrated in Figs. 1a and 1b. Some more examples of this character will be shown in the following although the levelling data are not necessarily complete.

The levelling net over the Noto Peninsula in the middle part of the Honshu Island is shown in Fig. 7 along with the location of the Nanao earthquake (1933, $M=6.0$). Figures 8 (a) and (b) show the vertical movements for the periods long before, immediately before, and after the earthquake along the levelling routes running on the east and west coasts of the peninsula, respectively. In the figures, it is tentatively assumed that the two bench marks at both ends of the route did not move in order to indicate the movement in relation to the earthquake. It is seen from these figures that the central part of the peninsula which had been

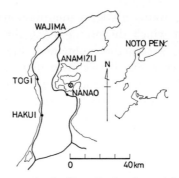

FIG. 7.   Levelling net in the Noto Peninsula, and the location of the epicenter of the Nanao earthquake (Sep. 21, 1933, M 6.0) (DAMBARA, 1973c).

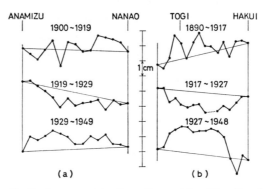

FIG. 8.   Vertical movements along the east levelling line (a), and along the west line (b) in the Noto Peninsula.

comparatively stable over a fairly long period of time subsided before the earthquake, and upheaved at the time of the earthquake.

Figure 9 is another example obtained for the earthquake of the central part of Gifu Prefecture (1969, $M=6.6$).   In this case, subsidence and upheaval are also obtained before and after the earthquake, respectively.   Although a subsidence before an earthquake is demonstrated for these two examples, there are many instances for which an upheaval is observed prior to an earthquake as shown in the case of the Niigata earthquake.

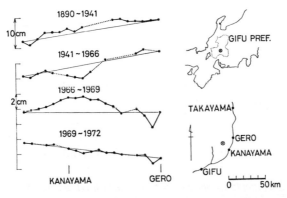

FIG. 9.   Levelling lines and vertical movements before and just after the earthquake in the central part of Gifu Prefecture (by data of the GSI).

Some of the abnormal upheavals do not end up with an earthquake. Such examples recently reported are, for instance, the abnormal upheaval in Kawasaki City near Tokyo and the one along the central part of the San Andreas fault to the north of Los Angeles in California, U.S.A.   An abnormal upheaval with a speed of about 10 mm/year was found by the Geographical Survey Institute in the southern part of Kawasaki City in 1974.   The area was nearly circular in shape with a diameter of about 10 km.   The high speed of upheaval continued over a period of 3–4 years.   This area had been subsiding since 1950 because of extraction of underground water.   After a regulation for restriction of pumping up underground water had been put in effect, however, the water level has been recovering rapidly.   When the anomalous uplift was noticed, many geophysical and geochemical observations were concentrated over the Kawasaki area, but no abnormal changes in any other elements of earthquake prediction were detected.   An earthquake of which the magnitude was expected to amount to 6 or there-about, if it should occur, did not occur after all, and the speed of up-heaval has gradually slowed down.

The anomalous upheaval along the San Andreas fault to the north of Los Angeles was reported by the U.S. Geological Survey in 1976. The anomalous area amounts to about 200×100 km², and the maximum upheaval reached about 25 cm during a period from 1961 to 1976. The area almost coincides with the rupture zone of the great earthquake in 1857. The uplift seems to have stopped growing around 1977, but the anomalous area is said to be expanding to the Arizona-California border.

In the normal case, the following sequence of crustal movement should be observable by geodetic means; normal crustal movement over a long period of time, anomalous movement over a period of several years before an earthquake, anomalous movement immediately prior to an earthquake, abrupt and large deformation accompanied by the earth-quake, transient movement just after the shock, and again slow and normal movement over a long period. This kind of geodetic informa-tion doubtless plays an important role in earthquake prediction.

5. 2 *Anomalous vertical movement accompanied by swarm earthquakes*

A swarm activity of earthquakes had been occurring in a small area near Ito City in the Izu Peninsula about 100 km south-west of Tokyo since February, 1930. The main seismic activities took place during a period from middle of February to early April, and also throughout the whole of May. Following a somewhat weak activity that revived in August, more intense swarm earthquakes hit an area 10 km west of Ito since November 7 culminating in the North-Izu earthquake (Nov. 26, $M=7.0$). The last swarm activity is usually regarded as the fore-shocks directly connected to the main shock. All the regions of these seismic activities are shown in Fig. 10.

During a series of these swarms, the Military Land Survey fre-quently carried out levelling surveys over the earthquake area at the request of the Earthquake Research Institute (ERI) of the University of Tokyo. A part of the result is shown in Fig. 11 (TSUBOI, 1933). It is clear from the figure that land upheaval had continued during the period throughout the seismic activities. The speed of the upheaval was very large for the period from March to November in 1930, the yearly speed having amounted to 190 mm/year.

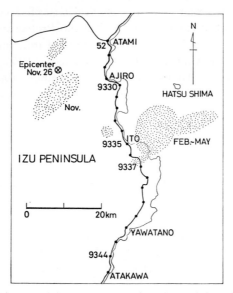

FIG. 10.   Clusters of the Ito swam earthquakes and the epicenter of the North-Izu earthquake of 1930 including its foreshocks and aftershocks.

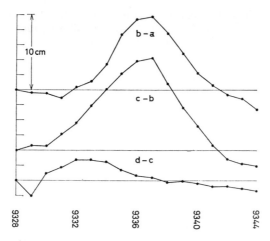

FIG. 11.   Vertical movements associated with the Ito swarm earthquakes. Periods of surveys are: a (1923–1924), b (Mar.–Apr., 1930), c (Nov., 1930), and d (Dec., 1930–Jan., 1931).

A similar high speed of land upheaval was observed at the time of the Matsushiro swarm earthquakes in Nagano Prefecture, Central Japan. The periods of active seismicity can be divided into the following three stages, that is from August, 1965 to February, 1966, from March to July, 1966, and from August, 1966 to December, 1966. Throughout all the stages of activity, frequent levelling surveys were carried out by the Geographical Survey Institute (GSI) and ERI. Figure 12 shows some of the results obtained by ERI (1970).

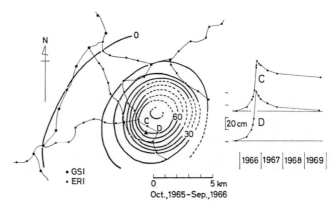

FIG. 12. Vertical movement associated with the Matsushiro swarm earthquakes. The left figure is in the period from Oct., 1965 to Sep., 1966. A triangle mark near a point C is Mt. Minakami.

The center of the upheaved area was located near Mt. Minakami, a Quaternary volcanic dome composed of dacite, while the ultimate upheaval is estimated as 80 cm or thereabout. An extremely large speed of upheaval was reported during the second and third stages of activity. The speed at these stages is so high that the vertical movement is hardly regarded as a precursory crustal movement. It may be that such a movement represents that corresponding to a coseismic deformation accompanied by a single large earthquake.

The two above-stated examples are vertical movements due only to a rather special type of fracturing of the crust. It is in general very difficult to forecast a coming large earthquake only by an anomalous vertical movement. An anomalous uplift similar to the one related to the Ito earthquakes has been found in the eastern part of the Izu Peninsula since 1975. Since October, 1975, the micro-earthquake observation network of ERI observed a series of swarm earthquakes occurring here and there in a limited area in the eastern part of the middle Izu Peninsula. The seismic area extends to the sea between the peninsula and the Izu-Oshima Island. On August 18, 1976, an earthquake of $M=5.4$ occurred at Kawazu Town in the area concerned. Its epicenter was located right in one of the swarm clusters whose activity had commenced since June of that year. These seismic activities are shown in Fig. 13.

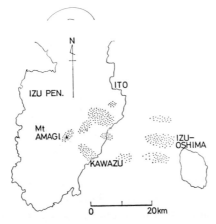

FIG. 13. Swarm earthquakes in the eastern Izu Peninsula in the period from Oct., 1975 to Dec., 1977 (ERI, 1977–1978).

On the other hand, GSI found an anomalous uplift in the northeast Izu Peninsula through a series of intensive levelling surveys. Meanwhile, a comparison of mean sea level between a tidal station at Ito and that of Aburatsubo about 40 km east of Ito, brought out very clearly that an anomalous movement had commenced in January–February, 1975. Figure 14 shows the vertical movement during the period from 1974–1975 to 1976.

As shown in Fig. 14, the area of the maximum upheaval is located in a mountainous area about 10 km west of Ito City, and does not coincide with the clusters of earthquake swarm. It appears that the maximum upheaval zone later shifted a little to the west, while the speed of upheaval rapidly decreased.

On January 14, 1978, a large earthquake of magnitude 7.0 occurred. The quake is called the Izu-Oshima Inshore earthquake. Considerable land deformations were found along the levelling route

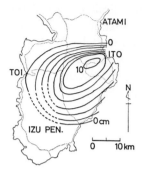

FIG. 14.   Abnormal upheaval in the eastern part of Izu Peninsula during the period from 1974–1975 to 1976 (GSI, 1976a).

that crosses a right-lateral fault which appeared at the time of the earthquake.   The aftershock area is divided into three parts, that is, the E-W line from Izu-Oshima to the east coast of Izu Peninsula, the ES-WN line from the coast toward inland, and the NE-SW line in the central peninsula.   Although the former two zones overlap fairly well the area of swarm cluster, they do not coincide with the anomalously upheaved area.   It is not at all clear whether or not the anomalous uplift has or had any relation to the series of swarm earthquakes and to the Izu-Oshima Inshore earthquake.   The fact that the uplift is still there after these events might suggest another possibility of having an earthquake in the northern portion of the peninsula.

5. 3   Vertical movements and the plate tectonics

It has now become widely believed by geoscientists that the recently-developed plate-tectonics can account for the generation of great earthquakes along an island arc.   According to the plate-tectonics, the fact that the Pacific side of the Japan Islands is gradually displacing landwards and, at the same time, tends to subside can be explained by the subduction of the Pacific plate in the northern part and of the Philippine Sea plate in the central part of Japan, respectively.

The plate motions in a north-west direction compresses the land crust giving rise to accumulation of crustal strain.   When the strain exceeds its ultimate value, rupture takes place at the interface between the land and sea plates causing a rebound of the land plate which has been compressed and pulled down.   At that time, large oceanward and upward displacements of the earth's crust are observed as have often been brought to light by means of geodetic observation. Such a compression-rupture-rebound process as the cause of great earthquakes in front of the Japan arc was first discussed by TAKEUCHI and KANAMORI (1968), and MOGI (1970).

The above-stated process is supported by the fact that step-like topography is found along the coast of the Boso and Miura Peninsulas. Similar topography is also found on the Kii Peninsula and Muroto Cape. Such a topography reflects sudden uplifts of seashore at the times of great earthquakes off the Pacific coast. Repetitions of levelling surveys also brought out very clearly the rebound of the crust in the southern part of Kii Peninsula and near Muroto Cape associated with a great earthquake. It is noticed, however, that the subsidence that takes place over a long period of time after an earthquake does not compensate the coseismic uplift. It seems likely that about 1/3 or so of the uplift is left permanently. Although TAKEUCHI (1972) tried to explain such an overall upward movement by an oscillation theory with a frictional force, an assumption that the land crust upheaves slowly in association with the landward progression of the oceanic plate seems to provide a more simple and plausible explanation of such a gradual upheaval of land.

Figure 15 shows the subsidence of the bench mark at Aburatsubo tidal station near the extremity of the Miura Peninsula after the 1923 Kanto earthquake (DAMBARA, 1973b). The movement is estimated on the assumption that a bench mark at the northern part of the Miura Peninsula has not been subjected to any change in height. Such an assumption seems to be supported by a long range crustal movement over the area concerned. The bench mark at Aburatsubo upheaved by 1.38 m at the time of the earthquake. Although some fluctuations due to unknown causes from the general movement have sometimes

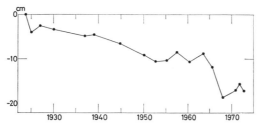

FIG. 15. Slow subsidence at Aburatsubo tidal station in 1923–1972 obtained by periodically repeated levelling surveys.

been observed, the overall vertical movement of the extremity of the peninsula is a subsidence with an almost constant speed of about 3 mm/year. If it is assumed that the subsidence continues at that rate over a long period of time, a total subsidence amounting to 1 m will be reached within a period of 350 years or so.

It is doubtless very important for earthquake prediction to know the mechanism of earthquake generation. Although the plate tectonics seems to be successful to some extent in accounting for the mechanism of generation of large earthquakes occurring along the Pacific Ocean side of the Japan Islands, the theory is only qualitative, and may merely be regarded as a first approximation to the truth. It is not at all clear, as yet, how the mechanism of earthquake generation along the Japan Sea coast and inland of the Japan Islands can be explained.

5. 4 Vertical movement in the Tokai area

The Tokai area facing the Pacific Ocean in the middle part of the Honshu Island has been one of the areas which has repeatedly suffered from great earthquakes throughout the historical time. It is feared by many Japanese seismologists that the area might again be hit by a destructive earthquake in the near future. It seems that we have every reason to expect a great earthquake there judging from the accumulation of crustal strain which is caused by the north-westward motion of an oceanic plate called the Philippine Sea plate extending to the south of the area concerned.

The levelling net over the area was founded around 1885. The net which is probably useful for predicting the coming earthquake in the Tokai area is shown in Fig. 16. The dense net near Omaezaki Point was newly established after nominating the area to a special observation area. In the same figure, the locations of tidal stations that are registered at the Coastal Movements Data Center in GSI are also shown in reference to Section 6.

The levelling surveys over the area concerned may be assigned to five epochs of 1889, 1929, 1951, 1966, and 1972 with some epoch reduction. The vertical movements during the three periods of 1889–1929, 1929–1951, and 1951–1966–1972 are shown in Figs. 17, 18, and 19, respectively. The surveys in 1966 and 1972 are combined because there is no large rate-change of movement between the two periods of 1951–1966 and 1966–1972.

In Fig. 17, the movements accompanied by the 1923 Kanto earth-

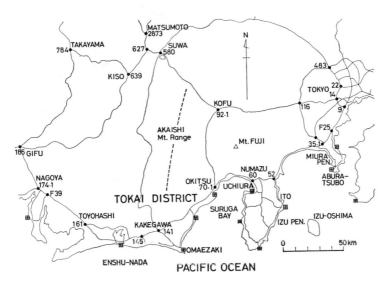

FIG. 16. A southern part of the levelling net in the Chubu District, and locations of main tidal stations.

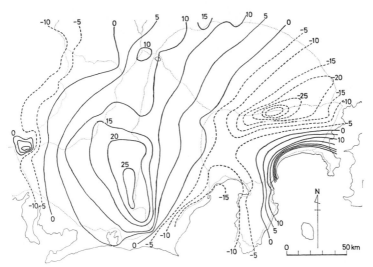

FIG. 17. Vertical movement in the southern Chubu District during the period from 1889 to 1929 in units of cm.

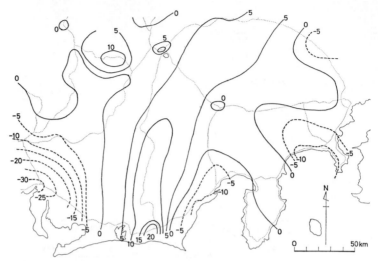

FIG. 18.   Vertical movement in the southern Chubu District during the period from 1929 to 1951 in units of cm.

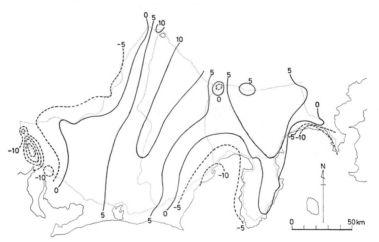

FIG. 19.   Vertical movement in the southern Chubu District during the period from 1951 to 1972 in units of cm, where the result of survey around 1966 is taken in.

quake, and the 1891 Nobi earthquake are seen in the eastern and the western parts, respectively. Apart from these movements, two conspicuous movements are found in the central part of the Tokai area. The first is the subsidence area along the western coast of the Suruga

Bay, the maximum speed of subsidence amounting to 4 mm/year. The second is the upheaval zone in the southern part of the Akaishi mountain range to the west-north of the subsiding area. The speed of these recent crustal movements is roughly twice as high as that of the Quaternary movement as revealed by geological investigation. Such a discrepancy may be caused by repetitions of slow subsidence over a long period of time and imperfect rebound of upheaval accompanied by a great earthquake in the area concerned.

The movements pointed out in the above are clearly seen in Figs. 18 and 19. The total movement during a period of 83 years from 1889 to 1972 indicates a subsidence of 35 cm along the western coast of the Suruga Bay, and an upheaval of 50 cm in the southern part of the Akaishi mountain range. If these movements are interpreted as a tilting motion in an SE-NW direction, the tilting in this direction is about 3.4″ in total, and its speed amounts to 0.041″/year. Such a tilting movement is not very much different from the one for a normal stage of crustal movement. Should the movement in the future tend to deviate from the normal movement, it can surely be said to be abnormal. The recent speed-up of subsidence at the western coast of Suruga Bay, especially in the area near Omaezaki, will be stated later.

6. Vertical Movement by Tide Observation

Changes in the sea level observed at a tidal station are nothing but those in the height of land relative to the sea level. Accordingly, if changes in the sea level by oceanographic and meteorological origins are eliminated, the remaining changes must reflect the vertical movements of the earth's crust. The accuracy of finding a crustal movement by means of tidal observation is entirely controlled by that of eliminating astronomical tides having various periods, seasonal variations of meteorological origin, and disturbances due to oceanographical conditions such as a mass of cold water and the like.

A number of methods for correcting disturbances of meteorological origin have been put forward by some investigators. Changes due to atmospheric pressure can readily be corrected by making use of a coefficient of 10 mm/mb. However, the influence of temperature change of sea-water is somewhat complicated, because it is the synthetic effect of temperatures at various depths. On top of these effects, the influences of change in ocean currents and of a mass of cold water can hardly be eliminated with much accuracy. It is almost hopelessly difficult to eliminate all the effects of meteorological and oceanographic origin from the results of tidal observation at a single station.

Periodic variations of sea level due to the tide-generating force can usually be eliminated by calculating yearly, monthly and daily means although the accuracy of elimination is high for a long period variation and low for a short period one.

Another method of eliminating noise in the sea level is to take the difference in the mean sea levels between two adjacent tidal stations. In this case, it is assumed that meteorological and oceanographical disturbances affect the sea levels at the two stations to more or less the same extent.

6. 1 Yearly mean sea level

By taking a yearly mean of the sea level, almost all astronomical tides of short period, and seasonal variations of meteorological origin can be more or less eliminated. But long-period tide such as that having a period of 18.6 years, and the eustatic uplift of sea level which is said to be caused by the melting of glaciers, cannot be eliminated. Attention should be paid to these remaining changes in the mean sea level when a long-range movement of the sea level is discussed.

It has been estimated that the standard deviation of a single yearly mean sea level from that smoothed over a long period of time amounts to about ±4 cm. Accordingly, a tidal observation over a period of several or more years are required in order to obtain a clear statement of the trend of the vertical movement of the earth's crust.

Changes in the yearly mean sea levels at Aburatsubo at the tip of the Miura Peninsula, and at Uchiura at the northern head of the Suruga Bay are plotted in Fig. 20. The distance between these two sta-

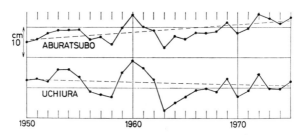

Fig. 20.   Changes in yearly mean sea level at Aburatsubo and Uchiura.

tions is about 80 km. It is seen that the vertical movements at both the stations tended to change with an almost constant rate during the 25-year period from 1950 to 1975. The estimated speed of movement is about -2.6 mm/year at Aburatsubo, and about $+0.8$ mm/year at Uchiura, respectively. From the results of levelling survey, it is already known that the subsidence speed at Aburatsubo is about -3 mm/year (Fig. 15), and the change in height at Uchiura is nearly zero. If the change in the mean sea level at Uchiura is assumed to be very small, the change at Aburatsubo relative to Uchiura becomes $-2.6-(+0.8)=-3.4$ mm/year, which coincides very well with the results of levelling survey.

Although some disturbances of short range that deviate from the general trend are seen in Fig. 20, it is difficult to assume that the crustal movement is similar at the two stations of which the distance is fairly large. It is natural, therefore, to presume that these disturbances must have been caused by some particular sea conditions. For example, a considerably large disturbance for the period from 1958 to 1963 is seen in both changes in Fig. 20. The irregular change during this period was noticed at many tidal stations along the coast from the Shikoku Island to the South Kanto District. It has become customary to explain the phenomenon by some change in the ocean current of Kuroshio (the Japan Current), and also by the mass of cold water that appeared off the Pacific coast over a fairly long period.

In conclusion, it may be said that the figures for a vertical movement obtained from changes in the yearly mean sea levels is fairly reliable when the period treated is long enough, i.e., 10 years or longer. But, no short-range change can be regarded as a true crustal movement until various checks have been completed. Long-range changes in height at a tidal station can also be used for checking the results of levelling survey.

6. 2 Difference in yearly mean sea level between two stations

As changes in sea level of meteorological and oceanographic origin usually appear over a certain sea region, the disturbances can be approximately eliminated by taking the difference in the mean sea level between two stations under favourable cases. The standard deviation of each value is reduced to about ± 2 cm after such an operation. The change is, of course, relative, and L_A-L_B means the change in ground height at B-station relative to that at A-station.

Figure 21 shows the changes in yearly mean sea level at Omaezaki and Uchiura stations located in the Tokai area, both of which are operated by JMA, and the changes in the difference L_U–L_O. It is seen in the figure that the fairly large disturbance originally observed for L_U and L_O vanishes by taking the difference as can be seen from L_U–L_O, and that the subsidence at Omaezaki has been accelerating since around 1974. Such an increase in the speed of crustal movement has also been found by a repetition of levelling surveys as shown in Fig. 22. As survey intervals of levelling are usually so long, the year of commencement of abnormal subsidence cannot be exactly determined. Although the speed-up of subsidence appears to be abnormal, no corresponding abnormal changes in other geophysical elements are found.

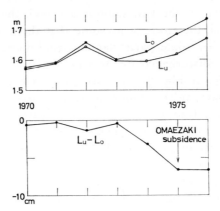

FIG. 21. Changes in yearly mean sea level at Omaezaki (L_O) and Uchiura (L_U), and change of L_U–L_O.

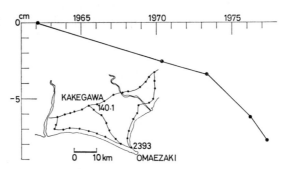

FIG. 22. Change in height of BM2593 at Omaezaki relative to BM140.1 at Kakegawa City by levelling survey (GSI, 1977).

There remains a possibility that the abnormal subsidence is caused by a fluctuation in the slow subsidence of long range which is not necessarily connected to earthquake occurrence.

6. 3   Difference in monthly mean sea level

A seasonal variation having an amplitude of about ±10–30 cm is usually superposed on a curve of monthly mean sea level. This variation is not exactly periodic, and its amplitude and phase vary year by year. Such seasonal variations are demonstrated in Fig. 23 for which the sea level data observed at Omaezaki and Uchiura are used again. Judging from the large fluctuations of the sea level due to seasonal variation, it seems almost impossible to obtain the change in the vertical movement of land relative to the sea level on a time scale less than a year on the basis of monthly mean sea level at a station. It may be observed, however, similar amplitude and phase are obtained even for seasonal variation of monthly mean sea level at stations of which the mutual distance is not very large. The differences in the monthly mean sea level between Omaezaki and Uchiura is shown in Fig. 24 for 1973–1976. It is seen that the abnormal subsidence at Omaezaki which has already been shown in Figs. 21 and 22, occurred some time in February–March, 1974. The year of commencement had already been estimated by making use of the yearly mean sea level. It is now demonstrated that even the month of commencement can thus be approximately determined.

The recent abnormal upheaval in the eastern part of the Izu Peninsula was stated in Subsection 5. 2. In order to find the time of commencement of this upheaval more exactly, differences in the monthly

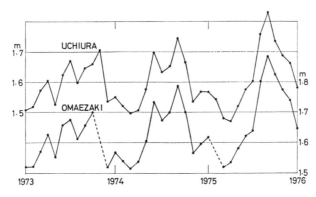

FIG. 23.   Changes in the monthly mean sea level at Omaezaki and Uchiura.

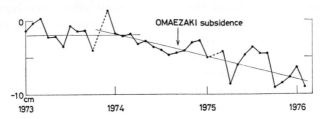

FIG. 24.　Change of difference between monthly mean sea levels at Uchiura and Omaezaki.

mean sea level between that at Aburatsubo and Ito are shown in Fig. 25.　The Ito tidal station is located at a point near the boundary of the abnormal upheaval area, so that the land deformation there is smaller than that in the maximum upheaval zone.　On the other hand, Aburatsubo has been subsiding slowly since the 1923 Kanto earthquake. Strictly speaking, the change in height at Ito relative to Aburatsubo should be corrected for the slow subsidence at Aburatsubo.　It is sufficient, however, to look at the simple differences in order to detect the commencement time of anomalous land deformation.　From Fig. 25, it is found that the upheaval at Ito commenced some time around January–February, 1975, and that after an upheaval of several centimeters which had been continuing until around March, 1976, the speed of upheaval has decreased rapidly.

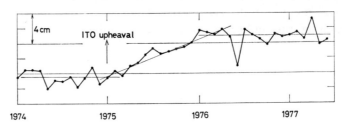

FIG. 25.　Change of difference between monthly mean sea levels at Aburatsubo and Ito.

6. 4　Short range change in sea level

It is no easy matter to detect an abnormal movement of the crust on the basis of daily mean sea level, because a daily mean barely eliminates semidiurnal and diurnal tides.　There are, however, several instances for which an abrupt and large change in height relative to the sea level accompanied by an earthquake are successfully detected. Nevertheless it is necessary to take the difference in sea level between two stations even for a short-range change in a manner similar to those in the previous subsections. An example is shown in Fig. 26, which was obtained after the Off Nemuro Peninsula earthquake (1973, $M=7.4$). In this figure, the subsidence by about 10 cm at Hanasaki relative to Kushiro is clearly seen.

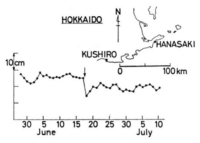

FIG. 26. Change of difference between daily mean sea levels at Kushiro and Hanasaki during the time of the Off Nemuro Peninsula earthquake of 1973 (GSI, 1974a).

A number of examples of anomalous sea retreats immediately before an earthquake are known in Japan. The precursor times of these sea retreats ΔT, are shown in Table 2.

TABLE 2. Time elapsed from sea retreat to earthquake, ΔT.

Earthquake	Year	M	ΔT	Location
Ajigasawa	1793	6.9	8 hrs	Aomori Prefecture
Ogi, Sado Is.	1802	6.6	4 hrs	Niigata Prefecture
Hamada	1872	7.1	0.3 hrs	Shimane Prefecture
Tango	1927	7.5	2.5 hrs	Kyoto Prefecture

Such sudden sea retreats must have been caused either by a local uplift of seashore or by a crustal depression at the sea bottom. Fairly large foreshocks are said to have occurred before the main shock of the Ajigasawa (several hours before), Ogi (4 or 5 hours before), and Hamada (several hours before) earthquakes. There is a possibility that these sea retreats were caused by a certain pre-slip deformation accompanied by a large foreshock. It is interesting to note that all the cited retreats occurred along the Japan Sea coast where the amplitude of daily tide is less than about 50 cm.

An ordinary tide gauge of Kelvin type records daily tide with a reduced scale of 1/20, and with a drum speed of 1–2 cm/hour. Supposing that the amplitude of daily tide is 150 cm, the recorded amplitude would, thus, be 75 mm. A tide gauge of digital type has a resolving power similar to that of the Kelvin type. Accordingly, the resolving power is in general 5–6 mm in amplitude and several minutes in time, so that a change in sea level of about 10 cm can be detected.

As a result of the above discussion, it is suggested that a precursory change in sea level exceeding 10 cm and continuing a few hours or more can be detected, should it actually occur prior to an earthquake. A workable system of imminent earthquake prediction based on tide gauge observation would be something like the following. Tidal records taken at a number of stations are to be telemetered to a center on a real-time basis. Differences in sea level between pairs of all the stations are to be automatically processed, so that real-time changes in sea level, 10-minute mean sea level say, at any station relative to other stations can be watched constantly. With such a system, a premonitory land deformation relative to sea level of the above-cited extent may certainly be detected.

Such a system of imminent prediction should be supported by a reliable prediction of long range, not only based on tidal observation but also on all other geophysical elements of various disciplines.

7. Horizontal Movement by Triangulation or Trilateration

Horizontal displacement of a triangulation station can be deduced by comparing the results of two successive triangulation surveys. The displacement of a triangular station thus disclosed for the period between the two surveys is customarily expressed by a displacement vector in the horizontal plane or its northward and eastward components. The displacement also enables us to calculate the horizontal strain within a triangle formed by any combination of three stations. In this case, principal strains (contraction and extension) and the maximum shearing strain are usually calculated.

The distance between two neighbouring first-order triangulation stations is usually about 40–50 km. Triangulation observation is carried out by making use of a theodolite of high precision which even under certain severe conditions, assures a high accuracy of measurement. The length of a base-line used is measured with a wire-scale made of invar. Since 1970, the length between two stations at appropriate locations, which is directly measured with an ODMD, is used in place of the classical base-line. The overall accuracy of the first-order triangulation is estimated as about 3×10^{-6} at the utmost, that is, about 10 cm for a length of 40 km.

Horizontal movement of the crust in Japan usually having a speed amounting to 10 mm/year or less under normal conditions, a 10-year accumulation of movement is at least necessary if it is to be detected by a first-order triangulation. A resurvey with a shorter time-interval would hardly be of any use except for horizontal movement having an anomalously high speed.

The first-order triangulation covering the whole of Japan was carried out twice in the past during 1883–1915 and 1949–1967. In addition to these surveys, many resurveys were carried out over limited areas struck by a destructive earthquake in order to recover the proper position of a triangulation station. As a result of these resurveys, many examples of crustal movement associated with a large earthquake were brought to light. These data certainly provide an extremely useful set of data for studying earthquakes and their prediction.

As a part of the national project of earthquake prediction in Japan a systematic precise trilateration work with an ODMD has been planned by GSI since 1973. The first- and second-order triangulation stations about 6,000 in number are to be surveyed under the plan. In this case the side-length for a trilateration net becomes about 10 km. The relative accuracy of this new survey is expected to be 2×10^{-6} or so, that is, 1–2 cm for a regular side length. Accordingly, the workable time-interval for successive surveys can be shortened from 10 to 3–5 years. It is anticipated that the high density of stations will enable us to detect an abnormal area which might become the seat of an earthquake of magnitude 6–7 in the future.

The amount of strain accumulated over a certain area during a period of several decades can be obtained from the horizontal movement obtained by comparing the new result of precise trilateration with that of former triangulation. The speed of accumulation of strain will also be obtained from repetitions of precise trilateration with a short time-interval. By using these quantities, occurrence time of a coming earthquake may be estimated, at least probabilistically, on the assumption that the ultimate strain of the earth's crust to rupture is 50 or 100 micro-strain (1×10^{-6}).

7. 1   False displacement vectors

When horizontal displacements of stations are to be calculated, the simplest assumption is to assume that one station and one azimuth have been subjected to no change throughout the period of two triangulations.   If this assumption is really correct, no trouble occurs, and correct displacement vectors will be obtained.   It is feared, however, that the selection of one fixed station and one fixed azimuth among many stations usually depends on personal intuition.   For example, stations in remote mountains, or at the outskirts of damaged area by an earthquake are often apt to be selected.   There is, however, no objective guarantee that the assumption is correct.

When the fixed station and azimuth, as mentioned above, involve implicit movements, it is known that the assumption gives rise to some trouble.   Figure 27 shows, in a schematic triangulation net, the appearance of false displacement vectors which are caused when the fixed station involves an implicit displacement, ΔX, in the X-component.   Directions of these false vectors become radial, and their amounts of displacement are proportional to the radial distance from the fixed station.   If ΔX is negative, the direction converges to the fixed station. In the case of ΔY, a shape of displacements similar to that of ΔX is obtained.   Such a false displacement will hereafter be called the false signal caused by the scale factor, because ΔX means change of side length.

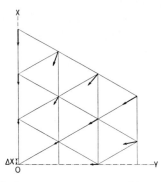

FIG. 27.   False displacement vectors caused by an implicit displacement, ΔX.

Figure 28 shows the appearance of false displacement vectors which are caused when the fixed azimuth involves an implicit rotation, ΔA. Their directions are perpendicular to the line connecting a station to

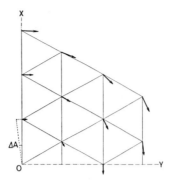

FIG. 28.   False displacement vectors caused by an implicit rotation of azimuth, $\varDelta A$.

the fixed station, and amounts of displacement are again proportional to the radial distance.   Such a false displacement will hereafter be called the false signal caused by the rotation factor.

An apparent displacement of a station is usually given as a resultant vector which is the result of combining true and false signals.   Even if one assumes a fixed station and a fixed azimuth, fairly correct displacements can only be obtained when the true signals are large enough as in the case of deformation accompanied by a great earthquake.   For a normal stage of crustal movement, however, false signals, which are of the same order as the true signals, must be removed from the apparent displacement in order to get a real displacement.

In order to process the survey results most reasonably in relation to the displacement of each station, MITTERMAYER (1972) presented a method of readjustment that relies on a condition for which $\varSigma V^2=0$, where V is the displacement vector.   A condition of $\varSigma V=0$, is sometimes used, too.   These conditions seem to work very well for the results of a survey over an area of inactive crustal movement for which the least squares method may be applied.   Such methods of analysis seem, however, to have no physical meaning for obtaining actual horizontal movements over an area of active crust because the displacements of individual stations cannot be treated as accidentally determined quantities.

The horizontal displacements associated with the North-Izu earthquake of 1930 were discussed by TSUBOI (1933), and were recalculated by SATO (1973).   The latter result is shown in Fig. 29.   In his calculation, three stations at the outskirts of the area concerned are fixed.   The

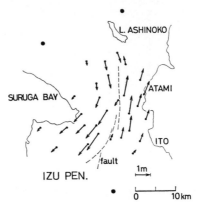

FIG. 29.   Horizontal displacements in the northern part of the Izu Peninsula associated with the North-Izu earthquake of 1930.   Large circles are fixed stations.

large displacement vectors clearly show a movement of the left-lateral fault which appeared at the time of the earthquake.

Forty-three years after the earthquake, a trilateration survey supplemented by an angle observation was carried out by GSI.   Figure 30 is the vector representation of the displacements during the period from 1933 to 1973, in which station No. 1 (see Fig. 31) and the azimuth connecting No. 1 to No. 13 are fixed.   It is highly probable that false signals are mixed up with true signals in this example.

When there are some stations of which the true displacements are believed to be negligibly small, their displacements must represent false signals.   By a method of finding such false signals which is tentatively

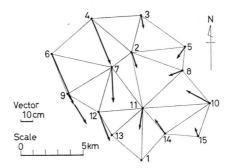

FIG. 30.   Horizontal displacements in the period from 1933 to 1973 in the northern part of the Izu Peninsula (GSI, 1974b).

called a method of signal coincidence (DAMBARA, 1976a), the result is readjusted as shown in Fig. 31. It is seen that the displacements are large in the eastern area, and their directions are nearly the same as those of the movements at the time of the 1930 earthquake.

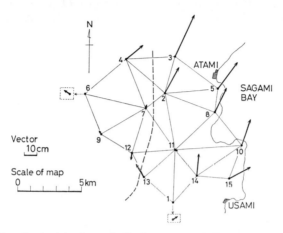

FIG. 31. Revised horizontal displacements of Fig. 30. Vectors at stations No. 1 and No. 6 are enlarged by 10.

It would not be unreasonable that, should the displacements as discussed here continue further, a left-lateral fault would again occur along the previous fault line as illustrated in Fig. 32. As stated in Section 2, two earthquakes occurred in this area with a time-interval of 1,089 years. The left-lateral fault which appeared in 1930 is associated with coseismic displacements of 2–3 m. As the speed of displacement during 1933–1973 is estimated as about 5 mm/year, it is supposed that the displacement will reach 2 m within a time of 400 years or thereabout.

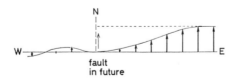

FIG. 32. Accumulation of displacement, and fault movement in the future.

7.2   Calculation of horizontal strain

Horizontal strain can be estimated from displacements of triangulation stations, or directly from changes in the side-length of a triangle (cf., JAEGER, 1964).   Let us assume that a point $P(x, y)$ moves to $P'(x+u, y+v)$ and a nearby point $Q(x+x', y+y')$ to $Q'(x+u', y+v')$.   Then, displacement components u' and v', are given as follows:

$$
\left.
\begin{aligned}
u' &= \frac{\partial u}{\partial x} x' + \frac{\partial u}{\partial y} y' \\[2mm]
v' &= \frac{\partial v}{\partial x} x' + \frac{\partial v}{\partial y} y'
\end{aligned}
\right\} .
\tag{5}
$$

We may put

$$
\left.
\begin{aligned}
\varepsilon_x &= \frac{\partial u}{\partial x}, \qquad \varepsilon_y = \frac{\partial v}{\partial y} \\[2mm]
\gamma_{xy} &= \gamma_{yx} = \frac{\partial u}{\partial x} + \frac{\partial v}{\partial y}
\end{aligned}
\right\}
\tag{6}
$$

where ε_x and ε_y are linear strains in the x and y directions respectively, and $\gamma_{xy} = \gamma_{yx}$ is an angle modulus between two axes representing shearing strains.   Now a linear strain ε_α in any direction α is given by

$$
\varepsilon_\alpha = \varepsilon_x \cos^2\alpha + \gamma_{xy} \sin\alpha \cos\alpha + \varepsilon_y \sin^2\alpha.
\tag{7}
$$

In Eq. (7), ε_x, ε_y, and γ_{xy} are unknowns to be determined from actually abserved data.   When changes in length of the three sides of a triangle are observed, the three unknowns can be determined.   The principal strain, ε_1 and ε_2, and the azimuth of ε_1 from the x-axis, ϕ, are then calculated from the following relation:

$$
\left.
\begin{aligned}
\varepsilon_1 &= \frac{1}{2}(\varepsilon_x + \varepsilon_y) + \frac{1}{2}\{(\varepsilon_x - \varepsilon_y)^2 + \gamma_{xy}^2\}^{1/2} \\[2mm]
\varepsilon_2 &= \frac{1}{2}(\varepsilon_x + \varepsilon_y) - \frac{1}{2}\{(\varepsilon_x - \varepsilon_y)^2 + \gamma_{xy}^2\}^{1/2} \\[2mm]
\tan 2\phi &= \frac{\gamma_{xy}}{(\varepsilon_x - \varepsilon_y)}
\end{aligned}
\right\} .
\tag{8}
$$

ε_1 and ε_2 are the maximum and minimum linear strains, respectively, and they are perpendicular to each other.   The dilatation, $\varDelta$, and the maximum shearing strain, γ, are given by the following simple relations:

$$
\left.
\begin{aligned}
\varDelta &= \varepsilon_1 + \varepsilon_2 \\[2mm]
\gamma &= \varepsilon_2 - \varepsilon_1
\end{aligned}
\right.
\tag{9}
$$

All these horizontal strains are regarded as the values at the geo-metric center of the triangle in question.

Comparing the results of the first-order triangulation survey of a certain epoch to those of the last epoch, horizontal strains for each tri-angle of the net developed during the period between the two surveys can be estimated. Except areas considerably deformed in association with a great earthquake, a mean horizontal strain accumulated during the recent several decades usually amounts to 5–10 micro-strain. Ac-cordingly, the speed of strain accumulation is estimated 0.1–0.2 micro-strain per year, such a speed probably representing the one in a normal stage of crustal movement.

7. 3 Horizontal movement and the plate tectonics

Vertical crustal movement associated with the rebound of a land plate at the interface between the land and sea plates at the time of rupture was stated in Subsection 5. 3. Figure 33 shows the horizontal displacements of triangulation stations in the South Kanto area during a period from 1887 to 1925 (SATO, 1973). Although the first survey was carried out in 1887, the crustal movement as revealed in the figure are certainly associated with the 1923 Kanto earthquake. Such large oceanward displacements of triangulation station accompanied by the earthquake have already been pointed out by MUTO (1932).

Figure 34 shows the horizontal principal strains developed over the

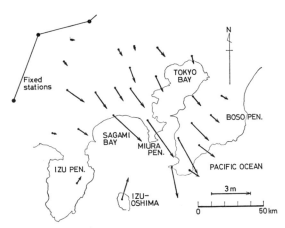

FIG. 33. Horizontal displacements in the South Kanto District associated with the Kanto earthquake of 1923.

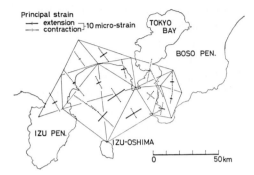

FIG. 34.   Horizontal principal strains in the South Kanto District in the period from 1925 to 1971.

same area during a period from 1925 to 1971 (GSI, 1972).   On comparing Fig. 34 to Fig. 33, it is noticeable that the large extension in a NW-SE direction at the time of the earthquake has tended to be compensated by the slow contraction taking place in almost the same direction after the earthquake.   This provides a typical example of a horizontal rebound.

As stated in Subsection 5. 4, the Tokai area is one of the targets of recurrence of great earthquakes.   Although there is no record of horizontal displacement accompanied by an earthquake in the past, movements similar to those in the South Kanto and the Nankai areas may well be expected.   Figure 35 shows the horizontal displacements in the Chubu District during the period from 1883–1897 to 1951–1960 (DAMBARA, 1976b).

The displacement amounting to about 2 m in a WNW-ESE direction, as can be seen in Fig. 35, seems likely to reflect a possible compression by the motion of the oceanic plate.   The last destructive earthquake off the Pacific coast of this area was the one in 1854 of magnitude 8.4.   If large oceanward displacements occurred at the time of the earthquake, a slow landward contraction might have been continuing during a 40-year period preceding the epoch of the survey.   It is surmised, therefore, that the total amount of compressional displacement developed since 1854 may well exceed 2 m.

In estimating the horizontal strain of a triangle from displacement measurements, one must be cautious of the combination of the three stations.   Even if displacements are large at all stations, the strain is not large when displacements are nearly equal to one another at the

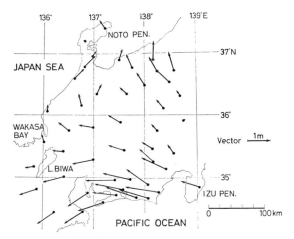

FIG. 35. Horizontal displacements in the Chubu District during the period from 1883–1897 to 1951–1960.

stations in almost the same direction as shown in Fig. 36. Then, such a combination of stations A, B, and D as illustrated in the figure is required. The third station, D, is selected in the barrier zone against the compressional force. Where is the barrier zone? Vector expression of horizontal displacement at respective triangulation stations will be helpful to this aim, if vectors are correctly calculated. Thus the strain expression should be used after careful consideration of its physical meaning.

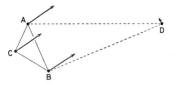

FIG. 36. Relation between displacements and strain in a triangle of three stations.

Figure 37 shows the triangulation net constructed with the first- and auxiliary first-order triangulation stations over the Suruga Bay in the eastern part of Shizuoka Prefecture. The triangulation was carried out in 1884. Some of the side lengths of the same net was resurveyed with an ODMD in 1973 and 1977 by GSI.

Figure 38 (upper) shows the changes in the horizontal strain during

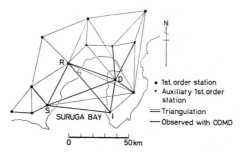

FIG. 37. The first order triangulation net over the Suruga Bay (GSI, 1977).

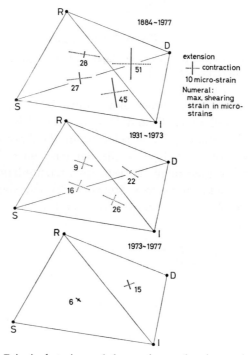

FIG. 38. Principal strains and the maximum shearing strains of triangles over the Suruga Bay in 1884–1977 (upper), 1931–1973 (middle), and 1973–1977 (lower) (GSI, 1977).

the 93-year period from 1884 to 1977 (GSI, 1977). Accumulation of moderately large amount of strain is seen in the area concerned. It shows a contraction in an E-W direction. The mean maximum shearing strain is about 40 micro-strain, and the speed of accumulation is estimated as 0.4 micro-strain per year.

A part of the same net was revised in 1931 after the North-Izu earthquake of 1930. By comparing the result for 1973 with that for 1931, the middle figure of Fig. 38 is obtained. The mean maximum shearing strain of the triangles common to those of 1884–1977 is about 20 micro-strain, and the strain rate is estimated as 0.4 micro-strain per year, the value coinciding well with that during the period 1884–1977. It may therefore be concluded that the result of triangulation in its early days is fairly reliable as far as the first-order triangulation survey is concerned.

In order to see the recent rate of strain accumulation, the result in 1977 is compared with that in 1973 as shown in Fig. 38 (lower). Both results are based on a survey with an ODMD, so that a high accuracy can be expected. The tendency of contraction in an E-S direction is unchanged, but its speed is only about 0.1 micro-strain per year. The apparent diminution of speed may be caused by the displacement at the triangulation station in the Izu Peninsula which is very close to the fault which appeared at the time of the 1974 Off Izu Peninsula earthquake. It may be premature, however, to say anything definite about the diminution of the contraction rate because the 4-year time-interval of successive surveys is too short even for accurate observations with an ODMD.

7. 4 Regular precise trilateration

The precise trilateration by GSI under the national programme has now been in progress over many seismologically important areas in Japan. At present, accumulated strain can only be obtained by comparing the new result with that of an earlier triangulation, so that to obtain the up-to-date speed of strain accumulation is a problem for the future. Indeed, many interesting data are accumulating, while some puzzling problems that cannot be explained properly are also being presented.

A part of the results of the precise trilateration in the Tokai area is shown in Fig. 39. The left figure illustrates the maximum shearing strains over the newly organized triangles during the period from 1885 to 1974. The early surveys of the first- and the second-order triangulations are readjusted for the newly combined net with an appropriate number of fixed stations. The precise trilaterations in this area were carried out in 1974 and 1977. The right figure of Fig. 39 shows nu-

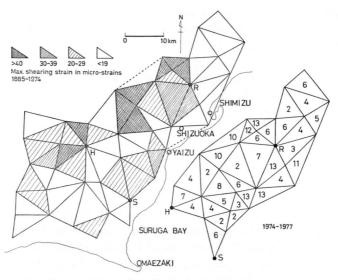

FIG. 39.   The maximum shearing strains in the Tokai area during the period from about 1885 to 1974 (left), and from 1974 to 1977 (right) (GSI, 1978).

merically the maximum shearing strain developed during these three years.

In Fig. 39, we see several triangles which are characterized by considerably large strains in an area between the coast near Shizuoka City and the southern part of the Akaishi mountain range.   The strains are approaching the limit of fracture of the crust provided the limit of about 50 micro-strain by RIKITAKE (1976) is correct.   The recent rate of strain accumulation in the triangles concerned is also comparatively large.   The large amount of strain, however, seems likely to be partly due to the low accuracy of the second-order triangulation of the early days.   Apart from these peculiar triangles, it is seen that overall strains are of the order of 20 micro-strain, and that the principal axis of contraction is in a NW-SE direction.

8.   Horizontal Strain by Base-Line Observation

In the case of the Japanese first-order triangulation in its early days, a number of base-line nets were established at places with a distance between them of about 200 km.   The base-line having a length of 3–5 km was accurately measured by making use of invar-wire scales.

The relative accuracy was estimated as about 2×10^{-6}. On the basis of such lengths of a base-line, a side-length of the first-order triangulation net is determined through a quadrilateral net of triangulation, i.e. the so-called base-line net.

Special mention should be made of the base-line measurement of high accuracy which was applied to the rhombic base-line net with 100 m side lines prepared in the premises of the Tokyo Astronomical Observatory. The net was set up originally to study a certain relation between the latitude variation and deformation of the earth's crust. Although this net is small in its extent, a remarkable change in dilatation was found at the time of the Kanto earthquake. The idea of using a rhombic base-line net for monitoring precursory crustal movements in relation to earthquake prediction has been succeeded by the GSI observation of a quadrilateral base-line net with 3–5 km side lines. The side length is nowadays measured with an ODMD.

Radiant type base-lines (see Subsection 8.2) have also been established by ERI and other universities. The length is measured with an ODMD. Such a study aims mostly at observing "the recent crustal movement" along an active fault. These base-lines can be used for obtaining horizontal strain provided at least three lines in different directions are available.

8. 1 *Some results of base-line net observation*

The above mentioned base-line net of rhombic shape set up on the premises of the Tokyo Astronomical Observatory in the suburbs of Tokyo was measured 32 times with invar-wire scales during a period from 1916 to 1958. Changes in the linear strain of each side line and dilatation of the rhombus as revealed by earlier measurements were discussed by TSUBOI (1933). The change in dilatation is shown in Fig. 40 in which an abrupt change amounting to about 30 micro-strain that took place at the time of the 1923 Kanto earthquake is markedly observed. It seems likely that an irregularity even in a small scale baseline net observed several years before the quake may suggest a possibility of detecting a precursor change. There are some ambiguities, however, that the irregularity might be caused by some imperfect correction of the influence of atmospheric temperature on the length of invar-wire.

It is natural to think that the larger the net is, the larger the change in the side-length of the net, if any. Development of the optical dis-

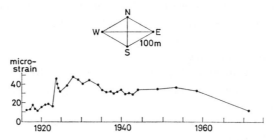

FIG. 40. Change of dilatation of the rhombic base-line net in the premises of the Tokyo Astronomical Observatory.

tance measuring device also makes it possible to utilize a method of measuring long distances which is sufficiently accurate. Under the circumstances, quadrilateral base-line nets with side lines of 3–5 km have become fashionable in recent years in place of small scale rhombic base-line nets. About 20 quadrilateral base-line nets have now been set up in Japan. Some of the nets have been combined with the new project of precise trilateration which was described in the previous section.

Figure 41 shows the quadrilateral base-line net near Pt. Omaezaki at the southern extremity of the Tokai area, and the change in distance between Takatenjin (T) and Sakabe-Mura (S) stations. The value for

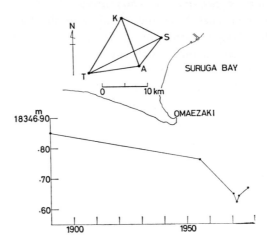

FIG. 41. The Omaezaki quadrilateral base-line net where T (Takatenjin), S (Sakabe-Mura), K (Kikugawa-Mura), and A (Akatsuchihara), and the change in side-length between T and S (GSI, 1974c).

1890 is obtained by a first-order triangulation, so that it is deduced from a rather indirect measurement.   But it seems that the accuracy of measurement is so high that the data can be taken into account in the present discussion, because the distance concerned is that of a side line of a base-line net.   The values after 1970 are the results of measurement with an ODMD.

Granting that the 1890 value is reliable, the distance between stations T and S tends to have become shortened since 1890, the shortening in terms of linear strain amounting to 14 micro-strain in about 80 years.   The tendency during the last 7 years, however, is opposite to that for the previous long period.   Is this tendency correct?   Or, is it a mere fluctuation of the gradual contraction over a long period?   Such fluctuations have already been noticed for the vertical movement by frequent repetition of levelling surveys at Aburatsubo (Fig. 15).   In order to see the changes of other base-lines, all the observed data after 1970 are listed in Table 3.

TABLE 3.   Observed distances of base-lines and their changes during the respective period.

	I Oct., 1970	II Nov., 1971	III Oct., 1973	IV Apr., 1977	II–I	III–II	IV–III	IV–I
	m	m	m	m	mm	mm	mm	mm
T–K	14416.276	.276	.282	.303	0	+ 6	+21	+27
T–S	18346.647	.618	.641	.665	−29	+23	+24	+18
T–A	11460.009	.001	.004	.022	− 8	+ 3	+18	+13
S–K	9981.062	.037	.058	.051	−25	+21	− 7	−11
S–A	8030.764	.751	.765	.772	−13	+14	+ 7	+ 8
A–K	11091.760	.763	.766	.764	+ 3	+ 3	− 2	+ 4

The trilateration stations are T (Takatenjin), S (Sakabe-Mura), K (Kikugawa-Mura), and A (Akatsuchihara), respectively.

As can be seen in Table 3, the distances do not always change in a linear manner with regard to time.   A change amounting to 20 mm for a side length of 1.8 km corresponds to a linear strain of 1.1×10^{-6}, which is just comparable to the accuracy of ODMD measurement. Accordingly, the conclusion at present is that no large rate of strain accumulation in any direction of this region is found by the quadrilateral nets.   In order to get at a much more correct conclusion, however, further surveys are necessary in the future.

The result of observation at the Toyohashi quadrilateral base-line net near Shinshiro City in Aichi Prefecture is shown in Fig. 42. The observations were carried out in March, 1969, January, 1970, and November, 1975. Only the change during 1970–1975 is shown in the figure, because the change during 1969–1970 is so small that it gives only a marginal conclusion judging from the accuracy of ODMD.

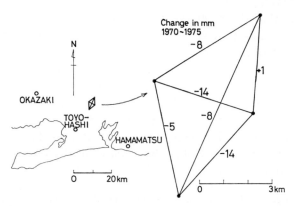

FIG. 42.　The Toyohashi quadrilateral base-line net in Aichi Prefecture, and changes in distance of side lines during the period from 1970 to 1975 (GSI, 1976b).

From the changes in distance during 1970–1975, the maximum shearing strain of 3–4 micro-strain is obtained. Such an amount of strain accumulation during 5 years is more or less the same as the normal value in Japan. The fact that the Toyohashi base-line net does not indicate large strain accumulation has an important bearing on earthquake prediction because the net is located in the hinterland of the Tokai area (which is one of the suspected areas for a future earthquake of large magnitude).

8. 2　Radiant base-lines

We sometimes construct a set of several base-lines in a radial manner from an origin station. Such lines are called radiant base-lines. Radiant base-lines have been set up at about 20 locations in Japan, which are mainly along an active fault, since 1966.

Figure 43 shows the Yahiko radiant base-lines and the changes in linear strain throughout the observations that were carried out 5 times during the period from 1969 to 1977 (ERI, 1978). Yahiko is located at a point about 40 km south-west of Niigata City. The base-lines are

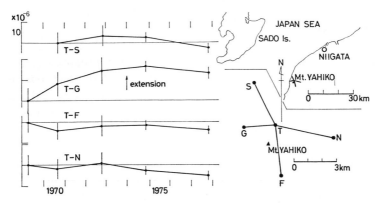

FIG. 43. The Yahiko radiant base-lines, and change of the linear strain of each line.

spread over a mountainous area of Tertiary formation. As the observation covers a period of nearly 10 years with measurements at several epochs, the data must be fairly reliable. The line T-G shows a slight increase of strain linear in time, the speed of which amounts to about 2 micro-strain per year. The speed is considerably larger than that deduced from the result of the first-order triangulation. It is suspected that the station G is located in an area which is under the influence of the eastward tilting of Yahiko mountain mass and of the local activity of micro-earthquakes.

Some radiant base-lines in the Izu Peninsula showed moderately large changes at the time of the 1974 Off Izu Peninsula earthquake and the 1978 Izu-Oshima Inshore earthquake. But we find as yet no evidence of precursory change from such base-lines.

In the U.S.A., there is a report of precursor change that the rate and sense of movement along the San Andreas fault changed in 1964 presumably at the time of the Corralitos earthquake (MEADE, 1966). Horizontal creep movements are not usually observed along an active fault in Japan in the normal stage of crustal movement. Granting the rebound theory of earthquake generation, the zone along a fault should be locked, and pre-earthquake movements would be observed over the outskirt areas. The configuration of radiant base-lines set up so far is not always satisfactory. It may be more preferable to set an origin station along a fault line, and to set up many stations at points having various distances from the origin on a nearly linear direction which is perpendicular to the fault line.

9. Conclusion

Under the title of geodesy and earthquake prediction, the basic principle of earthquake prediction by the geodetic methods is first stated. The principle is very simple, that is, any physical quantity which shows abrupt and large change at the time of an earthquake will be useful for earthquake prediction ,because such a quantity posseses the potentiality of showing an abnormal change even before the shock. Next, the problem of " how to predict the three elements, i.e. rupture zone, magnitude and occurrence time " is discussed. Generally speaking, geodetic work is effective for prediction of rupture zone and magnitude. As geodetic surveys are essentially intermittent, only long-range prediction of occurrence time is possible by a geodetic method. Prediction of short range is certainly an extremely difficult problem, because the very moment of fracture is subjected to a law of probability. As for long-range prediction of occurrence time, a number of examples of geodetic survey and other geophysical observations present promises for expecting an actual success towards our ultimate aim.   But the empirical relation between premonitory effect and occurrence time is as yet unsatisfactory especially for earthquakes of large magnitude.

Theoretical approach to the mechanism of earthquake generation based on various models is also important along with observational approach to the problem of earthquake prediction.   Differences in physical conditions between small and large earthquakes, and between interplate and intra-plate earthquakes should be clarified.

Many examples of success in earthquake prediction have been reported from the U.S.S.R., the People's Republic of China, and the U.S.A.   Although the importance of anomalous animal behaviour is stressed in China, the manner of aiming at earthquake prediction by means of instrumental observation of earthquake precursors are more or less the same for the four nations.   It is regrettable, however, that the history of systematic observation in relation to earthquake prediction is so short in Japan that only very little can be discussed about the possibility of actual prediction.

It is at present most urgent to accumulate basic data for earthquake prediction by a nation-wide effort.   In Sections 5, 6, 7, and 8, outlines of various geodetic methods that may be useful for obtaining such data are reviewed along with their accuracies.

REFERENCES

The following abbreviations are used in the references. *CCEP, Coordinating Committee for Earthquake Prediction*; *ERI, Earthquake Research Institute, the University of Tokyo*; *GSI, Geographical Survey Institute*; *GSJ, Geodetic Society of Japan*; *JMA, Japan Meteorological Agency*.

DAMBARA, T., Vertical movements of the earth's crust in relation to the Matsushiro earthquake, *J. GSJ*, **12**, 18–45, 1966 (in Japanese).

DAMBARA, T., Crustal movements before, at and after the Niigata earthquake, *Rep. CCEP*, **9**, 93–96, 1973a (in Japanese).

DAMBARA, T., Vertical movement of the crust at Aburatsubo, *J. GSJ*, **19**, 22–33, 1973b (in Japanese).

DAMBARA, T., Vertical movements before some inland earthquakes in Japan, presented at U.S.-Japan conference, earthquake prediction and control, 1973, Boulder, Colorado, 1973c.

DAMBARA, T., Adjustment of horizontal displacements by a method of signal coincidence, *J. GSJ*, **22**, 1–9, 1976a (in Japanese).

DAMBARA, T., Horizontal movements in the Chubu District, *J. GSJ*, **22**, 10–16, 1976b (in Japanese).

DAMBARA, T., A revised relation between the area of the crustal deformation associated with an earthquake and its magnitude, *Rep. CCEP*, **21**, 167–169, 1978 (in Japanese).

ERI, Vertical movement accompanied by swarm earthquakes in Matsushiro area, *Rep. CCEP*, **2**, 34–40, 1970 (in Japanese).

ERI, Earthquake swarms in the eastern part of the Izu Peninsula, *Rep. CCEP*, **17**, 71–75; **18**, 42–46; **19**, 64–68, 1977–1978 (all in Japanese).

ERI, Geodimeter surveys at the Ojiya and Yahiko base-line networks in the Niigata Prefecture, *Rep. CCEP*, **19**, 109–113, 1978.

GSI, Recent crustal movement in South Kanto district (4), *Rep. CCEP*, **8**, 23–26, 1972 (in Japanese).

GSI, Crustal movements in eastern part of Hokkaido associated with Off Nemuro earthquake, 1973, *Rep. CCER*, **11**, 7–13, 1974a (in Japanese).

GSI, Precise strain measurement in Yamakita and Tanna regions, *Rep. CCEP*, **11**, 94–95, 1974b (in Japanese).

GSI, Deformation of Omaezaki rhombus, *Rep. CCEP*, **11**, 105–106, 1974c (in Japanese).

GSI, Crustal deformation in the central part of Izu Peninsula, *Rep. CCEP*, **16**, 82–87, 1976a (in Japanese).

GSI, Re-observation of Toyohashi rhombus line, *Rep. CCER*, **16**, 126–127, 1976b (in Japanese).

GSI, Crustal movements in the Tokai district, *Rep. CCEP*, **18**, 75–80, 1977 (in Japanese).

GSI, Horizontal strains in the Tokai district, *Rep. CCEP*, **20**, 166–171, 1978 (in Japanese).

JAEGER, J. C., *Elasticity, Fracture and Flow*, pp. 40–46, Methuen & Co., London, 1964.

JMA, Seismic activity at the Suruga Bay and off the Tokai District, *Rep. CCEP*, **17**, 105–108, 1977 (in Japanese).

MEADE, B. K., Geodetic surveys for crustal movement studies, *Proceedings Second U.S.-Japan Conference on Research Related to Eathquake Prediction Problems*, Tokyo, 1966.

MITTERMAYER, E., A generalisation of the least-square method for the adjustment of free networks, *Bull. Geod.*, **104**, 139–157, 1972.

MOGI, K., Recent horizontal deformation of the earth's crust and tectonic activity in Japan (1), *Bull. ERI*, **48**, 413–430, 1970.

MUTO, K., A study of displacements of triangulation points, *Bull. ERI*, **10**, 384–391, 1932.

RIKITAKE, T., *Earthquake Prediction (Development in Solid Earth Geophysics, 9)*, pp. 295–296, Elsevier Scientific Publishing Co., Amsterdam, 1976.

RIKITAKE, T., Classification of earthquake precursors, *Tectonophysics*, **54**, 293–309, 1979.

SATO, H., A study of horizontal movement of the earth crust associated with destructive earthquake in Japan, *Bull. GSI*, **19**, 89–130, 1973.

SCHOLZ, C. H., L. R. SYKES, and Y. P. AGGARWAL, Earthquake prediction: A physical basis, *Science*, **181**, 803–809, 1973.

TAKEUCHI, H., Crustal deformations and frictional forces, *J. Seismol. Soc. Japan*, **25**, 270–271, 1972 (in Japanese).

TAKEUCHI, H. and H. KANAMORI, Crustal deformations before and after great earthquakes and the mantle convection, *J. Seismol. Soc. Japan*, **21**, 317, 1968 (in Japanese).

TSUBOI, C., Investigation on the deformation of the earth's crust found by precise geodetic means, *Japan. J. Astron. and Geophys.* **10**, 93–248, 1933.

TSUBOKAWA, I., On relation between duration of crustal movement and magnitude of earthquake expected, *J. GSJ*, **15**, 75–88, 1969 (in Japanese).

TSUBOKAWA, I., On relation between duration of precursory geophysical phenomena and duration of crustal movement before earthquake, *J. GSJ*, **19**, 116–119, 1973 (in Japanese).

TSUBOKAWA, I., Y. OGAWA, and T. HAYASHI, Crustal movements before and after the Niigata earthquake, *J. GSJ*, **10**, 165–171, 1964.

WHITCOMB, J. H., J. D. GARMANY, and D. L. ANDERSON, Earthquake prediction: Variation of seismic velocities before the San Fernando Earthquake, *Science*, **180**, 632–635, 1973.

PREDICTION-ORIENTED SEISMOLOGY

Kazuo Hamada

National Research Center for Disaster Prevention, Niiharigun, Ibaraki, Japan

Abstract Seismological papers including materials helpful in predicting earthquakes were reviewed and classified into several subjects in prediction-oriented seismology which was constituted by the author. Seismicity patterns in the broad sense are classified into two patterns. Ones pertinent to long-term prediction are recurrence intervals of earthquakes, seismic gaps and seismicity quiescence, migration and long-term change in seismicity, correlation between interplate and intraplate earthquakes, etc. The others pertinent to short-term prediction are foreshocks, precursory swarms, ground rumbling and tremors, reorientation of compressional axes, acoustic emission, etc.

The new directions and prospects are concluded as follows;

(1) Deterministic precursors to an earthquake for a short-term prediction do not seem to exist so far. Research of the precursor with a kind of reliability that any phenomenon is a precursor to an earthquake, namely, the quantitative research, is obviously a new direction in the prediction-oriented research.

(2) Data collection and processing systems constitute an important branch of the short-term prediction problem. How to detect the precursor or its candidates within a short time prior to the earthquake occurrence is a practical, technological problem and an essential part of the whole prediction system.

(3) Model studies such as the dilatancy and dislocation models of earthquake occurrence including the precursor treatment are typical prediction-oriented seismological studies. Such fundamental research constitutes the frame work of earthquake prediction research and leads to new findings in the observation or research.

(4) New big projects such as the deep borehole measurements of crustal activities or the submarine-cable-type OBS observation are two of the special features of the present day national program, which are supported by both the scientific necessity and the social needs. That is, the progress in the present day prediction is strongly dependent on the national policy regarding the prediction.

1. Introduction

The topic of earthquake prediction is drawing more and more attention from the general public, and is now one of the most exciting challenges against nature for many geoscientists throughout the world. Seismology was born because peoples in many countries have repeatedly suffered from large/great earthquakes, especially in countries along the main seismic belts, like Japan or the western United States. Modern seismology has a history of about a century since instrumental observation of ground motion caused by an earthquake was started. We are now able to learn various aspects of seismology, from materials in most countries. The present effort of earthquake prediction, which is a national program in some countries, was started only in the middle 1960's. Therefore, the present problem, " prediction-oriented seismology " is not yet established but is at a developing stage, although there have already been many successes in predicting earthquakes in the broad sense in China, Japan, the United States, and the Soviet Union, etc. The goal to aim at is to be able to specify the location, the occurrence time, and the size of an impending earthquake with a certain reliability when issuing prediction imformation to the general public. The present text itself is an attempt to create prediction-oriented seismology, based on the review of previous investigations. The author lays stress on up-to-date Japanese works because recent prediction researches in other countries are treated elsewhere.

2. Investigation of Historical Documents of Large / Great Earthquakes

Small earthquakes occur very often in the seismic zones of the world. Medium size earthquakes still occur frequently, but large/great

TABLE 1. Earthquake magnitudes, energies, effects, and statistics (PRESS and SIEVER, 1974).

Characteristic effects of shallow shocks in populated areas	Approximate maginitude	Number of earthquakes per year	Energy (ergs)
Damage nearly total	≥ 8.0	0.1–0.2	$> 10^{25}$
Great damage	≥ 7.4	4	$\geq .4 \times 10^{24}$
Serious damage, rails bent	7.0–7.3	15	$0.04–0.2 \times 10^{24}$
Considerable damage to buildings	6.2–6.9	100	$0.5–23 \times 10^{21}$
Slight damage to buildings	5.5–6.1	500	$1–27 \times 10^{19}$
Felt by all	4.9–5.4	1,400	$3.6–57 \times 10^{17}$
Felt by many	4.3–4.8	4,800	$1.3–27 \times 10^{16}$
Felt by some	3.5–4.2	30,000	$1.6–76 \times 10^{15}$
Not felt but recorded	2.0–3.4	800,000	$4 \times 10^{10}–9 \times 10^{13}$

Source: Data from B. Gutenberg.

earthquakes seldom occur. Table 1 shows the world seismicity by the number of earthquakes per year for each magnitude range with characteristic effects in populated areas. From the viewpoint of mitigation of damage caused by earthquakes, large/great earthquakes should be predicted because they are the cause of most of the damage. However, as seen in Table 1, great earthquakes occur only once or twice in ten years all over the world. If we select a certain place, the frequency of great earthquakes is extremely small. Table 2 shows

TABLE 2. List of great earthquakes off Nankaido, the Pacific coast of Southwest Japan (USAMI, 1976).

Year	Month	Day	Epicenter		Magnitude
684	11	29	32.5°N,	134.0°E	8.4
887	8	26	33.0 ,	135.3	8.6
1099	2	22	33.0 ,	135.5	8.0
1361	8	3	33.0 ,	135.0	8.4
1605	2	3	33.0 ,	134.9	7.9
1707	10	28	33.2 ,	135.9	8.4
1854	12	24	33.0 ,	135.0	8.4
1946	12	21	33.0 ,	135.6	8.1

as an example, historically great earthquakes which have occurred off Nankaido, the Pacific coast of Southwest Japan. Seismological observations by modern instruments do not cover these shocks except for one, the last shock of 1946. Therefore, investigations of historical documents regarding the earthquakes necessarily play an important role for earthquake prediction research, especially for large/great earthquakes, although we can not avoid a certain amount of ambiguity about the documentation. Historical documents about earthquakes enable us to examine mainly space-time patterns of large/great earthquakes. There are two major difficulties, one is the regional degree of uniformity of earthquake data in the past, the other is the accuracy of hypocentral locations and magnitude estimated. In spite of the ambiguities of historical documents, such works are indeed useful and recent new findings about ground uplift along Sagami Bay, in the Tokai area, Japan are taken into account for the present long-term prediction there. Sometimes, the historical documents tell us unusual phenomena before the earthquake such as gorund uplift, ground cracks, appearance of new springs, lightening, unusual animal behaviors, and so on.

Regarding the past seismicities in various regions in the world have been made by many investigators (e.g. REID, 1910; MILNE, 1912; REID and TABER, 1919a, b, 1920; TABER, 1922; LOUDERBACK, 1947; TOCHER, 1959; ROBSON, 1964; ALLEN, 1968; ROTHÉ, 1969; LOMNITZ, 1970; GUDARZI, 1971; MCEVILLY and RAZANI, 1973; ALSINAWI and GHALIB, 1975). Examination works in Japan have also been made by many investigators (e.g. TAYAMA, 1904; ŌMORI, 1920; IMAMURA, 1928, 1929; MUSHA, 1941–1943, 1951, 1950–1953; USAMI, 1966, 1973, 1974, 1975). In China, catalog of historical earthquakes has been issued, covering a surprisingly long era of about 3,000 years (INSTITUTE OF GEOPHYSICS, ACADEMIA SINICA, 1976).

3. Patterns of Seismicity Pertinent to Long-Term Prediction

3. 1 Recurrence intervals of large/great earthquakes

Regularity about recurrence intervals of large/great earthquakes, if any, in a specific area is available for statistical prediction of earthquakes. The problem seems to be simple, but, actually, the conclusions reached by different investigators even for the same areas are different. The difference results from the different data selected and different methods of data processing. KAWASUMI (1970) claimed a 69-year periodicity of strong earthquakes in Tokyo and its vicinities, using historical data of earthquakes since 818 A.D. at Kamakura about 50 km south-southwest of Tokyo. Kamakura was the capital of Japan from 1180 to 1333. Therefore, historical documents on earthquakes there are abundant compared to other regions. He applied Fourier analysis to a time series of strong earthquakes, whose intensities were V or more on the Japan Meteorological Agency (JMA) scale (corresponding to VIII or more on the M.M. scale) and found a 69-year period with a standard deviation of 13.2 years. To check the result of this periodicity, Schuster's test was adopted. As a result, the probability of the existence of the 69-year period was calculated to be as high as 99.94%. There are, however, a number of criticisms about Kawasumi's 69-year period with respect to the data selected and adequency of statistical methods applied. SHIMAZAKI (1971, 1972b) pointed out that Schuster's test which Kawasumi adopted is not adequate to test the significance of the harmonic constituent having the maximum amplitude. Shimazaki adopted FISHER's (1950) test and found that the 69-year period was not significant. USAMI and HISAMOTO (1970, 1971) did similar works with the aim of finding the probability of a future earthquake with intensity V or more on the JMA scale in Tokyo and Kyoto, applying a method similar to Kawasumi's. A 36-year period was found from earthquakes from 818 to 1969 which gave an intensity V or more on the JMA scale (VIII or more on the M.M. scale) in Tokyo. A 38.5-year period was found for earthquakes from 599 to the present date which gave an intensity V or more in Kyoto. One of their interesting conclusions was that even a stationary random process was found to be acceptable.

There are many statistical investigations based on the theory of extreme values. According to SHAKAL and WILLIS (1972), the Alaska-Aleutian region from 155° to 167°W has an 80% probability of earthquake occurrence with magnitude 8 or more, the probability value being calculated from seismic data from 1930 to 1971.

CHEN and LIN (1973) examined 13 seismic zones in China based on the same statistics. They selected some zones of high probability of earthquake occurrence within 5 years up to 1975 as follows: (1) North Tien-Shen, estimated magnitude 6, calculated probability 0.8; (2) South Tien-Shen, 6–6.5, 0.7; (3) Tsinghai plateau, 6–6.5, 0.75; (4) Lanchou area, 7, 0.6; (5) West Yunan, 6.5, 0.75; (6) Chengtu area, 6, 0.6.

In California, LOMNITZ (1974) obtained the following main conclusions from statistics applied to the California earthquake. (1) The probability of occurrence of earthquakes with magnitude 8 or more was about 1% in any year. (2) The median return period was 70 years, half of all intervals being more than 70 years. (3) The risk of a Californian earthquake with magnitude 8 or more was about 9% in a decade, and about 39% in 50 years.

3. 2 Seismic gap and seismicity quiescence

The seismic gap is a positive phenomena for great earthquakes and plays an important role in long-term prediction. Some earthquakes have already been predicted successfully from this phenomena. This gap is very obvious for great earthquakes along the northwestern, northern, northeastern, and eastern margins of the Pacific Ocean.

At first, FEDOTOV (1965) pointed out " gaps " in seismic activity in Kamchatka, in the Kuril Islands, and in Northeast Japan regions. From examinations of the seismic activity of the northwestern Circum-Pacific seismic belt, FEDOTOV (1965), MOGI (1968a), KELLEHER (1970), and SYKES (1971) found that the gap in seismic activity had been successively filled within several tens of years with a series of great earthquakes without significant overlap of their rupture zones. In and around Japan, more detailed investigations were made by UTSU (1968, 1970a, 1972), MOGI (1968b, 1969a), and NAGUMO (1973). They found that some great earthquakes in and around Japan were preceded by seismically quiescence periods of several to a few tens of years. This notion was already suggested by ALLEN et al. (1965) for California earthquakes in the United States and by INOUE (1965) for the Niigata earthquake (M 7.5) of 1964 in Japan.

Similar gaps or seismicity quiescence were reported in many seismic zones in the world, e.g. in the Alaska and the Aleutian regions by SYKES (1971), South America by KELLEHER (1972), major parts of the Circum-Pacific seismic belt by KELLEHER (1972), KELLEHER et al. (1973), and KELLEHER and SAVINO (1975). Also the concept of seismicity quiescence was extended to smaller earthquakes with magnitude 6 or 7 by BOROVIK et al. (1971), OHTAKE (1976a), and OHTAKE et al. (1977a, b, 1978) for Mexico, Central America, and California in the United States.

Recently, MOGI (1976, 1978) proposed a classification of the seismic gap into two types, pointing out an ambiguity in the definition of the seismic gap. One is the seismic gap of the first kind and the other is of the second kind. The seismic gap of the first kind is the gap which was found by FEDOTOV (1965), MOGI (1968a), and SYKES (1971) and others. That is, this type of gap has been successively filled within several tens of years with a series of great earthquakes without significant overlap of their rupture zones. And this is typical for great earthquakes along the major plate boundaries. The seismic gap of the second kind is the seismical quiescence as a precursory phenomena to large or great earthquakes. In general, some seismic gaps of the first kind become seismic gaps of the second kind just prior to the occurrence of a major earthquake.

In the present text, the author has not classified the seismic gap according to the two types but when describing examples has merely used the definition adopted in the particular reference.

3. 2. 1 The Aleutians, southern Alaska, southeast Alaska, and offshore British Columbia

Aftershocks of shallow earthquakes larger than magnitude 7 in the Aleutians, southern Alaska, southeast Alaska, and offshore British Columbia from 1920 to 1970 were relocated by SYKES (1971) in an attempt to delineate the rupture zones of large earthquakes. Three prominent gaps in seismicity are delineated: one is in southeast Alaska; another in southern Alaska near the epicenters of the great earthquakes of 1899 and 1900; and one in the far western Aleutians. These gaps deserve high priority for study and instrumentation. Large earthquakes appear to be much more regular than smaller shocks in their distributions with respect to space, time, and size. Aftershock zones of events since 1930 that are larger than magnitude 7.8 are longer than 250 km and those less than 7.5 are shorter than 125 km. Major earthquakes $(7 \leq M \leq 7.8)$ tend to have relatively small aftershock zones and can be regarded as the " mortar between the bricks " of great $(M > 7.8)$ earthquakes. Great earthquakes in this zone tend to have aftershock zones several hundred kilometers long. They occur in parts of the arc that are tectonically simple. Aftershock zones of large earthquakes tend to abut without significant overlap even for rupture zones as long as 1,200 km. Nearly the entire Alaska-Aleutian zone from 145°W to 171°E has been ruptured since 1938 in a series of large earthquakes. The rupture zones of five large events appear to form a space-time sequence that progressed from 155°W in 1938 to 171°E in 1965. This sequence is much like the well-known westward progression of activity since 1939 along the North Anatolian fault. Figure 1 illustrates aftershock areas of major earthquakes in the Aleutian and the northeast Pacific (SYKES, 1971).

3. 2. 2 South America

KELLEHER (1972) has attempted to forecast likely locations for large shallow South American earthquakes in the near future by examining the past space-time pattern of occurrence of large earthquakes ($M \geq$

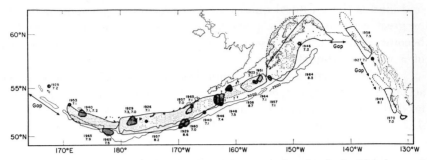

FIG. 1.   Aftershock areas of major and great earthquakes in the Aleutians and the northeast Pacific.   Dates and magnitudes are indicated.   Contours for 2,500 fm and 3,000 fm are shown in the Aleutian trench.   Question marks denote uncertainties in size of aftershock zones (SYKES, 1971).

7.7), the lateral extent of their rupture zones, and, where possible, the direction of rupture propagation.   Figure 2 illustrates estimated rupture zones in South America (KELLEHER, 1972), where the rupture zones of large shallow earthquakes generally abut and do not overlap.   Patterns of rupture propagation appear to follow certain trends.   These facts, plus the non-random behavior of the space-time history of seismic activity, provide consistencies that may permit prediction in a broad sense of future events.   By mapping the rupture zones of large earthquakes, it is possible to identify segments of the shallow seismic zone that have not ruptured for many decades.   Limited experience elsewhere indicates that these gaps between rupture zones tend to be filled by large-magnitude earthquakes.   In certain places it is possible to make approximate estimates of the time of occurrence of the next large earthquake.   For at least 300 or 400 years, the entire fault segment near the Central Valley province of central and southern Chile (about 32°–46°S) has fractured once each century with a general N-S progression of the several large earthquakes ($M \geq 8$).   Large earthquakes in this region have almost always occurred to the south of a previous large earthquake.   In addition, it is possible to infer a direction of rupturing for two large earthquakes in this century (1928 and 1960).   Both these earthquakes fractured southward away from the rupture zone of an earlier earthquake.   It would be consistent with these observations if a new series started about the end of this century near Valparaiso (33°S) and progressed southward.   In other sections of South America there are several extensive segments of the active seismic belt that have not

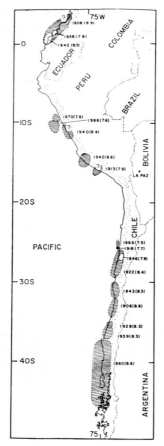

Fɪɢ. 2.   Hatched areas represent estimated rupture zones of large ($M > 7.7$)
South American earthquakes of this century.   Question marks indicate less
reliably determined zones.   Solid circle represents the epicenter of the 1918
Chilean earthquake.   Magnitudes are in parentheses.   Note the gaps between
rupture zones, particularly in the extensive region of northern Chile and SE
Peru (Kᴇʟʟᴇʜᴇʀ, 1972).

ruptured during this century.   In northern Chile, southernmost Peru
(17°–25°S) and South of Lima (12.5°–14°S), there is another significant
gap in recent activity.   Both these regions are probably areas of relatively high earthquake risk.   The northern Peru and southern Ecuador
region (about 9°–1°S) has also been relatively aseismic during this
century.   However, this region differs from the two previously mentioned gaps in that this coastal zone was a region of moderate seismicity

during historic times.   Perhaps aseismic creep is an unusually important factor in relieving tectonic strain along this particular segment of the shallow seismic zone.   Another possibility is that large shallow earthquakes in this region have an extremely long recurrence time.   Much of the shallow seismic zone of northern Ecuador and southwestern Columbia has ruptured twice during this century.   During large earthquakes in this region, the rupturing trends to be directed toward the north or NE.   The data for this region suggest that the area to the NE of the 1958 Columbia earthquake may be a region of relatively high earthquake risk (KELLEHER, 1972).

3. 2. 3   Eastern, northern, and northwestern margins of the Pacific from Chile to Japan, and the Caribbean loop

KELLEHER et al. (1973) have presented several possible criteria for forecasting the locations of large shallow earthquakes in the near future along major plate boundaries and for assigning a crudely determined rating to those forecasts.   These criteria, some of which have been proposed by other investigators, are based on the past space-time pattern of large earthquakes, the lateral extent of their rupture zones, and the direction of rupture propagation.   The criteria are applied in two stages. Application of the first set of these criteria to major plate boundaries along the eastern, northern, and northwestern margins of the Pacific from Chile to Japan and also to the Caribbean loop east of about 74°W results in the delineation of several areas of special seismic potential along each of the boundaries.   At present the validity of the criteria is not firmly established, and profound social changes based on these predictions are uncalled for, but the forecast presented here can, at the very least, serve as a guide in selecting areas for intensive study and instrumentation prior to the occurrence of a major earthquake.   In certain areas where additional information is available the subsequent application of a second set of supplementary criteria focuses special attention on certain of the areas delimited by the first set of criteria. Rupture zones of large shallow earthquakes as determined from aftershock locations, intensities, tsunami descriptions, and coastal uplift tend to cover plate boundaries without any significant overlap.   Because of this tendency to fill in the plate boundaries and because large earthquakes account for such a high percentage of the total seismic moment released in a seismic zone, segments of the zone that have not experienced a large earthquake recently are likely location for future shocks. Their results are summarized in Fig. 3 (KELLEHER et al., 1973).

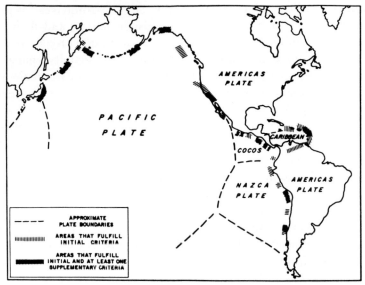

FIG. 3. A summary of segments of plate boundaries that fulfill the criteria of the study for identifying likely locations for large earthquakes of the near future. All areas that meet these criteria are designated regions of special seismic potential. Because of the scale, the areas indicated are approximate. See KELLEHER *et al.* (1973) for detailed descriptions and qualifications, especially concerning Middle America and the Caribbean. The occasional great normal fault earthquakes that occur seaward of trenches are in no way accounted for in this forecast, nor are intermediate and deep earthquakes and large shallow earthquakes in the interior of plates. Note also that only certain seismic zones are included in this study (KELLEHER *et al.*, 1973).

3. 2. 4 *The northwestern, northern, and eastern margins of the Pacific*

KELLEHER and SAVINO (1975) investigated the patterns of seismic activity before large strike-slip and thrust-type earthquakes for intervals as long as 45 years before the main shocks. The earthquakes chosen were the 1952 Kamchatka earthquake, the 1964 Alaska earthquake, three large earthquakes in western Canada-southeast Alaska and the 1960 Chile earthquake which occurred along the northwestern, northern, and eastern margins of the Pacific. Most of these events had rupture zones that extended hundreds of kilometers. An obvious and consistent features of these spatial distributions are the relatively aseismic character of extensive portions of the rupture zones until the time of the main shocks. These reduced levels of seismic activity extended to events even several magnitudes smaller than the main shocks and possible many magnitudes smaller. For most of the large earthquakes examined, rupture initiated in an area of moderate seismic activity and then propagated as far as hundreds of kilometers into adjacent quiet regions. There are some indications that the level of this prior activity, particularly in the vicinity of the epicenter, increased as the time of the main shocks approached. Thus one important conclusion is that gaps in seismicity for great earthquakes along major plate boundaries may also exist for smaller-magnitude activity. Further, such gaps may commonly remain so until the time of the principal shock. That is, if premonitory activity is associated with large earthquakes, such activity should be sought along the boundaries of a seismic gap rather than in the gap itself. Along the great transform fault system near western Canada and southeastern Alaska the nature of the seismicity varies noticeably with distance northwest of the Juan de Fuca spreading center. This changing seismic regime is similar to that of the San Andreas system proceeding north from the spreading centers in the Gulf of California (KELLEHER and SAVINO, 1975).

3. 2. 5   California, Central America, and Mexico

ALLEN et al. (1965) identified several areas in California where remarkably little seismic activity occurred between 1934 and 1963 despite abundant evidence of through-going Quaternary faults. They speculate that " holes " in activity along major active fault systems may be gradually filled in by future earthquakes. Large gaps in activity along the San Andreas, especially areas that have not been ruptured by large earthquakes in more than 100 years, should be regarded as likely places for future large earthquakes. BRUNE and ALLEN (1967) have suggested that one phenomenon which may ultimately be useful in estimating earthquake hazard is the apparent correlation between the zone of faulting during the great 1857 Fort Tejon earthquake and the zone of minimal seismic activity, both for microearthquakes and larger events. For example, sections of active faults characterized by occasional very large earthquakes may, in the intervening periods, be characterized by extremely low seismicity, possibly due to some " locking " mechanism.

OHTAKE et al. (1977a, b, 1978) have investigated patterns of seismic activity preceding the 15 largest earthquakes, whose magnitudes ranged from 5.2 to 7.5, which occurred in Mexico, Central America, and California since 1964. The pattern found in the cases studied, include features recognized by many previous investigators. In general, quiescence developed in the region of eventual shock, followed by resumption of seismic activity leading up to the main shock. They refer to the quiescent period as the α stage, and to the period of renewed activity as the β stage. The diameter of the region of anomalous seismicity ranged from 200 km to 300 km for the cases studied and the onset of quiescence (α stage) preceded the main shock by intervals ranging between 3 months and 5 years. Renewal of activity was detected in every case but one. The onset of β stage activity preceded the main shock by intervals ranging between 8 days and 2 years. As an exception the β stage activity was not detected for the magnitude 6.5 earthquake that occurred in northern Costa Rica in 1973. The results by OHTAKE *et al.* (1977a, b, 1978) suggest that resumption of activity in a previously quiescent zone may be one of the least ambiguous signals that failure is about to occur. A region in southern Mexico that may be in the quiescent stage of seismicity has been identified. This identification will be under future examination. A crutial point, of course, is whether precursory shocks to a large earthquake do occur, or not.

OHTAKE *et al.* (1978) tried to find a possible correlation between the total time of anomalous seismicity ($\alpha+\beta$), the time of the β stage, and linear dimensions of the zones of anomalous seismicity vs. main shock magnitudes. Results are illustrated in Figs. 4, 5, and 6, respectively. In Fig. 4, while the precursory times for the four California shocks fall reasonably close to the empirical curve developed by RIKITAKE (1976), the Central American and Mexican cases fall well below the curve, particularly, at high magnitude. Such scattering of data plotted is not unreasonable in view of the quite different tectonic regimes involved. Figure 5 shows a plot of β stage intervals vs. magnitude. The distribution of seismic stations relative to a particular region should have a strong influence on the detection of renewed activity and hence on the β value (OHTAKE *et al.*, 1978). The linear dimensions of the regions of anomalous seismicity vs. magnitude are illustrated in Fig. 6, and additional relationships of interest in this connection are shown for comparison: (1) the relationship by UTSU (1961) between magnitude and length of aftershock zones, (2) the relationship by DANBARA (1966) between magnitude and the diameter of the zone of crustal deformation detected by geodetic measurements related to the earthquake, and (3) an empirical relation between magnitude and the linear dimension of regions in which velocity changes were recognized (ANDERSON and WHITCOMB, 1975). These comparisons suggest that the zone of anomalous seis-

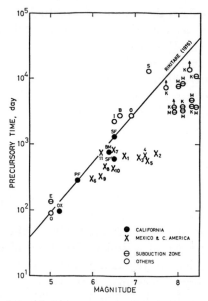

FIG. 4.   Time intervals between the onset of quiescence and the main shock
(α plus β).   Also shown is an empirical curve for various precursory phenomena developed by RIKITAKE (1976).   Full circles correspond to the California
earthquakes studied; OX=Oxnard, PF=Parkfield, BM=Borrego Mountain,
SF=San Fernando.   Open circles are data points from previous studies.   Data
points for the Mexican and Central American Shocks are marked by X's.
Subscripts indicate the case number given in OHTAKE et al. (1977b).   Sources
of data are indicated as follows: B=BOROVIK et al. (1971), E=ENGDAHL and
KISSLINGER (1977), K=KELLEHER and SAVINO (1975), I=ISHIDA and KANA-
MORI (1977), M=MOGI (1969a), O=OHTAKE (1970, 1976a), S=SU-SHAOSEIN
(1976) (OHTAKE et al., 1978).

micity is considerably wider than that in which other precursory phenomena are detected.

3. 2. 6   In and near Japan

UTSU (1970a, 1972) has reported the seismic activity of large earthquakes near Hokkaido, Japan and drew attention to the seismic gap off
Nemuro in Hokkaido at the meeting of the Coordinating Committee for
Earthquake Prediction (CCEP).   He mentioned that a possible earthquake might fill the gap, based on the results with respect to the seismic
gap.   Figure 7 shows the seismic gap off Nemuro (area C), Japan (UTSU,
1972).   Utsu's conclusions are as follows: (1) The recurrence interval
of great earthquakes ($M \geq 8$) in the region under consideration scatters

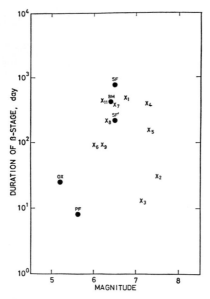

FIG. 5. Interval between the renewal of activity vs. magnitude. Symbols are as described in the caption for Fig. 4 (OHTAKE *et al.*, 1978).

from several tens of years to one hundred and several tens of years, although there is some ambiguity because of the incompleteness of the past historical data. (2) In five areas from A to F except for C, there were large earthquakes during a period of twenty years. (3) Although seismic activity is low in the area of a rupture zone of a future large earthquake, it is seldom that forecasting seismic activity begins there for several years to several tens of years prior to a large earthquake. The area C is an obvious gap in seismicity having no great earthquakes ($M \geq 8$) for the past 78 years. The magnitude of a possible earthquake was estimated to be 8.0 from the empirical formula by UTSU (1961). According to the GEOGRAPHICAL SURVEY INSTITUTE (1970a, b, 1971), significant crustal movement to the north-northwest direction and ground subsidence in eastern Hokkaido was found during the past seventy years. Such crustal movements are also an important indication of stress accumulation for an impending earthquake off Nemuro. From these crustal movements and the seismic gap, UTSU (1972) considered the probability of a great earthquake occurrence in the area C off Nemuro in Fig. 7, as being high although the occurrence time could not be predicted. The magnitude 7.4 earthquake of June 1973 originated

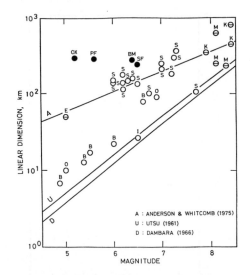

Fig. 6. Linear dimensions of the zones of anomalous seismic activity vs. magnitude of the main shock. Empirical relationships between: (A) magnitude and the linear dimension of zones in which P-wave velocity changes were detected prior to an earthquake (Anderson and Whitcomb, 1975); (U) magnitude vs. length of aftershock zone (Utsu, 1961); and (D) magnitude vs. diameter of regions of crustal deformation detected by geodetic measurements prior to earthquakes (Danbara, 1966). Barred circles are for earthquakes associated with subduction. Open circles correspond to intraplate earthquakes (Ohtake et al., 1978).

in this seismic gap and its aftershocks filled up this gap.

In the same area, Mogi (1977) reported as in Fig. 8 a series of large shallow earthquakes along the northern Japan-southeastern Kurile trench, where rupture zones of the shocks fill the gap in seismicity without significant overlap during the past 55 years from 1918 to 1973. The last event of 1973, Nemuro-Hanto-Oki earthquake, had been successfully predicted by Utsu (1972). The right figure in Fig. 8 illustrates Mogi's idea that the 1973 shock would have been predicted not only in its potential site but also the time of its occurrence. As seen in the figure, there is a clear relationship between the sequential number of the shocks and their occurrence times. Time intervals between the adjacent shocks are gradually and regularly reduced. A simple model in Fig. 9, which explains the regularity between the sequential number of the shocks and their occurrence times, is proposed by Mogi.

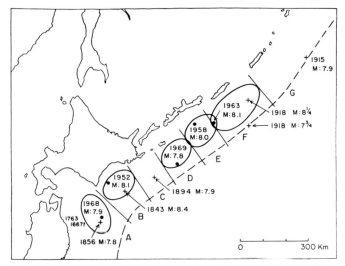

FIG. 7. Source regions of large earthquakes near Hokkaido, Japan (UTSU, 1972).

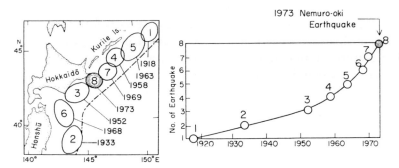

FIG. 8. Left: Locations of focal regions of large shallow earthquakes along the northern Japan-southeastern Kurile trench. The number of each earthquake shows the order of the occurrence time. Right: The earthquake number is shown as a function of time (MOGI, 1977).

Regularities in space-time pattern of recent seismicity in and around Japan have been discussed by MOGI (1969a). The recent seismicity in the Japanese region is represented by four great earthquakes or earthquake groups and associated foreshock and aftershock sequences. Each series shows the following characteritic pattern: in the period (several to twenty years) just prior to a great event, the area where the great event occurs is abnormally calm and the wide area surrounding this

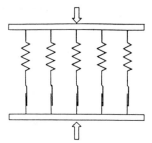

FIG. 9.   A simple model for the occurrence of large shallow earthquakes along the southeastern Kurile trench (MOGI, 1977).

calm area becomes markedly active (foreshock activity in a broad sense). With the great event, the above-mentioned doughnut pattern of the seismicity disappears, the area in and near the rupture zone of the great earthquake becomes active (aftershock activity in a broad sense). The area and the duration in which the foreshocks or aftershocks in a broad sense occur are very large in comparison with those in an ordinary sense.   Seismic activities are summarized by Mogi as follows:

1)   The 1952 Tokachi-Oki earthquake (M 8.1), off the southeast coast of Hokkaido, Japan.   During the nine years before the great event, no major shocks occurred in the focal region of the great event.   Another feature of importance is the increase of seismic activity in the wide area surrounding the focal region of the great event.   It should be noted that the pattern of a doughnut type of active area is a special feature in the period just before the great event.

2)   The 1944 Tonankai (M 8.0) and the 1946 Nankaido (M 8.1) earthquakes, off the south coast of the Kii Peninsula.   Since their focal regions abut and time interval between them was only two years, the above two great earthquakes are regarded as a group.   The seismic activity in the period 1910–1923 was not so high in Southwestern Japan. However, in the period just before the great earthquake (1924–Nov., 1944), the focal region of the great shocks was significantly calm and the surrounding area became very active.   The great shocks occurred to fill the gap of seismicity.   After the great shock occurrences, the doughnut pattern of seismicity disappeared and the active region was in and near the rupture zone of the great shocks.

3)   The 1933 Sanriku-Oki (M 8.3), the 1936 Kinkazan-Oki (M 7.7), and the 1938 Fukushima-Oki (M 7.7) earthquakes, off the east coast of Northeastern Japan.   These earthquakes occurred successively in the

six years from 1933 to 1938 and showed a clear migration feature. Because their rupture zones almost abut, these shocks are regarded as a group. Although a few shocks ($M \geq 6$) occurred before the great shock series in their rupture zones, the seismicity in their rupture zones was anomalously low compared to the adjacent regions which were very active. After the great shock series, seismicity patterns did not show any remarkable relation to the rupture zones of the great shocks.

4) The 1968 Tokachi-Oki earthquake (M 7.9) (8.2 Pasadena, 8.4 Palisades), off the east coast of Northeastern Japan. In the period 1961–1967 just prior to the great earthquake, the rupture zone was appreciably calm and wide areas surrounding it were markedly active.

The increased seismicity in a wide area surrounding the rupture zone of the great shock is, according to Mogi's criteria, a foreshock activity in a broad sense. Therefore, the 1943 Tottori earthquake (M 7.4), the 1941 Hyūganada earthquake (M 7.4) and even the 1927 Tango earthquake (M 7.5) are foreshocks of the Tonankai and Nankaido earthquakes, and the 1964 Niigata earthquake (M 7.5) and the 1962 Miyagiken-Hokubu earthquake are also foreshocks of the 1968 Tokachi-Oki earthquake in a broad sense.

Summarizing the past seismicity in and near Japan, MOGI (1968b, 1969a) has delineated areas where great or large earthquakes may occur in future.

UTSU (1974) has classified the regions of intense seismic activity of the Pacific coast of Japan into several types according to the space-time pattern of large earthquakes ($M \geq 6$). In type A regions, great earthquakes occur at fairly regular intervals. In type B regions, large carthquakes of magnitude up to about 7.5 have a tendency to occur in swarms. In type A′ regions, large earthquakes of magnitude up to 7.5 take place sporadically. The regions labelled as AB, A′B, A(B), (A)B, etc., are illustrated in Fig. 10 (UTSU, 1974). Type AB is a mixed type of A and B. Type A(B) is a type A with slight indication of the nature of type B. A remarkable seismicity gap is developed before a great earthquake in type A region. In regions of types B, (A)B, and AB, the absence of a seismicity gap does not secure the non-occurrence of a large earthquake in the near future. In regions B and (A)B, the probability of the successive occurrence of a large earthquake is high after the occurrence of a large earthquake. A seismic gap for great earthquakes found in regions of types A′, A′B, and B is not an indication of a future great earthquake.

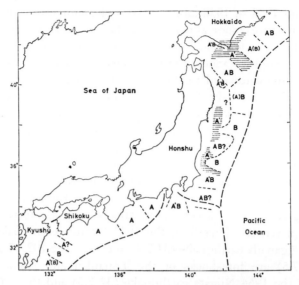

Fig. 10.　Division of the seismic zone off the Pacific coast of Japan according to the type of occurrence of large earthquakes (Utsu, 1974).

3. 2. 7　The Tokai area, Japan

The region along the Pacific coast in the central part of Japan (137°–139°E) is commonly called " the Tokai area."　Great earthquakes took place in the Pacific Ocean off this region in 1498, 1605, 1707, and 1854.　It is remarkable that the great earthquake in 1854 completely destroyed towns and villages in this region by the shaking itself and by the tsunami that followed.　The CCEP assigned this region as " an area of intensified observation " in early 1974.　That actually means a long-term prediction based on the recent results of observations and prediction research.　The main elements of such a predicative judgement consist of research on (1) great earthquakes along the Nankai trough with respect to their space-time pattern (e.g. Ando, 1975), (2) the seismicity gap in the ocean bed off the Tokai districts since 1926 (Sekiya and Tokunaga, 1974), and (3) land subsidence in the vicinity of Suruga Bay.　Figure 11 shows the seismicity gap off the Tokai districts noticed by the observational data of JMA (Sekiya and Tokunaga, 1974). Vertical movements in this area amounting to 30 cm between the years 1889 and 1967 are illustrated by contour lines in Fig. 12 (Geographical Survey Institute, 1974).　Recently GSI has conducted a trilateral survey (triangulation in part) in this area by means of geodimeters and

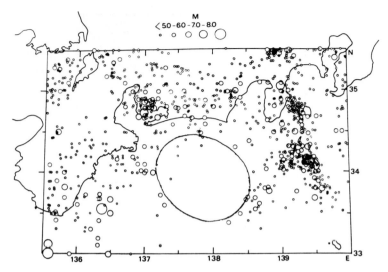

FIG. 11. Epicentral distribution for the period from 1926 to 1972 (SEKIYA and TOKUNAGA, 1974).

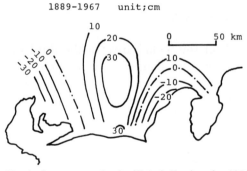

FIG. 12. Vertical movement in the Tokai districts for 1889–1967 (GEOGRAPHICAL SURVEY INSTITUTE, 1974).

compared the results with the triangulation survey conducted in 1885. This result showed that the average strain is 2×10^{-5} for the 85-year period up to 1974 and the direction of the principal axis of contraction is approximately NW-SE (GEOGRAPHICAL SURVEY INSTITUTE, 1976). Space-time distributions of great earthquakes along the Nankai trough off the Pacific coast of Southwest Japan are schematically shown in Fig. 13 (ANDO, 1975). During these 300 years, there are three cycles of great earthquakes. However, there has not been a great earthquake

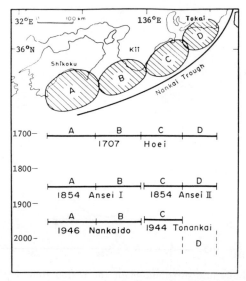

FIG. 13.   Sequence of major earthquakes along the Nankai trough since the 18th century.   A, B, C, and D represent the unit fault planes, respectively. The seismicity gap in D since the last cycle suggests the high potentiality of a future earthquake (ANDO, 1975).

in the region " D," off the Tokai area, in the third cycle.   The figure suggests that the potentially dangerous zone along the Nankai trough is the Tokai region.   These three factors, the seismicity gap, crustal deformations, and the space-time patterns of great earthquakes are consistent with each other with respect to the location of an impending earthquake.

Based on the data concerning earthquake faults, crustal movements and tsunamis associated with the Tonankai earthquake of 1944 and the Ansei-Tokai earthquake of 1854, and including the old historical descriptions of their tsunamis and crustal movements newly discovered, ISHIBASHI (1976) stressed the conclusion that the most potentially risky zone was Suruga Bay.   The main reasons were: the earthquake fault of Ansei-Tokai earthquake of 1854 extends to Suruga Bay but that of Tonankai earthquake of 1944 does not reach there, namely, there is still dangerous stress energy accumulated in the Suruga Bay area.

3. 3 Migration of seismic activity

Close correlation in seismic activity between adjacent regions is of interest in understanding the mechanisms of earthquake occurrence and supplies a useful clue for earthquake prediction. Of particular importance is the successive migration of seismicity. It is well known that the migration of seismicity from east to west over 800 km along the active fault happened in Anatolia in Turkey, principally in 1939 and the following years (RICHTER, 1958). Thereafter, several branches of migration have been reported in Japan, Alaska, Aleutian, Central America, South America, and the Philippine regions (MOGI, 1968a, c, 1969a, 1973a; KELLEHER, 1970, 1972; SYKES, 1971; OHTAKE et al., 1977b).

MOGI (1968a) suggested three migration branches of great shallow earthquakes ($M \geq 8.2$) in the world during the last thirty years. In the northwestern circum Pacific seismic belt, which is one of the three migration branches, the epicentral regions of recent great earthquakes were systematically displaced from Japan to Alaska. In this sequential occurrence of great earthquakes, the considered seismic belt was nearly continuously covered by aftershock areas of these earthquakes without any appreciable overlap of aftershock areas. This remarkable feature provide some important suggestions on the mechanism of earthquake generation. The migration branches are (a) the northwestern part of the circum Pacific seismic belt, (b) the Sumatra-Burma-Kansu seismic zone, and (c) the southeastern part of the circum Pacific seismic belt. In branches (a) and (b) the epicenters seem to have migrated from south to north (or northeast). Particularly, the migration feature in the Sumatra-Burma-Kansu-Baikal seismic zone is also found for smaller earthquakes ($7.6 < M < 8.2$). Branch (c), the southeastern part of the circum Pacific seismic belt, seems to suggest a southward migration with some irregularities. The migration velocities were 150 to 270 km per year. From the results mentioned above, the following process of earthquake occurrence is suggested. The strain energy is gradually and nearly uniformly stored in the seismic belt, and when the stored energy density increases to a critical value of the fracture strength of the earth's crust or upper mantle, a sudden great fracture, namely a great earthquake, takes place. Thus the results strongly support the fracture theory of earthquakes which was discussed in MOGI's (1967) paper. The global migration of recent great earthquakes leads to the suggestion that great earthquakes did not occur independently, but occurred on a global system in the period considered.

MOGI (1969a) has found a different type of migration of shallow earthquakes whose magnitudes were only 6 or more in and near the Japanese islands. The first example is an earthquake series, south to north, from the Kita-Izu earthquake of 1930 to the Miyagiken-Hokubu earthquake of 1962 as illustrated in Fig. 14. Their magnitudes were

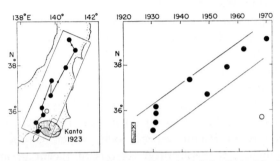

FIG. 14.   Migration of shallow earthquakes ($M \geq 6$) in Northeastern Japan (mainly from MOGI, 1969a).

greater than 6 and focal depths were less than 30 km. It is very interesting that the migration branch started from the epicentral region of the Kanto earthquake of 1923, which is the greatest one in this region during recent years. The second is a series of large earthquakes including the Sanriku-Oki earthquake of 1933. All large earthquakes with magnitude 7.5 or more in Northeastern Japan during the years from 1924 to 1951 occurred within the eight years interval bofore and after the Sanriku-Oki earthquake. Their rupture zones systematically moved from north to south as shown in Fig. 15. The third series is the earthquakes in the Hyūganada region and its adjacent area during the period from 1954 to 1968 as shown in Fig. 16. This series corresponds to the northern part of the seismic belt along the Ryūkyū (Nansei-Shotō) arc from southwest to northeast along the seismic belt. The recent Hyūganada earthquake of 1968 (M 7.5) is also belongs to this migration branch. MOGI (1973a) also pointed out a strong linear trend of migration of shallow earthquakes in the western seismic zone in the Philippines.

KELLEHER (1970) has examined space-time patterns on the distribution of major earthquakes in the Alaska-Aleutian seismic zone. The evidence suggests that major earthquakes of this zone tend to progress in time from east to west. Extrapolation of past trends indicates that a major Alaska-Aleutian earthquake may occur near 55°N, 158°W

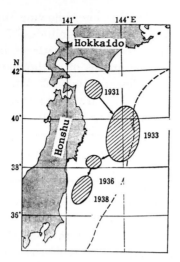

FIG. 15. Migration of large shallow earthquakes ($M \geq 7.5$) before and after the great Sanriku-Oki earthquake (MOGI, 1969a).

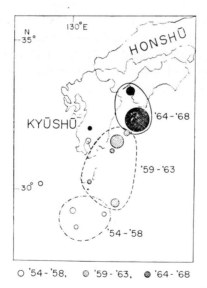

FIG. 16. Space distributions of focal regions of shallow earthquakes ($M \geq$ 6.0) which occurred in the Hyūganada region and its adjacent area during the successive periods (1954–1958), (1959–1963), and (1964–1968) (MOGI, 1969a).

between about 1974 and 1980.   The following three kinds of evidence indicate that earthquakes of about magnitude 7.7 and larger should be used to identify space-time earthquake patterns in the Alaska-Aleutian seismic belts: (1) Space-time graphs of earthquakes of about magnitude 7.7 and larger show strong linear trends, (2) Aftershock zones of successive large earthquakes ($M \geq 7.7$) are approximately adjacent, (3) The direction of fracture propagation is generally away from the focal zone of the previous adjacent large earthquakes, which suggests that the concentration of stress before the events was greater near the region of the adjacent earlier earthquake.

In the Alaska-Aleutian zone, SYKES (1971) revealed that the repture zone of large events appear to form a space-time sequence that progressed from 155°W in 1938 to 171°E in 1965 (see Fig. 1).

KELLEHER (1972) revealed a north-to-south migration of several large ($M \geq 8$) earthquakes along the Pacific coast of South America, near the Central Valley province of central and southern Chile (about 32°–46°S) during the past 300–400 years.

OHTAKE et al. (1977b) found a southerly migration of epicenters of moderatesized shallow earthquakes in the Nicaragua-Costa Rica-western extremity of Panama regions of Central America during the period December 1972 through December 1974.   A time versus distance plot of earthquakes suggested a linear trend in the migration of foci from the 1972 Managua earthquake.   The average rate of migration was 400 km/year.

3. 4   Long-term change in seismicity

The existence of active and inactive periods in the world's main seismic belts has been revealed by MOGI (1974a), based on space-time diagrams of the large shallow earthquakes throughout the world in the past 70-year period.   In these seismic belts which correspond to zones of convergence between the plates, even if the seismic belt is more than 10,000 km long, a number of large shallow earthquakes take place in succession throughout the belt within a certain active period.   This seismic activity seems to be periodic.   The world's main seismic belts of this type have nearly common active periods, except for the northern circum Pacific belt.   In other seismic zones, having triple junctions of plates or complex plate boundaries, the space-time pattern of large shallow earthquakes does not show such a regularity.   The activity of these belts is rather continuous.   These conclusions revealed that large shallow earthquakes are interdependent and coupled on a global scale. Such global scale interactions of large shallow earthquakes support the hypothesis of rigid-plate tectonics. A typical example by MOGI (1974a) is illustrated in Fig. 17. The figure shows the Alpide belt, including the Sunda arc, the arcuate structure in Burma, Himalaya, Iran, Anatolia, and the Mediterranean. This very long belt has a convergent boundary between the Eurasian and the Australian plates (LE PICHON, 1968). The figure shows the locations of large shallow earthquakes ($M \geq 7.7$) separated into four periods. The number of years for each period is shown in parenthesis. The solid circles are epicenters of earthquakes

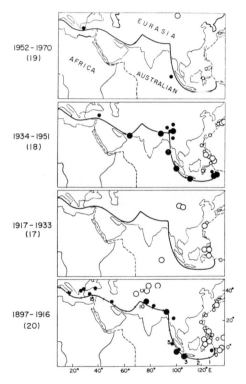

FIG. 17. Locations of large shallow earthquakes ($M \geq 7.7$), focal depth $<$ 100 km, in the Alpide belt during the periods 1952–1970, 1934–1951, 1917–1933, and 1897–1916. Large circles, $M \geq 8.0$; small circles, $7.7 \leq M < 8.0$. Solid and broken curves, boundaries between lithospheric plates; sold circles, earthquakes located within the belt with 1,500 km width along the Alpide belt. Earthquake data from DUDA (1965) and USCGS (1965–1972) (MOGI, 1974a).

located within the belt with 1,500 km width along the plate boundary and the open circles are those located outside the belt. The same result is also illustrated by the space-time diagram as shown in Fig. 18, where the occurrence time of earthquakes is plotted against distance measured from the eastern end of the considered belt along the belt. The distance is shown by numerals in 1,000 km unit beside the belt in Fig. 17. The figure shows that the seismic activity changes occur nearly simultaneously throughout the whole seismic belt of more than 15,000 km length and that the active or inactive periods have repeated periodically within the past seventy years.

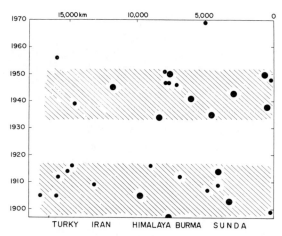

FIG. 18.   Space-time distribution of large shallow earthquakes ($M \geq 7.7$) in the Alpide belt including the Sunda arc.   The abscissa is the distance measured from the eastern end of the considered belt along the belt, which is indicated in 1,000 km in Fig. 17.   Data sources and symbols same as in Fig. 17 (MOGI, 1974a).

Another example is the northwestern, northern circum-Pacific seismic belt.   Seismicity along this belt changes markedly, that is, there have been rather more significant activity in the periods 1915–1933 and 1952–1969 than in the period 1934–1951.   Southwestern Japan, the boundary between the Eurasia and the Philippine plates, shows a different mode (MOGI, 1974a).

3. 5   Correlation between interplate and intraplate earthquakes

Along subducting plate boundaries, the underthrusting oceanic plate compresses the continental plate during the period between inter-plate earthquakes. The stress thus built up is released by a rebound of the continental plate associated with an interplate earthquake (e.g. MOGI, 1970; KANAMORI, 1972b; SHIMAZAKI, 1972a, 1974a, b). This cycle of loading and unloading by the oceanic plate probably triggers and suppresses the occurrence of intraplate earthquakes within the continental plate. Thus we expect a good correlation between intra-plate seismicity and the occurrence of interplate earthquakes. A good example of the correlation is found in Southwestern Japan, a region where one of the most complete historical earthquake documents are available (OZAWA, 1973; UTSU, 1974; SHIMAZAKI, 1976).

Figure 19 (UTSU, 1974) illustrates the epicenters of destructive earth-quakes listed in USAMI's (1966) table which occurred during the 1,000-years from 957 to 1956. For the period from 1870 to 1956, earthquakes of magnitude below 6.5 are excluded, because the frequency of destruc-

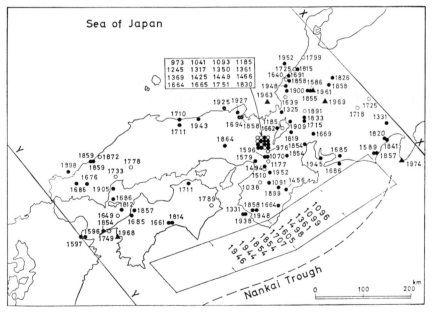

FIG. 19. Distribution of epicenters of destructive earthquakes in central and western Japan for the years from 957 to June 1974. Solid circles represent the events which occurred during the 50-year periods prior to and the 10-year periods after the great earthquakes along the Nankai trough. Open circles represent the events during the other periods through 1956. Triangles indi-cate the events between 1957 and June 1974 (UTSU, 1974).

tive earthquakes is remarkably increased in this period due to inclusion of events with minor damage. Solid circles represent the earthquakes which occurred within a 50-year period before and 10-year period after the occurrence of great earthquakes along the Nankai trough for the years indicated in the figure. Open circles denote earthquakes which occurred in other periods. There are 73 solid circles and 24 open circles. The total length of the periods for solid circles is 425 years and that for the open circles is 575 years. It can be seen that the rate of occurrence of destructive earthquakes during the former periods is about four times as high as that during the latter periods.

The triangles in the figure represent the destructive earthquakes of magnitude 6.5 and larger during the 17.5 years from 1957 to June 1974. The rate of occurrence in this period is still high. This may suggest the possibility of occurrence of a great earthquake off the Tokai area (UTSU, 1974).

MOGI (1969a, 1974b) has reported similar increased seismic activities in areas surrounding the focal regions of the 1944 Tonankai and the 1946 Nankaido earthquakes for about 20 years prior to these earthquakes. OZAWA (1973) has also stated that the destructive earthquakes in the Kinki district (Osaka, Kyoto, Hyōgo, Shiga, Nara, Wakayama, and Mie Prefectures) occurred within the period of about 30 years before the occurrence of the great earthquakes of Nankaido and Tonankai.

According to SHIMAZAKI (1976), over the last four hundred years the spatial variation of intraplate seismicity in Southwest Japan correlates well with the occurrence of great interplate earthquakes. For fifty years before an interplate earthquake the intraplate seismicity is highest along an inland belt. For ten years afterwards it falls off in this belt, but rises on both sides along the Philippine Sea and the Sea of Japan coasts. Then it becomes low and remains low throughout the whole region untill fifty to thirty years before the next interplate event, as shown by UTSU (1974). An intermittent underthrusting drag excerted by the Philippine Sea plate seems to control the intraplate seismicity, which partly takes up the relative plate motion as internal deformation. When a great interplate earthquake occurs, tectonic stress is released and seismic activity falls off in the central belt. The breaking of the plate boundary temporarily weakens the coupling between the two plates along the shallow part of the interface, which gently dips toward the Japan Sea coast. The decoupling causes stress concentration in the deeper part and results in increased seismic activity along the Japan Sea coast. The activity along the Philippine Sea coast may be interpreted as aftershock activity (SHIMAZAKI, 1976).

SHIMAZAKI (1978) has shown another example of the correlation in Tohoku, Northeast Japan, where the Pacific plate is subducting along the Japan trench. All of the active periods of intraplate earthquakes for the past 100 years from 1876 to 1975 in Tohoku, Northeast Japan correlate with the occurrence of great interplate earthquakes along the Japan trench. There were three seismically active periods before and after the three interplate earthquakes of 1896, 1938, and 1968. The seismically active periods are preceded by a period of quiescence lasting 10 to 15 years. One-fourth of intraplate earthquakes of magnitude 5.8 and above preceded interplate earthquakes by less than 5 years; a further one-half followed them by less than 10 years (Fig. 20). All foreshocks and aftershocks are excluded in these statistics. The duration of the high-seismic activity in Tohoku before interplate earthquakes is about one-tenth of that in Southwest Japan before interplate earthquakes along the Nankai trough occur. Sources of data used by SHIMAZAKI (1978) are: USAMI (1966) and the Seismological Bulletin of the JAPAN METEOROLOGICAL AGENCY (JMA) (1958, 1966, 1968). *The Magnitude Catalogue of Major Earthquakes which Occurred in the Vicinity of Japan (1885–1950)* (CENTRAL METEOROLOGICAL OBSERVATORY (CMO), 1952) was not used, because many earthquake magnitude listed in this catalog are one-half to two magnitude units larger than teleseismic magnitudes (MIYAMURA, 1977).

SENO (1977) presented the paper entitled " Interplate and intraplate earthquakes in the Tohoku and Hokkaido districts—a suggestion of an earthquake off the east part of Miyagi Prefecture— " at the fall meeting of the Seismological Society of Japan in November 1977. He examined the correlation between the interplate and intraplate earthquakes in the period from 1907 to 1976. In the examination, the areas are divided into several regions based on the rupture zone of the major interplate earthquakes and their direction of the slip vector as shown in Fig. 21. Most earthquakes of M 6 or larger occurred within 50 years before the interplate earthquake and 10 years after that in the past 100 years in each zone, except double circled shocks in the figure. This conclusion is similar to those of UTSU (1974) and SHIMAZAKI (1976, 1978).

Seno has claimed that an interplate earthquake off the east part of

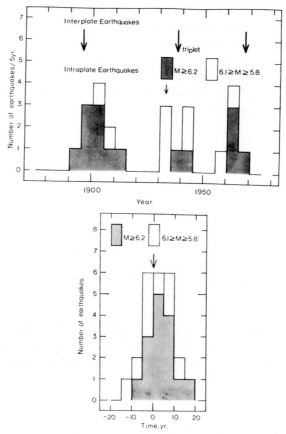

FIG. 20. Above: The number of intraplate earthquakes which occurred in Tohoku in each 5-year interval. The long arrows show the occurrence times of great earthquakes along the Japan trench. The short arrow indicates the occurrence time of the 1933 Sanriku earthquake. Below: Composite histogram of time intervals between the occurrence of each intraplate earthquake and the interplate earthquake which is closest in time to it. The minus sign indicates that the intraplate earthquake preceded the interplate earthquake (SHIMAZAKI, 1978).

Miyagi Prefecture might occur in the near future, since the double circled intraplate earthquakes have no corresponding interplate earthquake. Drawing attention to this area, he indicated that a possible earthquake is not inconsistent with other geophysical aspects such as historical earthquake activities, the seismicity gap, and crustal deformation based on geodetic survey. On June 12, 1978, an earthquake of

1907-1976

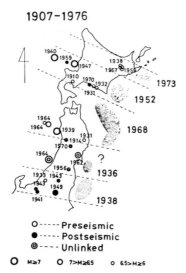

O----- Preseismic
●----- Postseismic
◎----- Unlinked

O M≥7   O 7>M≥65   o 65>M≥6

FIG. 21. Interplate and intraplate earthquakes in Tohoku and Hokkaido districts, Japan (SENO, 1977).

M 7.4 actually took place in the expected area off Miyagi Prefecture. Intensity distribution was quite similar to that of the 1897 Miyagiken-Kinkai earthquake. However, since the rupture zone of this shock was not sufficiently large enough, the area off Miyagi Prefecture is still regarded dangerous and newly assigned as an area for specific observation by the Coordinating Committee for Earthquake Prediction, Japan.

3. 6   Correlation between deep and shallow earthquakes

A relationship between shallow and deep seismicity was first found by MOGI (1973a) in island arc regions where the oceanic lithosphere penetrates beneath the arc. According to Mogi, there are two types of relationship: (1) A great shallow earthquake is preceded and sometimes followed by a marked increase of deep seismic activity in the same down-dip seismic zone perpendicular to the arc. This marked increase of deep seismicity is not accidental, but an essential precursory phenomenon. Such regularity is found in the Kurile-Kamchatka and the northern Japanese island arc regions, particularly near the ends of these two arcs and their junctions, where great shallow earthquakes frequently occur. (2) The seismic activity tends to progress in time from a shallow to a deep region within a descending lithosphere. This regularity is found in the Mariana and Tonga arcs, where the seismic activity is nearly continuous from the surface to the deep region and no great shallow earthquake in the period examined had occurred. The rate of migration was estimated to be about 50 km/year along the descending lithosphere.

3. 6. 1   Great shallow earthquakes and related deep and intermediate-depth seismic activity near Japan

Figure 22 shows the great shallow earthquake, the 1933 Sanriku-Oki earthquake (M 8.3) with its aftershock zone and those of deep earthquakes of magnitude 7 or larger which occurred in this region during the period from 1923 to 1943. The focal mechanism of the 1933 Sanriku-Oki earthquake is estimated to be a normal fault with a strike par-

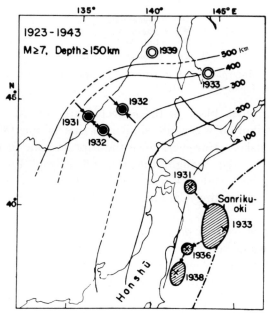

FIG. 22.   Locations of the deep and intermediate-depth earthquakes ($M \geq 7$) and the shallow earthquakes ($M \geq 7.7$) before and after the Sanriku-oki earthquake of 1933, from 1923 to 1943. Circles represent focal depths greater than 150 km, and filled and open circles denote deep events before and after the great shallow earthquake, respectively. Data from KATSUMATA (1970). Hatched areas represent the aftershock regions of shallow events. A converging pair of arrows indicates the horizontal projection of axis of maximum compression in focal mechanism solution after HONDA et al. (1967). The solid lines are contours of the depth of a deep seismic zone in km and the chain line marks the axis of a deep-sea trench (MOGI, 1973a).

allel to the trench axis (KANAMORI, 1971a; ICHIKAWA, 1971). During
the years prior to the 1933 event, three large deep earthquakes had occurred as seen in the figure. The horizontal direction of the maximum
compression from the focal plane solutions of these three shocks are also
illustrated by arrows in the figure (HONDA *et al.*, 1967). Figure 23 shows
the rupture zone of the 1952 Tokachi-Oki earthquake (*M* 8.1) and
related deep and intermediate-depth earthquakes of magnitude 6 or larger
for the period from 1943 to 1962. Earthquakes of a focal depth shallower than 150 km are excluded. Increases in deep seismic activity
are clearly seen prior to the 1952 great shock. Especially remarkable
is the large deep shock (*M* $7^3/_4$) that occurred two years before the 1952
event at 320-km depth.

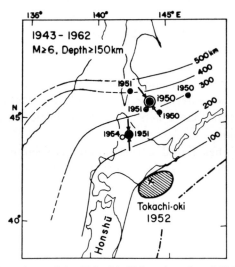

FIG. 23. Locations of the Tokachi-Oki earthquake of 1952 and the deep and
intermediate-depth earthquakes with magnitude 6 or larger during the period
from 1943 to 1962. Double circle, $M \geq 7$; large circle, $7 > M \geq 6.5$; small
circle, $6.5 > M \geq 6$. Data source and other symbols same as Fig. 22 (MOGI,
1973a).

MOGI (1973a) had found two other cases in the Japanese region.
One is the 1953 Bōsō-Oki earthquake (*M* 7.5) which occurred at the junction of the Japan trench, the Izu-Bonin trench, and the Sagami trough.
Along the Izu-Bonin trench seismic activity was markedly high at the
deep region and a number of deep shocks took place during the period
1935–1965. The other is the case of the 1968 Tokachi-Oki earthquake

(M 7.9) and the 1969 Hokkaido-Toho-Oki earthquake (M 7.8). These two earthquakes occurred 400–500 km apart along the Japan-Kurile trench system within only one and a half years. The deep seismic activity around Hokkaido obviously increased prior to these great shallow earthquakes. Similar phenomena are seen in the case of the 1923 Kamchatka earthquake (M 8.4) near the northern end of the Kurile-Kamchatka deep-sea trench.

3. 6. 2   *Migration of seismic activity from shallow to deep region within a deep seismic zone*

This type of migration has been found in deep seismic zones along the Mariana and Tonga trenches. An example found in the Tonga arc is illustrated in Fig. 24 (MOGI, 1973a). Although the reliability of data for the earlier period should be questioned, one or two trends in the early period are found, and this result leads to the suggestion of repetition of a shallow-to-deep migration process in this region.

The regularities of space-time patterns of shallow and deep earthquakes in island-arc regions are qualitatively discussed by Mogi from the viewpoint that the oceanic lithosphere is pushed downward under the island arc by an external force, probably by mantle convection and the descending lithosphere acts as a visco-elastic plate under gradual stress accumulation.

UTSU (1975) has found a correlation between intermediate earthquakes in Hida (the region near the City of Takayama, Central Japan) and shallow earthquakes in central Kanto (Fig. 25). Of 61 earthquakes of $M \geq 5.5$ which occurred in central Kanto during the 51 years from 1924 through 1974, 37 of these could be directly related with 16 earthquakes of $M \geq 5.0$ in the Hida area, the time difference between the two being less than one year. The probability that 37 or more earthquakes in central Kanto should have occurred during this period (total length is 14.46 years) is 1.5×10^{-7}, if the Kanto earthquakes are distributed randomly with respect to time. It seems improbable that this low probability is wholly attributable to effects other than the time correlation, such as the selection of data or the clustering of earthquakes, etc. The correlation suggests a mechanical connection between the two seismic regions, which are squeezed by the Pacific plate which underthrusts beneath the northeastern Japan arc.

3. 7   *Anomalous seismicity before a large earthquake*

Great earthquakes are a quick release of tectonic stress at any time

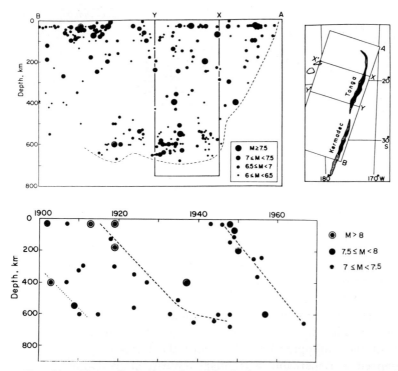

FIG. 24. Above: Projection of earthquake hypocenters onto a vertical plane parallel to the strike of the Tonga-Kermadec arc for the period from 1930 to 1965. Data from GUTENBERG and RICHTER (1954) and ROTHÉ (1969). Filled areas in the right figure indicate water depths greater than 9 km. Below: Focal depth of an earthquake plotted against the time of its occurrence in the segment (XY) of the Tonga arc during the period from 1900 to 1969. Data from DUDA (1965) and very recent data from USCGS Reports (MOGI, 1973a).

over a large area where stress had gradually accumulated during several tens or more years. Earthquake swarms, in contrast, are a gradual release of tectonic stress by minor fracturing. In general, the areas of swarms are relatively small and the total energy of a swarm is not large, although the number of shocks is large. That is, a swarm does not change the stress state significantly. These minor fracturing, earthquake swarms, might be an indication of local weakness in the underground material or stress concentration there. The activity of a swarm is probably controlled by the tectonic stress there. If there is an earthquake swarm in a certain tectonic province which also has a history of

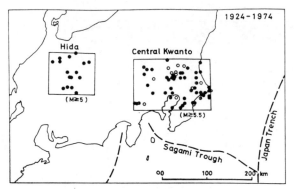

FIG. 25.   Epicenters of 16 earthquakes in Hida and 61 earthquakes in central Kanto.   Filled and half-filled circles in the central Kanto rectangle indicate the earthquakes which occurred within six months and within one year from one of the Hida earthquakes, respectively (UTSU, 1975).

great earthquakes, the swarm activity could by interpreted from the viewpoint of great earthquake prediction.

KANAMORI (1972a) has found significant relationships between earthquake swarms and great earthquakes in Southwestern Japan as shown in Fig. 26.   The area under consideration constitutes a junction of the Philippine Sea plate with the Eurasian plate, and has experienced a remarkable earthquake swarm at Wakayama in the Kii Peninsula and the following great earthquakes have occurred since 1880: the 1891 Mino-Owari earthquake (M 8.4), the 1923 Kanto earthquake (M 8.3), the 1944 Tonankai earthquake (M 8.3), the 1946 Nankaido earthquake (M 8.4), and the 1953 Bōsō-Oki earthquake (M 8.3). Figure 27 illustrates the annual number of earthquakes experienced at Wakayama from 1880 to 1970 and the dates of great earthquakes with magnitudes greater than 8 (NAKAMURA, 1964).   The next two features may clearly be understood from the figure; (1) two pronounced maxima in the number of felt earthquakes at the times of the 1923 Kanto earthquake and the 1953 Bōsō-Oki earthquake, (2) a sudden fall-off in the seismic activity at the end of 1944, the time of the 1944 Tonankai earthquake.   The Mino-Owari earthquake, on the other hand, did not affect the swarm activity at all.   KANAMORI (1972b) revealed that the Tonankai and the Nankaido earthquakes represented a rebound which occurred at the boundary between the Philippine Sea plate and the continental lithosphere where the earthquake swarm at Wakayama took place. The slip vector of the Philippine Sea plate towards the continental litho-

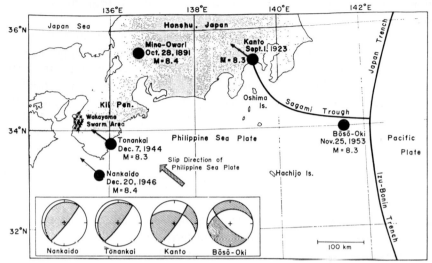

FIG. 26.   The northern part of the Philippine Sea plate.   Epicenters of all the earthquakes with magnitude larger than 8 that occurred in this region since 1880 are shown.   The magnitude data are taken from RICHTER (1958) except for the Mino-Owari earthquake for which the value listed in USAMI (1966) is used.   The arrows attached to the Kanto, Tonankai, and Nankaido earthquakes show the direction of the slip vector of the foot-wall side.   The mechanism diagrams show the lower half of the focal sphere in stereographic projection (ICHIKAWA, 1971; KANAMORI, 1971b, 1972b).   The compression quadrants are hatched.   The swarm earthquakes near Wakayama are shown by crosses (KANAMORI, 1972a).

sphere which was analyzed from the Nankaido and Tonankai earthquakes is shown in Fig. 26.   KANAMORI (1971b) and ANDO (1971) analyzed the source mechanism of the Kanto earthquake that indicated a relative movements between the two plates, the Philippine Sea plate and the continental lithosphere with the boundary of the Sagami trough.   The relative slip vector of the southern block, the Philippine Sea plate is also shown in Fig. 26.   The number of earthquakes of the swarm at Wakayama in the Kii Peninsula increased about four years before the 1923 Kanto earthquake and about five years before the 1953 Bōsō-Oki earthquake.   That is, swarm activity at Wakayama may provide a practical tool for predicting an impending great earthquake in the Kanto district and the adjacent.

One of the points of these phenomena is the precursory increase in the number of earthquake swarms and which leads to the suggestion of a creep-like continental movement at a depth prior to the occurrence of a great earthquake.

Kanamori proposed the following model as an explanation of this phenomenon. Several independent crustal blocks are opposing the motion of the Philippine Sea plate; each block is under stress from the motion of the Philippine Sea plate through a friction-coupled interface where great earthquakes are supposed to occur. When a great earthquake occurs at the boundary between the plate and the crustal block that includes the swarm area, the stress in this block is released and the swarm activity falls off. On the contrary, the stresses in this block increases when a great earthquake occurs, as a rebound, at the boundary between the Philippine Sea plate and one of the other crustal blocks, because such a rebound removes the force opposing the motion of the Philippine Sea plate, propels its motion, and therefore increases the stress in the block that includes the swarm area.

Seismic activities in the central part of Japan have been studied by many Japanese investigators (e.g. IMAMURA, 1928, 1929; NASU, 1973). IMAMURA (1929) had already pointed out that the 1703 Genroku earthquake and the 1923 Kanto earthquake had special importance and he called these shocks non-local great earthquakes. The cycle of the seismic activity that Imamura claimed is that the high seismicity period of local large earthquakes lasting 70–80 years before the non-local great earthquake is followed by an aftershock period of 1–2 years, the seismic activity then gradually decreases until the next active period. The recurrence time of great earthquakes in the southern part of Central Japan was estimated to be 100–150 years by IMAMURA (1928). This outline of seismic activity seems to be still correct even at the present time.

High seismic activities before the 1923 Kanto earthquake since 1884 were, according to NASU (1973), as follows.

1) Oct. 15, 1884 earthquake ($6.0 < M < 7.0$) damage in Tokyo and Yokohama;

2) Jun. 20, 1894 earthquake (M 7.5) (another name, the Iwatsuki earthquake) damage in Tokyo, Kanagawa, and Saitama;

3) Mar. 13, 1909 earthquake (M 7.3) damage in Yokohama;

4) Nov. 12–16, 1915 earthquake swarm near the Bōsō Peninsula;

5) Dec. 8, 1921 earthquake (M 7.1) and 44 aftershocks damage in Ryūgasaki 50 km east-northeast of the center of Tokyo;

6) Apr. 26, 1922 earthquake (M 6.9) damage in Yokohama, Yoko-

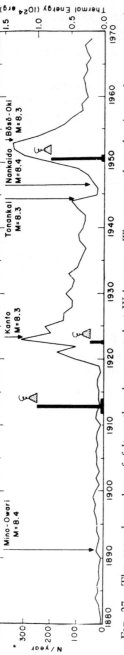

FIG. 27. The annual number of felt earthquakes counted at Wakayama. The arrows show the times of occurrence of great earthquakes ($M>8$) in this region. The vertical bars indicate the thermal energy emitted by eruptions of Oshima island (NAKAMURA, 1964).

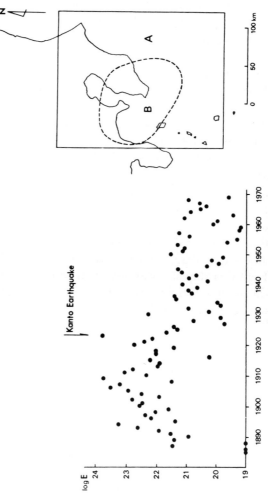

FIG. 28. Annual energy release in units of ergs in region A before and after the 1923 Kanto earthquake (SUYEHIRO and SEKIYA, 1972).

suka, Tokyo, and Kisarazu;

7) Jan. 14, 1923 near Mitsukaido earthquake (M 6.3) (southwest part of Ibaraki Prefecture) damage in Tokyo;

8) Jun. 2, 1923 off Hitachi earthquake (M 7.5) 250 km northeast of Tokyo.

Recently, SUYEHIRO and SEKIYA (1972) evaluated the seismic wave energy released from the south Kanto region before and after the 1923 Kanto earthquake. The area under consideration is shown in Fig. 28. Yearly values of seismic wave energy released from region A (34°–36°N, 139°–141°E except the region B of the dashed circle) are plotted from 1885 to 1965 in the figure. The yearly energy released from the region A before the Kanto earthquake is two orders of magnitude larger than that after the earthquake as seen in the figure. This indicates the unusual high seismic activity prior to the Kanto earthquake. The area B surrounded by the dashed circle in the figure was the main aftershock zone and pronounced crustal deformation was associated with the main shock there. The yearly energy of seismic waves released from region B was of the order of 10^{20} erg during the period of a few years up to 1921, however, it exceeded 10^{22} erg in 1922. The energy ratio, that of region B/that of region A, were estimated to be ≈ 0.002 in 1921 and ≈ 1 in 1922. That is, the seismic activity in the focal zone of the Kanto earthquake was relatively quite low before the shock, however, the seismic activity was extremely concentrated in the focal zone in the one year immediately prior to the occurrence of the Kanto earthquake. There was a kind of doughnut pattern of seismicity before the great earthquakes as pointed out by MOGI (1969a). If such anomalous phenomena concerning seismic activity exist, the present modern instrumentation covering the south Kanto region will definitely detect such an anomalous phenomena.

SEKIYA (1976, 1977) has reported the existence of anomalous seismic activities, different from the ordinary foreshocks, preceding ten major shallow earthquakes which occurred in and around Japan. The definition of this anomalous seismic activities is not described explicitly but illustrated in the figure where the cumulative number of earthquakes in a limited zone including the epicenter of the major shallow earthquakes are plotted against the elapsed time. The anomalous seismic activities were identified from these figures. The relation between earthquake magnitude M and the precursor time T in days, which is the time interval from the beginning of the anomalous seismic activity to the occurrence of the major shock, is given by $\log T = 0.77M - 1.65$. This relationship is similar to other precursor time-earthquake magnitude relations derived from other phenomena (e.g. TSUBOKAWA, 1969; WHITCOMB et al., 1973; SCHOLZ et al., 1973; RIKITAKE, 1976). Some of the anomalous seismic activities are followed by periods of quiescence. Such anomalous seismic activities may be interpreted as a seismicity before a period of seismicity quiescence just before the major earthquake. Sekiya's anomalous seismic activities are indeed newly-founded precursory phenomena. And further accumulation of observational data including that of microearthquakes would be of value. Table 3 is a list of ten earthquakes preceded by anomalous seismic activities taken from SEKIYA (1977).

TABLE 3. List of major earthquakes preceded by anomalous seismicities (SEKIYA, 1977).

Year	Earthquake	M	H (km)	Precursor duration	Remarks
Sept. 1, 1923	Kanto	7.9	10–20	82 years	2nd precursor duration, about 34 years
June 28, 1948	Fukui	7.3	20	19 years and 3 months	
Apr. 30, 1962	Northern Miyagi	6.5	0	5 years and 11 months	
Apr. 20, 1965	Shizuoka	6.1	20	4 years	
Oct. 1, 1965	Near Matsushiro	4.3	0	51 days	2nd precursor duration, 3 days
Feb. 21, 1968	Ebino	6.1	0	2 years and 4 months	2nd precursor duration, 3 months
Sept. 9, 1969	Central Gifu	6.6	0	7 years and 5 months	
May 9, 1974	Izuhanto-Oki	6.9	10	10 years and 4 months	
Nov. 16, 1974	Near Choshi	6.1	40	3 years and 5 months	
Oct. 17, 1975	Near Ebino	4.1	0	18 days	

3.8 Other seismicity patterns

Nation-wide observations by all kinds of instruments are not realistic, even though they are useful for earthquake prediction, because of limitation of manpower and cost.   The idea of test-fields in the Japanese earthquake prediction program is one of the practical ways in which various instruments and research items may be tested and for training in earthquake prediction.   The Yamazaki fault region in the southwest part of Japan was selected as a test-field, because the region is geologically and geophysically well characterized.   Various research and observations have been carried out in cooperation with several universities (OIKE and KISHIMOTO, 1977).   The Yamazaki fault is a typical left-lateral strikeslip fault in the southwestern part of Japan of which the average slip velocity is estimated to be 1 mm/year from geological and topographical observations (MATSUDA, 1976) and there have been no great earthquakes along the fault.   A seismographic network for microearthquakes has been operated since 1964 by a research group of Kyoto University and detailed space-time patterns of the seismic activity have been studied in comparison with other geological and geophysical data.   There were four peaks of seismic activity in the Yamazaki region in 1961, 1965, 1969, and 1973.   From such recent periodicity of every four years, a peak in seismicity of 1977 could be predicted.   Data obtained by an array observation of extensometers crossing the fractured zone show that movements are related to seismic activities.   Significant movements of the fractured zone were associated with isolated heavy rainfalls with a few days time delay after the rainfall.   Also the monthly distribution of the number of microearthquakes for these ten years has clear peaks in May and September in the Yamazaki region, which is also closely related to the monthly distribution of the precipitation of rain there.   That is, the rainfall seems to be a trigger of the earthquake occurrence (OIKE and KISHIMOTO, 1977).

OIKE (1977a) examined the seasonal frequency distribution of 196 shallow earthquakes of magnitude 4 or more in the southwestern part of Japan in the period from 1961 to 1972.   The results were similar to the case of microearthquakes in the Yamazaki fault region, that is, the peaks were in the spring to summer of the rainy season and in September of the typhoon season.   The number of rainfalls of which the precipitation is greater than a certain value has a peak 3–5 days before an earthquake.   The probability of rainfall of 30 mm/day or more within a week before an earthquake ($M{\geq}4$) was calculated to be 0.25 under the assumption of random rainfall occurrence.   The probability of rainfall of 30 mm/day or more within a week before an earthquake ($M{\geq}4$) was actually 0.45.   The trigger effect of rainfalls seems to exist when the variation of precipitation is large, rather than when the absolute value itself is large.

An earthquake of magnitude 4.0 occurred on September 30, 1977 in an area of seismicity quiescence along the Yamazaki fault. According to OIKE (1977b), this earthquake was successfully predicted based on the detailed pattern of seismicity which is closely related to the rainfall.

A pattern of seismic activity near Wakayama City in the northwestern part of the Kii Peninsula, Central Japan has been investigated in detail by a research group of the Earthquake Research Institute, Tokyo University (MIZOUE et al., 1977a, b). The characteristic of the seismic activity is that the epicenters are distributed in a linear direction which runs in a NE-SW direction through Wakayama City. It is thought to correspond to a hidden fault. The size of the swarm area is approximately 20 km in diameter. The area is further divided into three parts, two typical swarm areas and one seismically quiescent area which is located between the former two areas. Migration of seismicity among the three areas was investigated in detail. According to MIZOUE et al. (1977a), based on the regularity which was found from the long-term microearthquake observation, the earthquake (M 4.7) on August 7, 1977 was able to be predicted.

Regional seismicities have been examined by WESSON and ELLSWORTH (1973) to determine the nature of seismicity in the years preceding the several moderate earthquakes in California: 1952 Kern County (M 7.7), 1963 Watsonville (M 5.4), 1964 and 1967 Corralitos (M 5.0, 5.3), 1966 Parkfield-Cholame (M 5.1, 5.5), 1968 Borrego Mountain (M 6.4), 1969 Santa Rosa (M 5.6, 5.7), and 1971 San Fernando (M 6.4). In each instance, the moderate earthquakes occurred in areas characterized by a relatively high level of small earthquake activity. In most cases the preceding activity in the area immediately surrounding the epicenter was high relative to other segments of the fault zone or other nearby fault zones.

4. Patterns of Seismicity Pertinent to Short-Term Prediction

4. 1 Foreshocks, precursory swarms, ground rumbling, and tremors

The technical terminology " foreshock " has been extensively used to date. Usually it means earthquakes that occur near the epicenter of the following major earthquake just before the major event. However, there is no exact definition about foreshocks based on an appropriate scientific explanation. That is, strictly speaking, the definition is merely sensory and qualitative. For instance, Ishida and Kanamori (1977, 1978a) have found five earthquakes from 1969 to 1970 located within a small area around the main shock, the 1971 San Fernando earthquake, California, which are considered foreshocks. Until their discovery, however, no foreshock was recognized before the San Fernando earthquake. Their results suggest that detailed analysis of wave forms, spectra, and mechanisms can provide a powerful diagnostic method for identifying a foreshock sequence. Anyway, foreshocks have been mainly identified based on their space-time characteristics, cosidering those of the main shock. The physical nature of foreshocks is still an important research topic.

Foreshocks seldom occur before the major event in genral. According to Mogi (1963b), among 1,500 earthquakes of magnitude 4 or larger which occurred in and around Japan during the period from 1926 to 1961, only 64 earthquakes (4% of all) were accompanied with foreshocks whose time intervals up to the main shock were from a few minutes to a few months. Table 4 shows the number of main shocks

TABLE 4.   Number of earthquakes preceded by foreshocks.

Magnitude	Number of earthquakes with foreshocks	
	After Rikitake (1976)	After Mogi (1963b)
$M \geq 8$	6	1
$8 > M \geq 7$	15	6
$7 > M \geq 6$	24	17
$6 > M \geq 5$	19	27
$5 > M \geq 4$	17	12
$4 > M \geq 3$	7	

that were preceded by foreshocks, being classified by their magnitudes. Data were taken from Table 15–XIII of the paper by Rikitake (1976) and Mogi (1963b). Although Mogi pointed out that only 4% of major earthquakes were preceded by foreshocks, the percentage did depend on the magnitude of the main shock. Assuming that the Gutenberg-Richter's b value is equal to 1.0, which is actually close to the world average, the number of earthquakes increases one order as their magnitude decreases by 1. Therefore, as seen in the table, the percentage of the foreshock increases when the magnitude of the main shock becomes larger.

If we consider the earthquake with $M \geq 5$, the percentage of foreshocks is at least a few times that for $M \geq 4$, and for the earthquake with $M \geq 6$, that percentage increases probably several times more. Furthermore, by the use of modern high sensitive instrumentation for seismic observation, the coverage of earthquakes has been improved. That is the detection capability of foreshocks, even though they are small ones, is extremely high in comparison with that in the past.

An unusual earthquake swarm began to occur around the town of Matsushiro (36°34′N, 138°13′E), Nagano Prefecture, the central part of Japan in August 1965 and lasted over three years, accompanying a certain number of strong shocks and some remarkable crustal movement. The largest magnitude of these earthquakes was 5.3, which caused a minor damage in adjacent villages and towns. The total energy released as a seismic wave reached 1.6×10^{21} erg, which corresponds to the energy of a magnitude 6.3 shock (e.g. EARTHQUAKE RESEARCH INSTITUTE, 1966, 1967; HAGIWARA and IWATA, 1968; HAMADA, 1968; ICHIKAWA, 1969; HAGIWARA, 1973; OHTAKE, 1976b).

In the course of expanding swarm activity, precursory activities of ultra-microearthquake ($M < 1$) preceded larger earthquake swarms including strong shocks (approximately M 5 or larger) by several months. And the ultra-microearthquake activity had been watched as a forecast the following larger earthquake activity. This forecasting was actually made known to the local people, which was a kind of a prediction in a broad sense. That is, ultra-microearthquakes were observed in the Sakai village since August 1966, followed by felt earthquakes which included a strong shock of M 5.0 on 16 January 1967. In the Azuma village, ultra-microearthquakes began to occur seven months before the strong shock. Similar phenomena were observed four months before in the Kōshoku-Tokura-Kamiyamada region and five months before in Sanada Town. Six examples are listed in Table 15–XIII of the paper by RIKITAKE (1976).

There have been many cases where precursory foreshocks, earthquake swarms, ground rumblings, or tremors were observed by instruments or merely slightly felt by the inhabitants before the main shock (e.g. KAMINUMA et al., 1973). Among 60 destructive earthquakes which have occurred in and around Japan during the 100 years up to 1972, KAMINUMA et al. (1973) reported that 29 of these were preceded by such precursory phenomena.

Another report by JONES and MOLNAR (1976) states that 44% of all shallow large earthquakes in the world from 1550 to 1973 were preceded by foreshocks.

At present, unfortunately, we have no established method to distinguish such precursory phenomena. However, it encourages us to investigate an adequate method to distinguish these phenomena since about half of the major earthquakes have been preceded by some foreshock-like phenomena.

4. 2 Characteristics of foreshocks

ŌMORI (1910) and IMAMURA (1913, 1915) had already investigated foreshocks in and near Japan and pointed out that several of the earthquakes preceded by foreshocks had a tendency to occur in some limited regions. However, after that, there were no systematic investigations until MOGI's (1963b) work. Based on experimental investigations (MOGI, 1963a), Mogi presented three types of frequency-time curves of fractures and also showed that natural earthquakes could be classified into these three types. They are shown in Fig. 29. The type 1 appears only at regions of homogeneous structure caused by a uniformly applied stress. The third type appears at remarkably fractured regions or it also may be caused by concentrated application of stress (e.g. in intrusive stress by viscous magma). The second type is an intermediate one between the first and the third types. This pattern appears at regions of moderately heterogeneous structure or it may be caused by a stress which is not uniform.

The epicenters of the earthquakes preceded by foreshocks are dis-

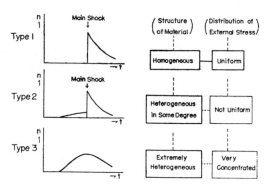

FIG. 29. Three types of the successive occurrence of the elastic shocks accompanying fractures and their relations to the structures of materials and the applied stresses (MOGI, 1963b).

tributed in the following limited regions (MOGI, 1963b): (1) Izu-Nagano-Kyoto region, (2) Miyoshi-Hamada region (north part of Hiroshima Prefecture), (3) Ohita-Kumamoto region, and (4) Hyūganada region as illustrated in Fig. 30. It is noteworthy that the occurrence of fore-shocks is not accidental but has a general tendency to be limited in a certain region. Such regionality is attributed to the geological and geophysical structure of the Japanese islands. Mogi claims that ob-servations of foreshocks may give a clue for the prediction of major earthquakes at least in certain regions where the normal sequence of foreshocks is expected.

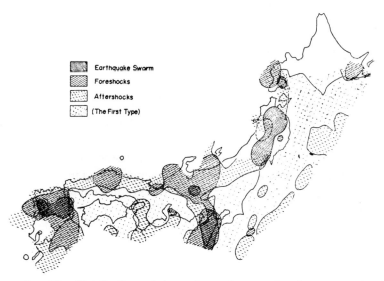

FIG. 30. Distributions of the regions where earthquake swarms, foreshocks and aftershocks occur with high probability (MOGI, 1963b).

The localities of earthquake swarms are also illustrated in Fig. 30. The swarms are also distributed into certain regions. The spatial dis-tribution of the swarms is nearly identical to that of the foreshocks, and to that of the volcanic regions. The swarms occur at remarkably fractured regions. That is, as the stress gradually increases, a high stress concentration appears around numerous cracks and faults, so local fractures begin to occur even under a low stress. Therefore, it is difficult for a single large shock to occur. This pattern of occurrence is also caused by an extremely concentrated stress like intrusion of magma in the volcanic regions. The spatial distribution of the regions fractured to some degree, which correspond to the types 1 and 2, is similar to the geographical distributions of Quaternary volcanoes and volcanic regions in Neogene, Tertiary, and Quaternary periods.

As referred to in Subsection 3. 2, UTSU (1974) has investigated space-time patterns of large earthquakes and divided the region into several types as illustrated in Fig. 10.   Regarding the nature of foreshocks, in type A regions, foreshock activities are usually weak for great earthquakes. However, if they occur, they will be easily detected because of a pronounced seismicity quiescence developed before a great earthquake. In regions of types B, (A)B, and A′B, swarms usually have low b values and it is not a criterion for foreshocks.

Since SUYEHIRO et al. (1964) had discovered an anomalously small b value for foreshocks and an ordinary value for aftershocks of the main shock in Matsushiro, Japan, attention has been drawn to the b value as one of the methods to discriminate the foreshocks from the earthquake swarm.   Suyehiro and his colleagues observed 25 foreshocks, a mainshock of magnitude 3.3, and 173 aftershocks with a high-sensitivity tripartite seismographic observation in Matsushiro, Japan in 1964. A substantial difference was found between foreshocks and aftershocks for Gutenberg-Richter's b value which gives the relationship between frequency of occurrence and earthquake magnitude.   The b values were only 0.35 for the foreschocks and 0.76 for the aftershocks.   A similar investigation has been carried out on the 1960 great Chilean earthquake by SUYEHIRO (1966).   Forty-five foreshocks and 250 aftershocks were observed in a 33-hour period before and in a 33-hour period after the main shock.   The significant difference between foreshocks and aftershocks was also found, namely the b values were 0.55 for the foreshocks and 1.13 for the aftershocks.

Results of statistical investigations of the magnitude and time distributions of foreshocks of 62 cases in an area of Greece during the period 1914 through 1973 were reported by PAPAZACHOS (1975).   In the paper, evidence was presented that the b value has a smaller value before than after the main shock.   The mean value of b was 0.82 for foreshocks and 1.3 for aftershocks.   The difference in magnitude between the main shock and the largest foreshock seemed to be independent of the magnitude of the main shock.   The average difference was found equal to two magnitude units.

Recently, HAMADA (1978) has found an anomalously small m value of 1.55 for foreshocks and an ordinary m value of 1.78 for background seismicity and aftershocks of the 1978 Izu-Ōshima-Kinkai earthquake (M 7.0), where m is called the Ishimoto-Iida's coefficient and $m-1=b$ generally holds good. The data were from 584 earthquakes during the period October 1977 to January 13, 1978 which is considered the period of background seismicity, 56 foreshocks during 4.5 hours just before the main shock, and 1289 aftershocks for the period from February to April 1978. After a statistical test on the significance of the m value, the m value was confirmed anomalously small for the foreshocks with a confidence level of 99%.

Similar b values for foreshocks and aftershocks before and after a main rupture in heterogeneous rock speciments were reported based on rock-breaking experiments (e.g. MOGI, 1962a, b, 1967, 1969b, 1973b; SCHOLZ, 1968). Table 15–XIII in the paper by RIKITAKE (1976) gives 11 other examples regarding the b-value change before and after the main earthquake, which are taken from BUFE (1970), SCHOLZ et al. (1973), WYSS and LEE (1973), and FIEDLER (1974).

Attention should be given to the fact that the b value for fore-shocks is not always smaller than that for aftershocks or background seismicity. Conversely, earthquake sequences having a small b value are not always foreshocks. An earthquake swarm along the trench near Japan had small b values (UTSU, 1970b, 1971). Therefore, to use the b value as a practical tool to discriminate the foreshock, a systematic investigation would be required to reveal the degree of efficiency of the tool.

UTSU (1978) has presented a simple method to distinguish fore-shocks from swarms. The method is based on the three largest earth-quakes in magnitude and their time sequences. Data were from shallow earthquake series which occurred in and around Japan for the period from 1926 to 1964, of which the maximum magnitude was 5 or more for the period from 1926 to 1977 and 4.5 or more for the period from 1965 to 1977. There were 245 such earthquake series having a certain space-time condition that was adopted by Utsu. The condition is that the occurrence time falls within two weeks before and after the largest earthquake and the epicentral location is within 0.5° (or 1.0°, if the maximum magnitude is larger than 7) in both latitude and longitude from that of the largest earthquake. Therefore, earthquake sequences selected here do not include precursory earthquakes different from the ordinary foreshocks (e.g. SEKIYA, 1977; EVISON, 1977). Among the 245 series, only 13 cases, about 5%, were classified as foreshock series.

TABLE 5. Numbers of earthquake swarms and foreshock sequences classified according to M_1-M_2, M_1-M_3, and the time sequence of shocks of M_1 and M_2 (Utsu, 1978).

M_1-M_3	0.0	0.1	0.2	0.3	0.4	0.5	0.6
M_3 unknown	9	10 / 5	3 / 11	3 / 12	2. ③ / 9 ①	5 ② / 9	3 ② / 4
≥ 0.7	3	2 / 8	2 / 4	3 / 4	2 ② / 7	1 / 9	4 ① / 5
0.6		1		3 / 4	2 / 5	1 / 4	1
0.5		1 / 3	1 / 2	1 / 1	2	2	
0.4	3	2 / 1	2 / 2	2 / 1	3 / 1		
0.3	3	2 / 2	6 / 3	3 / 3			
0.2	5	1 / 2	4 / 1				
0.1	6	1 / 2					

Legend:
- Number of swarms
- Number of foreshock sequences
- s_1 f_1 — M_1 follows M_2
- s_2 f_2 — M_1 precedes M_2

$M_1 - M_2$

They are listed in Table 5, where M_1, M_2, and M_3 are the first, the second, and the third maximum magnitudes of the earthquake series. The numerals on the left correspond to the number of swarms and on the right, the foreshock series for each square in the table. The upper numerals mean that M_1 follows M_2 and lower numerals mean that M_1 precedes M_2 in the square. If we select the earthquake series which satisfies the conditions of the upper cases surrounded by the solid lines in the table, that is the conditions of $0.4 \leq M_1-M_2 \leq 0.6$ and $M_1-M_3 \geq 0.7$ and M_1 follows M_2 are satisfied, there are 27 earthquake series including 10 foreshock series, about 37% of the 27 events. The above three conditions can be applied during the occurrence of an earthquake series. As 32 earthquake series satisfy the above three conditions during the occurrence of an earthquake series, in this case, $10/32 = 31\%$ is the probability of foreshocks. Such simple test of Utsu's to distinguish foreshocks from swarms have practicality because of their simplicity and estimated probability of about 30%. Large magnitude difference like $M_1-M_2 \geq 0.4$ and $M_1-M_3 \geq 0.7$ among the three largest shocks in the earthquake series are qualitatively consistent with small b values.

4. 3   Reorientation of compressional axes

Reorientation of compressional axes, that is the focal mechanism change of microearthquakes prior to the occurrence of three major earthquakes (M 5.5–6.0), had been reported, and occurred during the period 1963 through 1969 in the Garm region in the Soviet Union (NERSESOV et al., 1973). The focal mechanism that changed prior to the main shock was similar to that of the main shock. The time of this change ranged from 3 to 6 months before the main shock. SIMBIREVA (1971) also reported the focal mechanism change of microearthquakes before the main shocks with magnitude 4.2–5.5 in the region of the Naryn River Basin, where the time of the mechanism change ranged from 1.5–2 months before the main shock. In this region earthquakes have a tendency to occur near the boundary between areas which are divided by the focal mechanism of earthquakes of background seismicity. Examples are illustrated in Fig. 31. This phenomenon is one of the creditable precursors that are known. However, not many similar phenomena have been reported in other countries, and there has not been such creditable results in Japan until the present time.

As stated in Subsection 3. 7, disastrous earthquakes had frequently occurred in the Kanto district in Japan before the 1923 Kanto earthquake. ISHIBASHI (1975) examined the focal mechanism of four earthquakes among these disastrous earthquakes, for which hypocentral locations were relatively well determined. His focal plane solutions of the earthquakes located in the southwest part of Ibaraki Prefecture (36°N, 140°E) showed a reverse fault with a SW-NE direction of the principal compression axis. On the other hand, ICHIKAWA (1966) and KAWASAKI and KATSUMATA (1975) have investigated the focal mechanism in the Kanto district during these 30 years. According to their results, earthquake faults are the reverse with NW-SE or E-W directions of the principal compression axis and they are stable in the southwest part of Ibaraki Prefecture. KAWASAKI and KATSUMATA (1975) classified the focal mechanism of the earthquake with depths deeper than 30 km in the Kanto district into three types; (I) E-W compression axis, reverse faults, (II) NE-SW compression axis, normal faults, (III) N-S compression axis, reverse faults. Figure 32 illustrates the regional distribution of the type (I) by open circles and the types (II) and (III) by solid circles, where there is an obvious boundary between the type (I) and the types (II)+(III) as shown by a solid line. ISHIBASHI (1975) interprets these results as follows. Earthquakes in the southwest part of Ibaraki

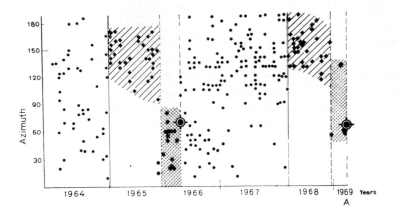

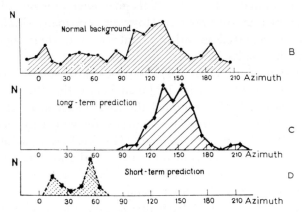

FIG. 31.　The reorientation process.　A: Distribution of azimuths of compressional axes of weak earthquakes with time.　B: Distribution of azimuths of the stress axes for normal background weak earthquakes.　C: Those used in long-term prediction.　D: Those used in short-term prediction.　Depth range of weak earthquakes=1–15 km (SADOVSKY *et al.*, 1972).

Prefecture occur under the compressional stresses from the southeast side (the Japan trench side) and the southwest side (the Sagami trough side).　The stress from the southeast side is ordinarily dominant there, however, when stress accumulates along the Sagami trough, that is when a large or great earthquake along the Sagami trough is imminent, the stress from the southwest side becomes dominant and at that time seismic activity becomes high.　Therefore, the direction of the compres-

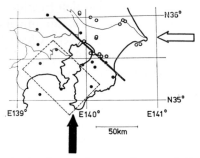

FIG. 32. The distribution of the tectonic stress infered from the earthquake mechanism solution in the Kanto area for the period 1951 through 1970. Open (solid) arrow indicates the compression axis derived from the focal mechanism of earthquakes shown by open (solid) circles (KAWASAKI and KATSUMATA, 1975). The broken square is a fault plane of the 1923 Kanto earthquake which was calculated from the crustal movements (KANAMORI and ANDO, 1973).

sional axis in the southwest part of Ibaraki Prefecture could be changed, caused by stress accumulation along the Sagami trough. The focal mechanism change may play an important role in the future for prediction of large or great earthquakes along the Sagami trough (ISHIBASHI, 1975).

The focal mechanisms of several earthquakes that occurred before and after the 1930 Kita-Izu earthquake (M 7.0, 35°N, 139°E) in the Izu Peninsula in Japan show another possible reorientation of the principal compression axis. That is, the focal plane solutions of the two earthquakes in 1930 before the 1930 Kita-Izu earthquake indicated a N60°–70°W compression axis, the 1930 Kita-Izu earthquake showed a N36°W compression axis, and four earthquakes from 1934 to 1974 showed approximately a N-S compression axis (JAPAN METEOROLOGICAL AGENCY, 1975).

The author would expect such a precursory reorientation of the principal compression axis especially in the Kanto and Tokai region of Japan, where the stress is a reflection of mutual movements of the Eurasian, the Pacific and the Philippine Sea plates, rather than within a single plate.

Change in the direction of the principal compression axis can be attributed mainly to two reasons. One is an abrupt release of a great amount of stress energy along the fault plane by a large or great earthquake and the other is a gradual release of a relatively small amount of stress energy by a creep or creep-like deformation in a certain limited zone, such as a zone along the fault of an impending earthquake or along boundaries between the crustal blocks which leads to some stress concentration.   The latter case probably takes place at almost critical stress level.   And the precursory stress state change must be intimately related to this.

GUPTA (1975) presented another possible explanation for the re-orientation of stress axes by assuming a systematic vertical migration of foreshock activity within a region of thrust tectonics.   When there was an uplift or upward buckling in the epicentral region, he qualitatively discussed idealized variations with respect to depth in the tectonic maximum compressive stress and other two orthogonal stresses and showed that the reorientation of stress axes could be explained by a systematic vertical migration of seismic activity.

4. 4   Acoustic emission, micro-fracture, and spectrum change of seismic waves

Research on acoustic emission, micro-fracture, and spectrum change of seismic waves are also related to earthquake prediction.   However, not very many papers relating to earthquake prediction from this view-point have been reported.   From the Soviet Union, the spectrum change of seismic waves before a major earthquake has been reported. Pronounced high frequency seismic waves were observed about four months before an earthquake of M 4.8 and occurred near the Garm region in April 1966 (NERSESOV et al., 1973; SADOVSKY and NERSESOV, 1974).   Conversely, a decrese of high frequency components of P and S waves of small earthquakes (M 3.0–3.5) was observed before the oc-currence of earthquakes of M 6 or more in the southern Kurile Islands (FEDOTOV et al., 1972).   Also, TSUJIURA (1976) reported that the domi-nant frequencies of seismic waves of foreshocks for the 1976 Yamanashi-ken-Tōbu earthquake (M 5.5, 35.5°N, 139°E) and the 1976 Kawazu earthquake (M 3.0, east coast of the Izu Peninsula in the central part of Japan) were lower than those of their aftershocks.

Recently, ISHIDA and KANAMORI (1980) found a significantly higher frequency of dominant S waves of foreshocks for the 1952 Kern County earthquake (M 7.7) in California, United States, as illustrated in Fig. 33. Namely, dominant frequencies of S waves of earthquakes occurred near the epicenter of the main shock during 1934–1949 ranging from 1.0 to 2.0 Hz, while those from 2.0 to 4.0 Hz were observed for foreshocks

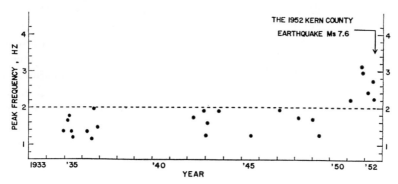

FIG. 33. Change of dominant frequencies of S waves of earthquakes occurred near the epicenter of the 1952 Kern County earthquake. The magnitude range was from 2.7 to 4.0 (ISHIDA and KANAMORI, 1980).

during 1951–July 21, 1952. The magnitude range of earthquakes used in their analysis was from 2.7 to 4.0 for both periods. They also found a significantly higher frequency for foreshocks for the 1971 San Fernando earthquake (M 6.5) being similar to those of the Kern County earthquake.

There are some papers reporting spectral analyses of seismic waves in Japan. The results have been used for research purposes such as the source process, fault parameters, or attenuation of seismic waves in the crust/mantle, etc. However, precursory time variation of the spectra before and after the major earthquake are not yet well established.

Acoustic emission which is high-frequency seismic waves generated by micro-fractures is expected to contribute to earthquake prediction. They are, however, not yet practically used in seismological field observation. It has been noted that a high-frequency geophone with a frequency range from 20 or 30 to 400 or 500 Hz detected intense high frequency seismic waves in the Garm region in 1950, where the epicentral distance was 15 km and the depth was 15 km. Remarkable activity lasted for some minutes about 2 hours prior to the occurrence of the main shock (ANTSYFEROV, 1968). As stated in Subsection 4. 2, there are many descriptions in the historical documents about ground rumblings or tremors before the major earthquakes in Japan (e.g. KAMINUMA *et al.*, 1973).

Acoustic emission (AE) is also studied in other fields. For landslide prediction, micro-fractures which emit AE are considered an indicator of the creep preceding the major landslide. To check the validity of the landslide prediction, the method of monitoring the micro-fracture is thought to be the most effective method for detecting precursory AE (e.g. WATARI and ITAGAKI, 1975). Regarding prediction and control of a rock burst, AE is also hopefully a precursory phenomena. NIWA *et al.* (1976), for instance, concluded that rockburst can be predicted if continuous monitoring of AE is possible. And actually the AE detector is used for rock burst prediction in mines. Although landslide and rock burst are different phenomena from natural earthquakes, application of such techniques to earthquake prediction should be investigated.

5.   Deep Borehole Observation of Microearthquakes and Ground Tilt

5. 1   *Introduction*

Tokyo is the most important place in Japan, because of the city's functions as the political, economic, and cultural center of the Japanese people. Unfortunately, Tokyo has repeatedly suffered considerable damage from earthquakes. In the second 5-year schedule (fiscal year 1969–1973) of the earthquake prediction program in Japan, to overcome the artificial noise at the surface, deep borehole measurements of crustal activities, which are expected to supply important data for earthquake prediction, were planned in and near Tokyo. The National Research Center for Disaster Prevention (NRCDP) had planned to carry out three deep borehole observations in and near Tokyo. The first place selected was in the city of Iwatsuki, about 30 km north of the center of Tokyo. Starting in 1970, the construction of a 3,500-m depth observatory was completed at the end of 1972. Observations of microearthquakes and crustal movements have been carried out since May 1973. The second one, the Shimohsa observatory of 2,300-m depth, was completed at Shonan Town about 30 km east-northeast from the center of Tokyo at the end of 1977. Preliminary observation of microearthquakes and ground tilt has been carried out in 1978. Observational data from these observatories are telemetered through the telephone line to the NRCDP in Ibaraki Prefecture about 60 km northeast of the center of Tokyo, where observational data of other seismometers and tiltmeters covering the south Kanto and Tokai areas are gathered together. The third one, the Fuchu observatory is now under construction and will be completed by the end of the fiscal year 1979 (TAKA-

HASHI and HAMADA, 1975; HAMADA *et al.*, 1979).

5. 2   *Performance of the deep borehole observatories*

Performances of the deep borehole observatories, the Iwatsuki and Shimohsa observatories are summarized in Table 6. Both observatories have a three-components set of velocity seismometers, a three-components set of accelerometers, a two-components set of tiltmeters, and two sets of thermometers within a sensor vessel which is rigidly fixed at the bottom of the borehole. The sensor vessel can be lifted up if necessary, and can be reset at the bottom again. Accelerometers are expected to supply important records for engineering seismology and enable us to study the effect of the outermost layers of the ground by using both seismograms at the surface and at the basement. Such records are valuable especially for seismic waves of a long period. The borehole tiltmeter is the only instrument that can detect the ground tilt in a flat region which is covered by soft and thick layers like the Kanto plain in Japan. The borehole tiltmeter has been developed to the point where the annual drift, that may include the effect of the setting at the bottom of the hole, has been reduced to several arc seconds/year. The present borehole tiltmeter is suitable for short-term crustal movements over at least a one or two months period.

The Iwatsuki deep borehole observatory can detect earthquakes with magnitude greater than 2 (or 3) within a 100 (or 200) km radius from the observatory. Namely, almost all earthquakes in and near Japan reported by Japan Meteorological Agency and local earthquakes in and near the Kanto district reported from the Earthquake Research Institude are detected at the Iwatsuki observatory.

Reduction of background noise with respect to depth was examined by YAMAMOTO *et al.* (1975) and YAMAMOTO and TAKAHASHI (1976). The conclusions were that the noise level at the bottom was fairly stable and the spectral amplitude of background noise at the bottom of the hole was about 1/30–1/1,000 compared with that at the surface. These ratios mainly depend on frequency range and surface conditions. There was an obvious frequency dependent reduction of background noise in the deep borehole. Noise reduction is more effective for high-frequency waves than low-frequency ones. One instance showed that relative spectral amplitude at 3,500-m depth to that of the surface is 0.06 at 1 Hz, but, it is only 0.01 at 10 Hz.

TABLE 6. Performance of the deep borehole observatories.

Station	Sensor	Sensitivity and constants				Overall sensitivity and ranges recordable
Iwatsuki 35°55′33″N 135°44′17″E 3,502 m below sea level 85.6°C at the bottom	A three-components set of velocity seismometers	X 1.71 v/kine	$Fo*$=0.98 Hz		h=0.65	Magnification of 1,000,000 at 10 Hz 2 μkine–10 mkine for 1–25 Hz
		Y 1.65	1.00		0.65	
		Z 1.70	1.01		0.65	
	A three-components set of forced balance accelerometers	X 16.1 v/G	Fo=50 Hz			5 mgal–30 gal for 0–30 Hz
		Y ditto				
		Z ditto				
X; N4°W Y; W4°S Z; vertical	A two-components set of tiltmeters	X 100 mv/arc sec				0.01″–5″ adjustable within ±3°
		Y ditto				
	Two sets of thermometers	0.2 v/°C				80°C–90°C
Shimohsa 35°47′36″N 140°01′26″E 2,277 m below sea level 61.0°C at the bottom	A three-components set of velocity seismometers	X 1.89 v/kine	Fo=1.00 Hz		h=0.62	Magnification of 570,000 at 10 Hz 5 μkine–70 mkine for 1–25 Hz
		Y 1.97	1.00		0.62	
		Z 1.43	0.98		0.62	
	A three-components set of forced balance accelerometers	X 16.1 v/G	Fo=50 Hz			5 mgal–30 gal for 0–30 Hz
		Y ditto				
		Z ditto				
X; W11°S Y; N11°W Z; vertical	A two-components set of tiltmeters	X 102 mv/arc sec				0.01″–5″ adjustable within ±3°
		Y 100				
	Two sets of thermometers	0.2 v/°C				56°C–68°C

* Fo; Natural frequency

6.   Submarine Observation of the Earth's Crust Activities

6. 1   *Submarine observation by ocean bottom seismometers*

The first development of the ocean bottom seismograph (OBS) goes back about 40 years ago in the United States.   Thereafter, the OBS has been rapidly developed since the 1960's by the United States, the Soviet Union, Japan, etc.   The main reason for the OBS development seems to be for the detection of underground nuclear explosion rather than for seismological or geophysical interests.   From the viewpoint of geophysical interests, however, the OBS has been quickly improved since the late 1960's, accompanied with the development of the new global tectonics " ocean floor spreading."   At present, however, the amount of observation by the OBS is still quite small compared with the conventional seismic observation on land.   A research group at Lamont Observatory of Columbia University in the United States had already developed the first submarine-cable-type OBS which operated at 400-m depth, and 180 km off the California coast in 1965, but the observation was discontinued.   There is no permanent OBS station now in operation, except the submarine-cable-type OBS of the JMA which was started in August 1978 off the Tokai area, Pacific coast of the central part of Japan.   Also almost all observations by OBS are for short period observation, being one month at the longest.   There are three types of OBSs, (1) anchored-buoy type, (2) free-fall-pop-up type, and (3) submarine-cable type.   The first anchored-buoy type OBSs were often used until a recent date.   However, the free-fall-pop-up type OBSs have been improved and now favored because of their mobility and convenience for setting and withdrawal of records.   Recently, the scale of the OBS observation has been expanded and the network observation consisting of some 20–30 OBSs are carried out in cooperation with the seismometrical network on land nearby, although it is only in a certain area and for a limited period.   The present status of OBS observation has been described in some papers (e.g. Asada and Shimamura, 1974; Shimamura and Asada, 1978; Kasahara, 1978).

The necessity for development and improvement of the OBS are stated in the Japanese earthquake prediction program as being a powerful method for detecting the earth's crust activities beneath the ocean around Japan, especially in the area of the Japan trench, the Sagami trough, and the Nankai trough along the Pacific and the Philippine Sea coast of Japan which is the zone where historically great earthquakes have caused great damage in Japan.   This is more strongly emphasized in the 4th 5-year schedule of the Japanese earthquake prediction program (fiscal year 1979–1983).   Seismic observations at the ocean bottom where the focal zone of disasterous earthquakes is located are expected to contribute to the following research and monitoring items that are important for earthquake prediction.   (1) Characteristics of the seismic activities along the trench or trough; the hypocentral distribution, the focal mechanisms, the stress state, and the seismicity of the oceanic crust. (2) Research on the seismic gap or quiescence which are particularly reliable parameters for the prediction of an impending earthquake; precise and three-dimensional aspects of the seismicity gap or quiescence, and the existing time intervals of seismicity quiescence. (3) Detection of foreshocks of large/great earthquakes.   (4) Investigation of seismic wave velocity change in the oceanic crust as a precursory phenomena to an earthquake.   (5) Investigation of seismic-wave velocity structure beneath the ocean near the trench/trough; this contributes to the precise determination of the hypocentral locations and the origin times and to the understanding of the tectonic situation of the down-going oceanic plate and surrounding crust and upper mantle and provides a fundamental information about earthquakes along the trench/trough.   If we adopt the submarine-cable-type OBS for providing on line data, real time monitoring of seismic activities and related parameters, which are pertinent to earthquake prediction, could be possible using a rapid data processing and display system.   As stated above, after the 4-year development period, the JMA has started continuous observation by the submarine-cable-type OBS off the Tokai area, Pacific coast of the central part of Japan since August 1978.   The sensors of this OBS system consists of 4 three-components sets of seismographs and a tsunami meter and are connected with a single submarine cable.

6. 2   Other geophysical observations at the ocean bottom
From the viewpoint of earthquake prediction, observations of tilt, gravity, and terrestrial magnetism, etc., on the sea bed, where the rupture zone of great earthquakes is situated along the trench/trough, provide important geophysical information in addition to seismic activity. Continuous observation of tilt, gravity, and magnetism seems to be promising to detect the precursory phenomena immediately prior to a great earthquake in co-operation with various observations on land near-

by. However, at present an instrument for such observations has not been designed. Therefore, the confronting problems are development of the instrument for measurements of tilt, gravity, and magnetism and so on including the technical development for setting it at the sea bed. Japan has repeatedly suffered huge amount of damage caused by great earthquakes and its tsunamis along the Japan trench, the Suruga trough, and the Nankai trough. Therefore, the prediction of such great earthquakes has been stressed in Japan much more than in other countries (GEODESY COUNCIL, 1978). Therefore the research project for submarine measurements of the oceanic crustal activities was commenced in 1978 in co-operation with several organizations and progress is expected in the future.

7. Data Collection and Processing of Seismographic Data for Earthquake Prediction

Data collection systems of the seismographic network have been developed and improved, being accompanied with the development and dissemination of telemeters and data processing by the electronic computer system. Such a system ensures rationalization of the operating system for routine base works and security of high-quality observational data. Typical examples of the rationalized system are seen in the United States and Japan. In Japan, telemeters and data processing by means of an electronic computer system for seismographic networks were mainly adopted in the 3rd 5-year schedule (fiscal year 1974–1978) of the earthquake prediction program, later than the United States.

Remarkable advantages by adopting such telemeters and computers have been found in the Tohoku district, the northeastern Honshu island, Japan, where the seismographic stations for observation of microearthquakes by Tohoku University are located. A research group of Tohoku University (HASEGAWA et al., 1978) has found a double-planed structure of deep seismic zone over a wide area of more than 300×200 km in the Tohoku district, Japan. This prominent feature of the configuration of the deep seismic zone has been ascertained through the precise determination of the microearthquake hypocenters by using the data from the seismic network of Tohoku University. The two planes are nearly parallel to each other, the distance between the two planes being from 30 to 40 km. An explanation for the double-planed structure of the hypocentral distribution based on the physical properties of the down-going slab was given by TSUKAHARA (1977, 1980). Among other new findings supported by the high-quality seismographic observation by Tohoku University is that almost all the shallow microearthquakes in the land area occur only in the layer with a P-wave velocity of 6 km/sec which was determined from explosive seismic observations (TAKAGI et al., 1977). The advantages, which are represented by the seismographic networks in the Tohoku district, Japan, are based on modern data collection and processing of the first stage of the prediction system.

Although the earthquake prediction technology is not yet established, to put the progressing technology to practical use, a second stage of data collection and processing is required. The second stage of the system is a short-term, prediction-oriented, on line data collection and processing system which consist of seismic stations, telemeter networks, and an electronic computer system. This system constitutes an important technological portion of the whole prediction system from a social point of view. The designed function of the second stage system will include not only rationalization of the system including manpower and security of high-quality observational data, but also on line, real time monitoring of the seismic activities and related parameters in a certain area under consideration. Therefore, it includes essentially the automatic or semi-automatic readings on seismograms, that is readings of P, S phases, the polarity of the initial motion, the maximum amplitude, and so on. Real time monitoring items will be expanded in future from the simple to the complex ones. At the primary stage, the number of earthquakes at each station or in a certain area, the Gutenberg-Richter's b value in a certain area or the Ishimoto-Iida's coefficient m at each station, and the P-traveltime residuals at each station could be easily monitored, and at the advanced stage, space-time patterns of seismic activity, foreshocks, focal mechanism, and seismic-wave velocities could be quickly monitored for overall judgements of earthquake occurrence in the area under consideration.

The data gathering and processing system constitutes a vital technological part of the earthquake prediction system, especially for the detection of anomalous phenomena immediately prior to an earthquake. Concerning not only seismographic observation, but also other observations for earthquake prediction, the intra-structure for earthquake prediction including the data gathering and quick processing system is emphasized in the 4th 5-year schedule of the earthquake prediction program (fiscal year 1979–1983) of Japan (GEODESY COUNCIL, 1978). The system for data collection and processing, especially for seismographic observation and continuous observation of crustal movement, is advancing into the second stage type for real time monitoring of seismic activities and related parameters pertinent to earthquake prediction.

8. Plate Tectonics as a Background to Earthquake Occurrence

The lithosphere is a plate of solidified rock that is mounted on a partially molten asthenosphere. The plate is approximately 70 km thick beneath the oceans and 100 km thick beneath the continents. The continents rests on the plate and they move together. The lithosphere is created at mid-ocean ridges from a plume of partially molten rock which comes up to the ridge and then sinks back into the mantle in the zone of subduction, where partial melting of basaltic crust takes place and which feeds the volcanic chain. The marginal basin or sea, like the Sea of Japan, which is located behind the island-arc volcanic chain, separates the chain from a continent. The earthquakes in the world are mainly controlled by such plate motions. They occur along the oceanic ridges where the plate is created, in the transform fault zone where the plates pass each other, and in the subduction zone where the plate penetrates the mantle and it is there that deep earthquakes occur.

It is not necessary to describe the general concept of the plate tectonics here. The only pertinent points that should be mentioned here are the following. Several regularities in space-time patterns of earthquake occurrence have been confirmed with the purpose of earthquake prediction. They are introduced in Sections 3 and 4. It is very interesting to note that some of the regularities could be well understood based on present plate movements, and conversely the plate movements lead to the suggestion of some regularities concerning the pattern of seismic activity pertinent to earthquake prediction. These regularities of earthquake occurrence are, as a whole, consistent with rigid-plate tectonics. They are: (1) correlation between interplate and intraplate earthquakes found in Southwest and Northeast Japan; (2) correlation between deep and shallow earthquakes along or within the plate in Japan and other countries; (3) correlation between high activity of the earthquake swarm near Wakayama in the Kii Peninsula and two great shocks, the 1923 Kanto and the 1953 Bōsō-Oki earthquakes in and near Japan; and (4) anomalous high seismicity before the great earthquake in the Kanto region, Japan. That is, the plate tectonics are intimately related to some regularities regarding the space-time pattern of earthquake occurrences which are pertinent to, not short-term, but long-term prediction.

9. Models for Precursory Phenomena of an Earthquake

There are two types of models for precursory phenomena of an earthquake: one is the dilatancy models which are further divided into a dilatancy fluid diffusion model and a dry dilatancy model; the other is a creep dislocation model. The dilatancy fluid diffusion model was first proposed by NUR (1972) based on laboratory experiments. This type of model has been improved by SCHOLZ et al. (1973), WHITCOMB et al. (1973), ANDERSON and WHITCOMB (1973) and others. The dry dilatancy model has been developed by BRADY (1974), MOGI (1974b), STUART (1974) and others. The dilatancy model is the first attempt in earthquake prediction research to try to explain various precursory phenomena simultaneously, such as seismic activity, seismic-wave velocity change, ground uplift, release of Radon from ground water, and change of terrestrial electro-magnetic properties. Development of this model was obviously stimulated by observational results of seismic-wave velocity change first reported in the Garm region of the Soviet Union, and the model seemed to explain various precursory phenomena very well, at least in the early stage. However, further improvement of the model was more difficult than expected. One reason is probably that the most important data, that is, the seismic-wave velocity changes, were not easily confirmed except in some cases. Another reason might be the non-generality of the model.

On the other hand, slow fault movements along the extension of co-seismic fault planes or along the fault itself could be the cause of precursory crustal movements. A kind of pre-seismic creep before the occurrence of the great earthquake has already been considered by KANAMORI (1972a). KANAMORI and CIPAR (1974) pointed out a creep-like reverse fault movement of which the dislocation was estimated as large as 30 m only fifteen minutes before the 1960 great Chilean earthquake (M 8.3). MORTENSEN and JOHNSTON (1976) treated a similar model, a creep instability model in their paper regarding the 1974 Hollister earthquake. THATCHER (1976) tried to understand the Palmdale bulge in the Mojave Desert, California as an aseismic creep, drawing attention to its possible relation to thrust-type earthquakes like the 1952 Kern County and the 1971 San Fernando earthquakes. Recently, FUJII (1976, 1977) discussed a creep dislocation model as a possible model of earthquake precursor regarding the crustal movements. As Fujii stated, the creep dislocation model is not necessarily inconsistent with the dilatancy model. The creep dislocation at depth changes the position of the materials along the dislocation plane and furthermore changes the state of stress distribution around the plane, therefore, as a secondary phenomena, dilatancy might be caused by some stress concentration at shallow depth. A combination of the two might be realistic.

10. Concluding Remarks

The author has tried to sketch the important aspects of the prediction-oriented seismology by summarizing a number of papers which include some materials helpful in predicting earthquakes through Section 1 to 9. As stated at the beginning, the prediction-oriented seismology is, of course, not as yet, well established and the present descriptions are merely citations, one after another. Rearrangements of the materials considering the possible inter-relationships among the various precursory phenomena and their systematic descriptions require further investigation.

There is, however, an explicit, common understanding about the prediction, that is, earthquake prediction should specify the location, the size, and the occurrence time with some probability or reliability. Deterministic precursors to an earthquake for a short-term prediction do not seem to exist so far. Seismological, seismotectonic, or other geoscientific investigations provide many materials helpful in predicting earthquake, however, the greater part of them are qualitative. Research of the precursors with a kind of reliability that any phenomenon is a precursor to an earthquake, namely, the quantitative research, is obviously a new direction in the prediction-oriented research. As such a reliability cannot be derived from a theoretical investigation, a great deal of observational data selected for certain conditions and effort for the data treatment are needed to establish the reliability of the precursor.

Another new direction of the prediction-oriented seismology is how to detect the precursor or its candidates within a short time prior to the occurrence of the earthquake. This is a completely practical and technological problem. There are a few research papers which are directed to practical usage. Of course, as a base to such works, the nature, the characteristics, or the regularity of the precursor must be known. As mentioned in Section 7, data collection and processing system constitute an important branch of the short-term prediction problem.

Model studies of earthquake occurrence including the precursor treatment in Section 9 are one of the most typical prediction-oriented seismological investigative studies. The dilatancy model and the creep dislocation model are appraisable as one of the most valuable products of earthquake-prediction efforts in the field of fundamental research. Although the results of the model study can not be directly connected with the practical problem, such fundamental research constitutes the framework of earthquake prediction research and leads to new findings in the observation or research.

New big projects such as the deep borehole measurements of crustal activity or the submarine-cable-type OBS observation are two of the special features of the national program of the present day earthquake prediction as stated in Sections 5 and 6. Expensive projects of this kind are supported by both the scientific necessity and the social needs for earthquake prediction, not merely by any individual scientific interest. That is, the progress in the present day earthquake prediction program is strongly dependent on the national policy regarding the prediction.

REFERENCES

ALLEN, C. R., The tectonic environments of seismically active and inactive areas along the San Andreas fault system, *Stanford Univ. Publ. Geol. Sci.*, **11**, 70, 1968.

ALLEN, C. R., P. S. AMAND, D. F. RICHTER, and J. M. NORDQUIST, Relationship between seismicity and geological structure in the southern California region, *Bull. Seismol. Soc. Am.*, **55**, 753–798, 1965.

ALSINAWI, S. and H. A. A. GHALIB, Historical seismicity of Iraq, *Bull. Seismol. Soc. Am.*, **65**, 541–547, 1975.

ANDERSON, D. and J. WHITCOMB, The dilatancy-diffusion model of earthquake prediction, in *Proceedings of the Conference on Tectonic Problems of the San Andreas Fault System*, edited by R. L. Kovach and A. Nur, 417 pp., Stanford Univ. Press, Palo Alto, California, 1973.

ANDERSON, D. and J. WHITCOMB, Time-dependent seismology, *J. Geophys. Res.*, **80**, 1497–1503, 1975.

ANDO, M., A fault-origin model of the great Kanto earthquake of 1923 as deduced from geodetic data, *Bull. Earthq. Res. Inst., Univ. Tokyo*, **49**, 19–32, 1971.

ANDO, M., Possibility of a major earthquake in the Tokai district, Japan and its pre-estimated seismotectonic effects, *Tectonophysics*, **25**, 69–85, 1975.

ANTSYFEROV, M. S., On the possibilities of geoacoustic prediction of local earthquakes, Acta 3rd All-Union Symp. Seismic Regime, June 3–7, 1968, Science Press, Siberian Division, Transrated into English from Russian by D. B. Vitaliano, U.S. Geological Survey, 1969.

ASADA, T. and H. SHIMAMURA, Ocean bottom seismographs and a new geophysics, *Kagaku*, **44**, 278–285, 1974 (in Japanese).

BOROVIK, N., L. MISHARINA, and A. TRESKOV, On the possibility of strong earthquakes in Pribayakalia in the future, *Izv. Acad. Sci. USSR Phys. Solid Earth*, Engl. Transl., 13–16, 1971.

BRADY, B. T., Theory of earthquakes, 1. A scale independent theory of rock failure, *Pure Appl. Geophys.*, **112**, 701–725, 1974.

BRUNE, J. N. and C. R. ALLEN, A microearthquake survey of the San Andreas fault system in southern California, *Bull. Seismol. Soc. Am.*, **57**, 277–296, 1967.

BUFE, C. G., Frequency-magnitude variations during the 1970 Danvill earthquake swarm, *Earthq. Notes*, **41** (3), 3–7, 1970.

CHEN, P. S. and P. H. LIN, An application of statistical theory of extreme values to moderate and long interval earthquake prediction, *Acta Geophys. Sinica*, **16**, 6–24, 1973 (in Chinese).

DANBARA, T., Vertical movements of the earth's crust in relation to the Matsushiro earthquakes, *J. Geod. Soc. Jpn.*, **12**, 18–45, 1966 (in Japanese).

DUDA, S. J., Secular seismic energy release in the circum-Pacific belt, *Tectonophysics*, **2**, 409–452, 1965.

EARTHQUAKE RESEARCH INSTITUTE, *Research on the Matsushiro Earthquakes*, Earthq. Res. Inst., Univ. Tokyo, (1) (2), 1966, (3) (4) (5), 1967.

ENGDAHL, E. and C. KISSLINGER, Precursory phenomena to a magnitude 5 earthquake in the Central Aleutian islands, *EOS*, **58**, 433 (abstract), 1977.

EVISON, F. F., The precursory earthquake swarm, *Phys. Earth Planet. Inter.*, **15**, 19–23, 1977.

FEDOTOV, S. A., On regularities in distribution of strong earthquakes of Kamchatka, Kurile Island, and northeastern Japan, *Trudy Inst. Phys. Earth Acad. Sci., USSR*, **36**, 66–93, 1965 (in Russian).

FEDOTOV, S. A., A. A. GUSEV, and S. A. BOLDYREV, Progress of earthquake prediction in Kamchatka, *Tectonophysics*, **14**, 279–286, 1972.

FIEDLER, B. G., Local *b*-values related to seismicity, *Tectonophysics*, **23**, 277–282, 1974.

FISHER, R. A., Tests of significance in harmonic analysis, in *Contribution to Mathematical Statistics*, 16. 58a–16. 59a, John Wiley and Sons, Inc., New York, 1950.

FUJII, Y., Seismic crustal movement and associated gravity change, *J. Geod. Soc. Jpn.*, **22**, 308–310, 1976.

FUJII, Y., Pre-slip as forerunner of earthquake occurrence, *Proceedings Simposium Earthq. Pred.* (*1976*), 127–137, 1977 (in Japanese).

GEODESY COUNCIL (Japan), *Authorized Proposal of the 4th 5-year Schedule of Earthquake Prediction Program* (*fiscal year* 1979–1983) *of Japan*, 1978 (in Japanese).

GEOGRAPHICAL SURVEY INSTITUTE (Japan), Recent result of triangulation survey in Hokkaido, Japan, *Rep. Coor. Comm. Earthq. Pred.*, **2**, 3–5, 1970a (in Japanese).

GEOGRAPHICAL SURVEY INSTITUTE (Japan), Recent result of levelling in Hokkaido, Japan, *Rep. Coor. Comm. Earthq. Pred.*, **4**, 1–2, 1970b (in Japanese).

GEOGRAPHICAL SURVEY INSTITUTE (Japan), Recent result of levelling in eastern Hokkaido, Japan, *Rep. Coor. Comm. Earthq. Pred.*, **5**, 1–2, 1971 (in Japanese).

GEOGRAPHICAL SURVEY INSTITUTE (Japan), Vertical movements in Tokai District (2), *Rep. Coor. Comm. Earthq. Pred.*, **11**, 102–104, 1974 (in Japanese).

GEOGRAPHICAL SURVEY INSTITUTE (Japan), Horizontal strains in Tokai District, *Rep. Coor. Comm. Earthq. Pred.*, **15**, 103–105, 1976 (in Japanese).

GUDARZI, K. M., A statistical study of the seismicity of the Iranian Plateau, presented to the 15th General Assembly of the IUGG, Moscow, 1971.

GUPTA, I. N., Precursory reorientation of stress axes due to vertical migration of seismic activity?, *J. Geophys. Res.*, **80**, 272–273, 1975.

GUTENBERG, B. and C. F. RICHTER, *Seismicity of the Earth and Associated Phenomena*, 310 pp., Princeton Univ. Press, Princeton, N.J., 1954.

HAGIWARA, T., The development of earthquake prediction research in Japan, *Gakujitsu Geppo, Japan Soc. Promotion Sci.*, **26**, 2–32, 1973 (in Japanese).

HAGIWARA, T. and T. IWATA, Summary of the seismographic observation of Matsushiro swarm earthquakes, *Bull. Earthq. Res. Inst., Univ. Tokyo*, **46**, 485–515, 1968.

HAMADA, K., Ultra microearthquakes in the area around Matsushiro, *Bull. Eathq. Res. Inst.*, **46**, 271–318, 1968.

HAMADA, K., Anomalously small value of the Ishimoto-Iida's coefficient *m* for foreshocks of Izu-Oshima-Kinkai earthquake of January 14, 1978, *Rep. Coor. Comm. Earthq. Pred.*, **20**, 53–55, 1978 (in Japanese).

HAMADA, K., H. TAKAHASHI, M. TAKAHASHI, and H. SUZUKI, Deep borehole measurements of crustal activities around Tokyo, in *Terrestrial and Space*

Techniques in Earthquake Prediction Research, edited by Andreas Vogel, 600 pp., Frieder Vieweg & Sohn, Braunschweig / Wiesbaden, 1979.

HASEGAWA, A., N. UMINO, and A. TAKAGI, Double-planed structure of the deep seismic zone in the northeastern Japan arc, *Tectonophysics*, **47**, 43–58, 1978.

HONDA, H., A. MASATSUKA, and M. ICHIKAWA, On the mechanism of earthquakes and stress producing them in Japan and its vicinity (third paper), *Geophys. Mag.*, **33**, 271–279, 1967.

ICHIKAWA, M., Statistical investigation of mechanism of earthquakes occurring in and near Japan and some related problems, *J. Met. Res.*, **18**, 83–154, 1966 (in Japanese).

ICHIKAWA, M., Matsushiro earthquake swarm, *Geophys. Mag.*, **34**, 307–331, 1969.

ICHIKAWA, M., Reanalyses of mechanism of earthquakes which occurred in and near Japan and statistical studies on the nodal plane solutions obtained, *Geophys. Mag.*, **35**, 207–274, 1971.

IMAMURA, A., On the 1854 Ansci, the 1872 Hamada, the 1896 Rikuu earthquakes and others, *Rep. Imp. Earthq. Inv. Comm.*, **77**, 1913 (in Japanese).

IMAMURA, A., Examination report on the east Akita earthquake of 1915, *Rep. Imp. Earthq. Inv. Comm.*, **82**, 1–30, 1915 (in Japanese).

IMAMURA, A., On the seismic activity of central Japan, *Jap. J. Astr. Geophys.*, **6**, 119–137, 1928.

IMAMURA, A., Cycles of seismic activities in the Kanto and Kinki districts and precursory phenomena before large earthquakes, *Zisin, J. Seismol. Soc. Jpn.*, **1**, 4–16, 1929 (in Japanese).

INOUE, U., Seismic activity around the epicentral area of the Niigata earthquake, *Kenshin-Jiho*, **27**, 139–144, 1965 (in Japanese).

INSTITUTE OF GEOPHYSICS, ACADEMIA SINICA, Catalog of historical large earthquakes in China (780 B.C.–1973A.D., $M \geq 6$), Proceedings of the lectures by the seismological delegation of the People's Republic of China, pp. 67–83, Seismol. Soc. Japan, 1976 (in Japanese).

ISHIBASHI, K., Stress field in the Kanto district before the Kanto earthquake of 1923—Indication of large earthquakes along the Sagami trough—(abstract), Spring Meeting Seismol. Soc. Japan, p. 69, 1975 (in (Japanese).

ISHIBASHI, K., Re-investigation into a great earthquake that is expected in the Tokai district—on the Suruga Bay great earthquake (abstract), Fall Meeting Seismol. Soc. Japan, pp. 30–34, 1976 (in Japanese).

ISHIDA, M. and H. KANAMORI, The spatio-temporal variation of seismicity before the 1971 San Fernando earthquake, California, *Geophys. Res. Lett.*, **4**, 345–346, 1977.

ISHIDA, M. and H. KANAMORI, The foreshock activity of the 1971 San Fernando earthquake, California, *Bull. Seismol. Soc. Am.*, **68**, 1265–1279,

1978a.

ISHIDA, M. and H. KANAMORI, Temporal variation of seismicity and spectrum of small earthquakes preceding the 1952 Kern County, California, earthquake, *Bull. Seismol. Soc. Am.*, **70**, 509–527, 1980.

JAPAN METEOROLOGICAL AGENCY, Catalog of major earthquakes which occurred in and near Japan (1926–1956), *Seismol. Bull. Jpn. Met. Agency*, Supp. 2, 1–91, 1958.

JAPAN METEOROLOGICAL AGENCY, Catalog of major earthquakes which occurred in and near Japan (1957–1962), *Seismol. Bull. Jpn. Met. Agency*, Supp. 2, 1–49, 1966.

JAPAN METEOROLOGICAL AGENCY, Catalog of major earthquakes which occurred in and near Japan (1963–1967), *Seismol. Bull. Jpn. Met. Agency*, Supp. 3, 1–62, 1968.

JAPAN METEOROLOGICAL AGENCY, An investigation report of the 1974 Izu-Hanto-Oki earthquake, *Kenshin-Jiho*, **39**, 89–120, 1975 (in Japanese).

JONES, L. and P. MOLNAR, Frequency of foreshocks, *Nature*, **262**, 677–679, 1976.

KAMINUMA, K., T. IWATA, I. KAYANO, and M. OHTAKE, Summary of Scientific Data of Major Earthquakes in Japan 1872–1972, 136 pp., Earthq. Res. Inst., Univ. Tokyo, 1973 (in Japanese).

KANAMORI, H., Seismological evidence for a lithospheric normal faulting—The Sanriku earthquake of 1933, *Phys. Earth Planet. Inter.*, **4**, 289–300, 1971a.

KANAMORI, H., Faulting of the great Kanto earthquake of 1923 as revealed by seismological data, *Bull. Earthq. Res. Inst., Univ. Tokyo*, **49**, 13–18, 1971b.

KANAMORI, H., Relation between tectonic stress, great earthquakes and earthquake swarm, *Tectonophysics*, **14**, 1–12, 1972a.

KANAMORI, H., Tectonic implications of the 1944 Tonankai and the 1946 Nankaido earthquakes, *Phys. Earth Planet. Inter.*, **5**, 129–139, 1972b.

KANAMORI, H. and M. ANDO, Fault parameters of the great Kanto earthquake of 1923, in *Pub. 50th Anniversary Great Kanto Earthq., 1923*, pp. 89–101, Earthq. Res. Inst., Univ. Tokyo, 1973 (in Japanese).

KANAMORI, H. and J. J. CIPAR, Focal process of the great Chilean earthquake May 22, 1960, *Phys. Earth Planet. Inter.*, **9**, 128–136, 1974.

KASAHARA, J., Ocean bottom seismology undergoing a change, *Shizen (nature)*, Sep., 60–71, 1978 (in Japanese).

KATSUMATA, M., Seismicity and some related problems in and near Japanese islands, *Kenshin-Jiho (Q. J. Seismol.)*, **35**, 75–142, 1970 (in Japanese).

KAWASAKI, I. and M. KATSUMATA, Stress field at depths in the Kanto district (abstract), Spring Meeting Seismol. Soc. Japan, p. 24, 1975 (in Japanese).

KAWASUMI, H., Proofs of 69 years periodicity and imminence of destructive earthquake in southern Kwanto district and problems in the countermeasures thereof, *Chigaku Zasshi*, **79**, 115–138, 1970 (in Japanese).

KELLEHER, J., Space-time seismicity of the Alaska-Aleutian seismic zone, *J. Geophys. Res.*, **75**, 5745–5756, 1970.

KELLEHER, J., Rupture zones of large south American earthquakes and some predictions, *J. Geophys. Res.*, **77**, 2087–2103, 1972.

KELLEHER, J. and J. SAVINO, Distribution of seismicity before large strike slip and thrust-type earthquakes, *J. Geophys. Res.*, **80**, 260–271, 1975.

KELLEHER, J., L. SYKES, and J. OLIVER, Possible criteria for predicting earthquake locations and their application to major plate boundaries of the Pacific and the Caribbean, *J. Geophys. Res.*, **78**, 2545–2585, 1973.

LE PICHON, X., Sea-floor spreading and continental drift, *J. Geophys. Res.*, **73**, 3661–3697, 1968.

LOMNITZ, C., Major earthquakes and tsunamis in Chile during the period 1535–1955, *Geol. Rundsch.*, **59**, 938, 1970.

LOMNITZ, C., *Global Tectonics and Earthquake Risk*, 320 pp., Elsevier, Amsterdam, 1974.

LOUDERBACK, G., Central California earthquakes of the 1830's, *Bull. Seismol. Soc. Am.* **37**, 33, 1947.

MATSUDA, T., Active faults and earthquakes—geological investigation, *Chishitsugaku-ron-shu*, **12**, 15–32, 1976 (in Japanese).

MCEVILLY, T. V. and R. RAZANI, A preliminary report of Qir, Iran earthquake of April 10, 1972, *Bull. Seismol. Soc. Am.*, **63**, 339–354, 1973.

MILNE, J., *A Catalog of Destructive Earthquakes*, 92 pp., British Association for the Advanced of Science, London, 1912.

MIYAMURA, S., Magnitudes of Japanese earthquakes in 1904–1906, presented at Joint IASPEI/IAVCEI Scientific Assemblies, Darham, August, 1977.

MIZOUE, M., M. NAKAMURA, M., ISHIKETA, and N. SETO, Regularity and pattern of seismic activity in and near Wakayama, Japan (abstract), Fall Meeting Seismol. Soc. Japan, pp. 90–91, 1977a (in Japanese).

MIZOUE, M., M. NAKAMURA, M. ISHIKETA, and N. SETO, Earthquake prediction from microearthquake observation in the vicinity of Wakayama City, northwestern part of the Kii Peninsula, Central Japan, *Rep. Wakayama Obs., Earthq. Res. Inst., Univ. Tokyo*, **14**, 1–45, 1977b.

MOGI, K., Study of the elastic shocks caused by the fracture of heterogeneous materials and its relation to earthquake phenomena, *Bull. Earthq. Res. Inst., Univ. Tokyo*, **40**, 125–173, 1962a.

MOGI, K., Magnitude-frequency relation for elastic shocks accompanying fractures of various materials and some related problems in earthquakes (2nd paper), *Bull. Earthq. Res. Inst., Univ. Tokyo*, **40**, 831–853, 1962b.

MOGI, K., The fracture of a semi-infinite body caused by an inner stress origin and its relation to the earthquake phenomena (2nd paper)—The case of the materials having some heterogeneous structures—, *Bull. Earthq. Res. Inst.*, *Univ. Tokyo*, **41**, 595–614, 1963a.

MOGI, K., Some discussions on aftershocks, foreshocks and earthquake swarms —the fracture of a semi-infinite body caused by an inner stress origin and its relation to the earthquake phenomena, *Bull. Earthq. Res. Inst., Univ. Tokyo*, **41**, 615–658, 1963b.

MOGI, K., Earthquakes and fractures, *Tectonophysics*, **5**, 35–55, 1967.

MOGI, K., Sequential occurrence of recent great earthquakes, *J. Phys. Earth*, **16**, 30–36, 1968a.

MOGI, K., Some features of recent seismic activity in and near Japan (1), *Bull. Earthq. Res. Inst., Univ. Tokyo*, **46**, 1225–1236, 1968b.

MOGI, K., Migration of seismic activity, *Bull. Earthq. Res. Inst., Univ. Tokyo*, **46**, 53–74, 1968c.

MOGI, K., Some features of recent seismic activity in and near Japan (2), *Bull. Earthq. Res. Inst., Univ. Tokyo*, **47**, 395–417, 1969a.

MOGI, K., Rock breaking test, *Kagaku*, **39**, 95–102, 1969b (in Japanese).

MOGI, K., Recent horizontal deformation of the earth's crust and tectonic activity (1), *Bull. Earthq. Res. Inst., Univ. Tokyo*, **48**, 413–430, 1970.

MOGI, K., Relationship between shallow and deep seismicity in the western Pacific region, *Tectonophysics*, **17**, 1–22, 1973a.

MOGI, K., Rock fracture, in *Annual Review of Earth and Planetary Sciences*, edited by F. A. Donath, pp. 63–84, Annual Reviews, 1, Palo Alto, California, 1973b.

MOGI, K., Active periods in the world's chief seismic belts, *Tectonophysics*, **22**, 265–282, 1974a.

MOGI, K., Rock fracture and earthquake prediction, *J. Soc. Mat. Sci. Jpn.*, **23**, 320–331, 1974b (in Japanese).

MOGI, K., Theory of great earthquakes, *Suuri-Kagaku*, August, 46–52, 1976 (in Japanese).

MOGI, K., Seismic activity and earthquake prediction, *Proceedings Symposium Earthq. Pred. (1976)*, 203–214, 1977 (in Japanese).

MOGI, K., Seismic gap and migration of seismic activity (abstract), Fall Meeting Seismol. Soc. Japan, p. 27, 1978 (in Japanese).

MORTENSEN, C. E. and M. J. S. JOHNSTON, Anomalous tilt preceding the Hollister earthquake of November 28, 1974, *J. Geophys. Res.*, **81**, 3561–3566, 1976.

MUSHA, K., *Zote Dai Nihon Jishin Shiryo* (Revised Japanese historical records relevant to earthquakes), 1–3, Earthquake Disaster Prevention Coucil, 1941–1943 (in Japanese).

MUSHA, K., *Nihon Jishin Shiryo* (Japanese historical records relevant to earthquakes), 1019 pp., Mainichi Press, Tokyo, 1951 (in Japanese).

MUSHA, K., *Earthquake List in and near Japan*, 1950–1953 (in Japanese).

NAGUMO, S., Activation mode of great submarine earthquakes along the Japanese islands, in *Pub. 50th Anniversary Great Kanto Earthq., 1923*, pp. 237–291, Earthq. Res. Inst., Univ. Tokyo, 1973 (in Japanese).

NAKAMURA, K., Volcano-stratigraphic study of Oshima volcano, Izu, Japan, *Bull. Earthq. Res. Inst., Univ. Tokyo*, **42**, 649–728, 1964.

NASU, N., Seismic activity in the Kwanto district before and after the great Kwanto earthquake of 1923, in *Pub. 50th Anniversary Great Kanto Earthq., 1923*, pp. 21–40, Eathq. Res. Inst., Univ. Tokyo, 1973 (in Japanese).

NERSESOV, I. L., A. A. LUKK, V. S. PONOMAREV, T. G. RAUTIAN, B. G. RULEV, A. N. SEMENOV, and I. G. SIMBIREVA, Possibilities of earthquake prediction, exemplified by the Garm area of the Tazhik S.S.R., in *Earthquake Precursors*, edited by M. A. Sadovsky, I. L. Nersesov, and L. A. Latynina, 216 pp., Acad. Sci. USSR, Moscow, 1973 (Translated in English from Russian by D. B. Vitalino, U. S. Geological Survey, 1973).

NIWA, Y., S. KOBAYASHI, T. FUKUI, T. YANAGIDANI, and M. OTSU, Some considerations on acoustic emissions dues to rock burst, *Proceedings of The 10th Symposium on Rock Mechanics*, pp. 46–50, Com. Rock Mechanics, J.S.C.E., 1976 (in Japanese).

NUR, A., Dilatancy, pore fluids and premonitory variations of t_s/t_p travel times, *Bull. Seismol. Soc. Am.*, **62**, 1217–1222, 1972.

OHTAKE, M., Micro-structure of the seismic sequence related to a moderate earthquake, *Bull. Earthq. Res. Inst., Univ. Tokyo*, **48**, 1053–1067, 1970.

OHTAKE, M., Search for precursors of the 1974 Izu-Hanto-Oki earthquake, Japan, *Pure Appl. Geophys.*, **114**, 1083–1093, 1976a.

OHTAKE, M., Review of the Matsushiro earthquake swarm, *Kagaku*, **46**, 306–313, 1976b (in Japanese).

OHTAKE, M., T. MATSUMOTO, and G. V. LATHAM, Seismicity gap near Oaxaca, southern Mexico as a probable precursor to a large earthquake, *Pure Appl. Geophys.*, **115**, 375–385, 1977a.

OHTAKE, M., T. MATSUMOTO, and G. V. LATHAM, Temporal change in seismicity preceding some shallow earthquakes in Mexico and Central America, *Bull. Int. Inst. Seismol. and Earthq. Eng.*, **15**, 105–123, 1977b.

OHTAKE, M., T. MATSUMOTO, and G. V. LATHAM, Patterns of seismicity preceding earthquakes in Central America, Mexico, and California, *Proceedings USGS Symposium on Earthquake Prediction*, 1978.

OIKE, K., Relationship between rainfall and occurrence of shallow earthquakes (abstract), Fall Meeting Seismol. Soc. Japan, p. 94, 1977a (in Japanese).

OIKE, K., Consequence of predicted seismic activity along the Yamazaki fault, Japan (abstract), Fall Meeting Seismol. Soc. Japan, p. 88, 1977b (in Japanese).

OIKE, K. and Y. KISHIMOTO, The Yamazaki fault as a test field, *Proceedings*

Simposium Earthq. Pred. (1976), pp. 83–90, 1977 (in Japanese).

ŌMORI, F., Foreshocks of large earthquakes, *Rep. Imp. Earthq. Inv. Comm.*, **68**, ko, 31–38, 1910 (in Japanese).

ŌMORI, F., Catalog of earthquakes in and around Japan, *Rep. Imp. Earthq. Inv. Comm.*, **88**, otsu, 1–71, 1920 (in Japanese).

OZAWA, I., Forecast of occurrences of earthquakes in the northwestern part of the Kinki district, *Cont. Geophys. Inst., Kyoto Univ.*, **13**, 147–161, 1973.

PAPAZACHOS, B. C., Foreshocks and earthquake prediction, *Tectonophysics*, **28**, 213–223, 1975.

PRESS, F. and R. SIEVER, *Earth*, 945 pp., W. H. Freeman and Co., San Francisco, California, 1974.

REID, H. F., The California earthquake of April 18, 1906, *Rep. State Earthquake Inv. Commission*, Vol. 2, *The Mechanics of the Earthquake*, pp. 16–19, Carnegie Institute of Washington, Washington, D.C., 1910.

REID, H. F. and S. TABER, The Puerto Rico earthquake of 1918 with descriptions of earlier earthquakes, Rep. Earthq. Inv. Commision, Doc. 269, p. 53, U.S. House of Representatives, 66th Congress, 1st Session, 1919a.

REID, H. F. and S. TABER, The Puerto Rico earthquake of October-November 1918, *Bull. Seismol. Soc. Am.*, **9**, 95, 1919b.

REID, H. F. and S. TABER, The Virgin Islands earthquakes of 1867–1868, *Bull. Seismol. Soc. Am.*, **10**, 9, 1920.

RICHTER, C. F., *Elementary Seismology*, 768 pp., W. H. Freeman and Co., San Francisco, California, 1958.

RIKITAKE, T., *Earthquake Prediction*, 357 pp., Elsevier, Amsterdam, 1976.

ROBSON, G., An earthquake catalog for eastern Caribbean 1530–1960, *Bull. Seismol. Soc. Am.*, **54**, 875, 1964.

ROTHÉ, J. P., *The Seismicity of the Earth, 1953–1965*, 326 pp., UNESCO, Paris, 1969.

SADOVSKY, M. A. and I. L. NERSESOV, Forecasts of earthquakes on the basis of complex geophysical features, *Tectonophysics*, **23**, 247–255, 1974.

SADOVSKY, M. A., I. NERSESOV, S. NIGMATULLAEV, L. LATYNINA, A. LUKK, A. SEMENOV, I. SIMBIREVA, and V. ULOMOV, The processes preceding strong earthquakes in some regions of middle Asia, *Tectonophysics*, **14**, 295–307, 1972.

SCHOLZ, C. H., The frequency-magnitude relation of microfracturing in rock and its relation to earthquakes, *Bull. Seismol. Soc. Am.*, **58**, 399–415, 1968.

SCHOLZ, C., L. SYKES, and Y. AGGARWAL, Earthquake prediction: a physical basis, *Science*, **181**, 803–910, 1973.

SEKIYA, H., The seismicity preceding earthquakes and its significance to earthquake prediction, *Zisin, J. Seismol. Soc. Jpn.*, **29**, 199–312, 1976 (in Japanese).

SEKIYA, H., Anomalous seismic activity and earthquake prediction, *J. Phys.*

Earth, **25**, Suppl., S85–S93, 1977.

SEKIYA, H. and K. TOKUNAGA, On the seismicity near the sea of Enshu, *Rep. Coor. Comm. Earthq. Pred.*, **11**, 96–101, 1974 (in Japanese).

SENO, T., Inter-plate and intra-plate seismic activities in the Tohoku and Hokkaido regions (abstract), Fall Meeting Seismol. Soc. Japan, p. 120, 1977 (in Japanese).

SHAKAL, A. F. and D. E. WILLIS, Estimated earthquake probabilities in the north circum-Pacific area, *Bull. Seismol. Soc. Am.*, **62**, 1397–1410, 1972.

SHIMAMURA, H. and T. ASADA, Ocean bottom seismology, problem and perspective, Extra of *Kaiyo-Kagaku*, **1**, 204–218, 1978 (in Japanese).

SHIMAZAKI, K., Periodicity of earthquake occurrence, *Kagaku*, **41**, 688–689, 1971 (in Japanese).

SHIMAZAKI, K., Focal mechanism of a shock at the north western boundary of the Pacific plate: Extensional feature of the oceanic lithosphere and compressional feature of the continental lithosphere, *Phys. Earth Planet. Inter.*, **6**, 397–404, 1972a.

SHIMAZAKI, K., Hidden periodicities of destructive earthquakes at Tokyo, *Zisin, J. Seismol. Soc. Jpn.*, **25**, 24–32, 1972b (in Japanese).

SHIMAZAKI, K., Pre-seismic crustal deformation caused by an underthrusting oceanic plate in eastern Hokkaido, Japan, *Phys. Earth Planet. Inter.*, **8**, 148–157, 1974a.

SHIMAZAKI, K., Nemuro-Oki earthquake of June 17, 1973, a lithospheric rebound at the upper half of the interface, *Phys. Earth Planet. Inter.*, **9**, 314–327, 1974b.

SHIMAZAKI, K., Intra-plate seismicity and inter-plate earthquakes: historical activity in southwest Japan, *Tectonophysics*, **33**, 33–42, 1976.

SHIMAZAKI, K., Correlation between intraplate seismicity and interplate earthquakes in Tohoku, Northeastern Japan, *Bull. Seismol. Soc. Am.*, **68**, 181–192, 1978.

SIMBIREVA, I. G., Focal mechanism of weak earthquakes in the Naryn River Basin, in *Experimental Seismology*, edited by M. A. Sadovsky, 423 pp., Science Press, Moscow, 1971 (Translated in English from Russian by D. B. Vitaliano, U. S. Geological Survey, 1973).

STUART, W., Diffusionless dilatancy model for earthquake precursors, *Geophys. Res. Lett.*, **1**, 261–264, 1974.

SU-SHAOSEIN, Characteristics of seismic activity related to the Haicheng earthquake, *Proceedings of the Lectures by the Seismological Delegation of the People's Republic of China*, pp. 27–41, Seismol. Soc. Japan, 1976 (in Japanese).

SUYEHIRO, S., Difference between aftershocks and foreshocks in the relationship of magnitude to frequency of occurrence for the great Chilean earthquake of 1960, *Bull. Seismol. Soc. Am.*, **56**, 185–200, 1966.

SUYEHIRO, S. and H. SEKIYA, Foreshocks and earthquakes prediction, *Tectonophysics*, **14**, 219–225, 1972.

SUYEHIRO, S., T. ASADA, and M. OHTAKE, Foreshocks and aftershocks accompanying a perceptible earthquake in central Japan, *Pap. Met. Geophys.*, **15**, 71–88, 1964.

SYKES, L., Aftershock zones of great earthquakes, seismicity gaps, and earthquake prediction for Alaska and the Aleutians, *J. Geophys. Res.*, **76**, 8021–8041, 1971.

TABER, S., The seismic belt in the greater Antilles, *Bull. Seismol. Soc. Am.*, **12**, 199, 1922.

TAKAHASHI, H. and K. HAMADA, Deep borehole observation of the earth's crust activities around Tokyo—Introduction of the Iwatsuki Observatory, *Pure Appl., Geophys.*, **113**, 311–320, 1975.

TAKAGI, A., A. HASEGAWA, and N. UMINO, Seismic activities in the northeastern Japan arc, *J. Phys. Earth*, **25**, Suppl., S95–S104, 1977.

TAYAMA, M., Dai Nihon Jishin Shiryo (Japanese historical documents on earthquakes), *Imp. Earthq. Inv. Comm.*, **46**, ko, 1–606, otsu, 1–595, 1904 (in Japanese), Reprinted by Shibunkan, Kyoto, 1973 (in Japanese).

THATCHER, W., Episodic strain accumulation in southern California, *Science*, **194**, 691–695, 1976.

TOCHER, D., Seismic history of the San Francisco region, *Calif. Div. Mines Spec. Rep.*, **57**, 39, 1959.

TSUBOKAWA, I., On relation between duration of crustal movement and magnitude of earthquake expected, *J. Geod. Soc. Jpn.*, **15**, 75–88, 1969 (in Japanese).

TSUJIURA, H., Spectral analyses of foreshocks, mainshocks and aftershocks (abstract), Fall Meeting Seismol. Soc. Japan, p. 24, 1976 (in Japanese).

TSUKAHARA, H., Plasticity of a descending plate and doble seismic planes (abstract), Fall Meeting Seismol. Soc. Japan, p. 49, 1977 (in Japanese).

TSUKAHARA, H., Physical conditions for double seismic planes of the deep seismic zone, *J. Phys. Earth*, **28**, 1981 (in press).

USAMI, T., Descriptive table of major earthquakes in and near Japan which were accompanied by damage, *Bull. Earthq. Res. Inst., Univ. Tokyo*, **44**, 1571–1622, 1966 (in Japanese).

USAMI, T., Error estimation for epicenters of Japanese historical earthquakes, in *Pub. 50th Anniversary Great Kanto Earthq. 1923*, pp. 1–12, Earthq. Res. Inst., Univ. Tokyo, 1973 (in Japanese).

USAMI, T., Error estimation for epicenters of Japanese historical earthquakes, *Urgent Rep. Earthq. Res. Inst., Univ. Tokyo*, **12**, 1–29, 1974 (in Japanese).

USAMI, T., *Nihon Higai Zisin Soran* (Comprehensive Data of Disasterous Earthquakes in Japan), 327 pp., Univ. Tokyo Press, Tokyo, 1975 (in Japanese).

USAMI, T., Earthquake disaster, in *Earthquake*, pp. 3–40, Univ. Tokyo Press, Tokyo, 1976 (in Japanese).

USAMI, T. and S. HISAMOTO, Future probability of a coming earthquake with intensity V or more in the Tokyo area, *Bull. Earthq. Res. Inst., Univ. Tokyo*, **48**, 331–340, 1970 (in Japanese).

USAMI, T. and S. HISAMOTO, Future probability of a coming earthquake with intensity V or more in the Kyoto area, *Bull. Earthq. Res. Inst., Univ. Tokyo*, **49**, 115–125, 1971 (in Japanese).

USCGS, Preliminary determination of epicenters (1965–1972), USCGS, 1965–1972.

UTSU, T., A statistical study on the occurrence of aftershocks, *Geophys. Mag.*, **30**, 521–605, 1961.

UTSU, T., Seismic activity in Hokkaido and its vicinity, *Geophys. Bull. Hokkaido Univ.*, **20**, 51–75, 1968 (in Japanese).

UTSU, T., Seismic activity and seismic observation in Hokkaido in recent years, *Rep. Coor. Comm. Earthq. Pred.*, **2**, 1–2, 1970a (in Japanese).

UTSU, T., Aftershocks and earthquake statistics (II), *J. Fac. Sci. Hokkaido Univ., Ser. VII (Geophysics)*, **3**, 197–266, 1970b.

UTSU, T., Aftershocks and earthquake statistics (III), *J. Fac. Sci. Hokkaido Univ., Ser. VII (Geophysics)*, **3**, 379–441, 1971.

UTSU, T., Large earthquakes near Hokkaido and the expectancy of the occurrence of a large earthquake off Nemuro, *Rep. Coor. Comm. Earthq. Pred.*, **7**, 7–13, 1972 (in Japanese).

UTSU, T., Space-time pattern of large earthquakes occurring off the Pacific coast of the Japanese islands, *J. Phys. Earth*, **22**, 325–342, 1974.

UTSU, T., Correlation between shallow earthquakes in Kwanto region and intermediate earthquakes in Hida region, Central Japan, *Zisin, J. Seismol. Soc. Jpn.*, **28**, 303–312, 1975 (in Japanese).

UTSU, T., An investigation into the discrimination of foreshock sequences from earthquake swarm, *Zisin, J. Seismol. Soc. Jpn., Ser. 2*, **31**, 129–135, 1978 (in Japanese).

UTSU, T. and A. SEKI, A relation between the area of aftershock region and the energy of mainshock, *Zisin. J. Seismol. Soc. Jpn., Ser. 2*, **7**, 233–240, 1955 (in Japanese).

WATARI, M. and O. ITAGAKI, Laboratory experiments for land slide prediction methods by using acoustic emission, *Doboku Gijutsu Shiryo*, **17–2**, 3–8, 1975 (in Japanese).

WESSON, R. and W. ELLSWORTH, Seismicity preceding moderate earthquakes in California, *J. Geophys. Res.*, **78**, 8527–8546, 1973.

WHITCOMB, J. H., J. D. GARMANY, and D. L. ANDERSON, Earthquake prediction: Variation of seismic velocities before the San Fernando earthquake, *Science*, **181**, 632–635, 1973.

WYSS, M. and W. H. K. LEE, Time variations of the average earthquake magnitude in Central California, in *Proceedings of the Conferrence on Tectonic Problems on the San Andreas Fault System*, edited by R. L. Kovach and A. Nur, Stanford Univ. Publ., *Geol. Sci.*, **13**, 24–42, 1973.

YAMAMOTO, E. and M. TAKAHASHI, Depth distribution of background noises in the deep borehole at the Iwatsuki observatory (abstract), Fall Meeting Seismol. Soc. Japan, p. 95, 1976 (in Japanese).

YAMAMOTO, E., K. HAMADA, and K. KASAHARA, Background seismic-wave noise and elimination of noise transmitted through water in a borehole at the Iwatsuki observatory, *Zisin, J. Seismol. Soc. Jpn.*, **18**, 171–180, 1975 (in Japanese).

ELECTRIC AND MAGNETIC APPROACH TO EARTHQUAKE PREDICTION

Yoshimori HONKURA*

Earthquake Research Institute, Unviersity of Tokyo, Yayoi, Bunkyo-ku, Tokyo, Japan

Abstract Electric and magnetic phenomena associated with earthquake occurrences have been reported from time to time in various countries. Some of them are coseismic changes, but some anomalies appeared before earthquakes as though they were signals warning impending earthquakes. Such precursors contributed to successful predictions of large earthquakes, particularly in China.

Electric and magnetic anomalies reported so far can be classified into four types: the geomagnetic field (G), the electric (telluric) field (E), the ground resistivity associated with strain ($R1$), and the resistivity in and around a focal region ($R2$). For each one of the four classes precursory changes have been reported. However, the nature of precursor time, the time from the first appearance of an anomaly to an earthquake occurrence, is different from one class to another. $R1$ precursors have appeared some hours before earthquakes, that is at an imminent stage of prediction. Most of E precursors have appeared some days to some tens of days prior to earthquakes and, therefore, they seem to be useful at a short-term prediction stage. $R2$-type anomalies are characterized by a precursor time of tens to hundreds of days. They will be used most effectively at medium- to short-term prediction stages. G anomalies tend to appear at medium- to long-term stages, although

---

* Present address: Department of Applied Physics, Tokyo Institute of Technology, 2-12-1 Ookayama, Meguro-ku, Tokyo, Japan

nothing clear could be obtained because of the small amount of available data.

If an appropriate combination of the four types of measurements is considered and applied to field work, the contribution of electric and magnetic methods to earthquake prediction will increase.   It is certainly better to combine the results of electric and magnetic measurements with other elements of earthquake prediction.   In that case, we may approach to general nature of earthquake precursors.   Then information on the particular process going on in the crust in association with an impending earthquake may possibly be obtained.

1. Introduction

Anomalous changes in geophysical and geodetical parameters as well as unusual phenomena such as peculiar animal behaviors have been reported in association with earthquake occurrences (RIKITAKE, 1976a). Some of them are classified into electric and magnetic phenomena which will be treated in this article.   However, not all of the electrically and magnetically anomalous changes reported so far have been well understood and physical interpretations have yet to be pursued.

One of the most frequently reported anomalies is an anomalous change in the geomagnetic field at the time of an earthquake occurrence. Recent measurements based on modern magnetometers having a high accuracy revealed, however, that the unusually large anomalies reported so far are unlikely to be of precursory nature (RIKITAKE, 1968).   Modern theories also predict only slight changes in the geomagnetic field at least as far as precursory changes are concerned (e.g. NAGATA, 1969).   In Section 2 are described various aspects of recent earthquake prediction studies based on geomagnetic field changes.   First of all, some examples of coseismic and precursory changes will be introduced.   Then other studies including theoretical work on piezomagnetism will be explained with emphasis on their contributions to the understanding of anomalous phenomena associated with earthquakes.

Another important factor is a change in electric currents flowing in the earth (often called the telluric current).   However, telluric current measurements are highly disturbed by various agencies unrelated to an earthquake such as rainfall, temperature, and so on.   This is one of the reasons why reliable data are relatively few in spite of many reports on anomalous changes in telluric currents.   That some large earthquakes were quite successfully predicted in China where telluric current measurements contributed much to the predictions (NORITOMI, 1978) is an epoch-making event and a re-examination of telluric current anomalies should be pursued further.   With this in mind, Section 3 is assigned to electric current anomalies with emphasis on precursory changes without excessive attention to data reliability.   The phenomenon of earthquake lightning is also included in this section because of its origin, widely believed to be an electric discharge due to the electric field in the earth and atmosphere.

Recently much attention has been focussed on electrical resistivity measurements and many successful results have been obtained.   Studies of resistivity changes can be classified into two types.   One aims at detecting a very small change in ground resistivity associated with strain as observed by a highly sensitive resistivity variometer constructed by YAMAZAKI (1967, 1977).   This type of resistivity change will be explained in Section 4.   The purpose of the other is to observe a resistivity change of several to a few tens of percents in and around a focal region.   Anomalies obtained from the two measuring methods based on artificial electric currents and natural electric and magnetic fields will be treated in Section 5 as well as experimental studies of resistivity changes associated with dilatancy, stick-slip along a fault, and so on.

In view of the important contributions of electric and magnetic methods to earthquake prediction, an attempt will be made to examine the relation between precursor time and earthquake magnitude for electric and magnetic precursors.   An attempt is also made to clarify the characteristics of such relations for four types of anomalies as described in Sections 2, 3, 4, and 5, respectively.   The result of the attempt will be found in Section 6.

The final section, Section 7, will be assigned to a brief survey of problems in the described studies pertaining to earthquake prediction. Some of them may turn out to be promising and some may not.   It is obvious that promising ones should be intensively pursued and others should be re-examined.   It is the author's wish that this article will be of some help for this kind of purpose.

2. Tectonomagnetism

It has widely been believed that the magnetic field observed on the earth's surface would change at the time of earthquake occurrence.   In fact, there are many reports on geomagnetic changes associated with earthquakes (RIKITAKE, 1976a). The effect of mechanical stress on rock magnetism has also been brought to light as a physical basis for changes in the geomagnetic field. NAGATA (1969) reviewed observational, experimental, and theoretical aspects of geomagnetic changes with special reference to crustal tectonics, particularly earthquakes, and expressed this phenomenon by the term " tectonomagnetism."

2. 1 Changes in the geomagnetic field associated with earthquakes

The magnitude of a geomagnetic change reported in association with an earthquake has greatly decreased with time, reflecting improvements of measuring instruments and techniques (RIKITAKE, 1968, 1976a). The proton precession magnetometer has improved the accuracy of measurements up to 1 nT (gamma) or even less in the total intensity and the procedure of field work also has greatly been reduced. Thus, the use of this magnetometer has to a considerable extent increased the reliability of observed changes in the total intensity. From the viewpoint of earthquake prediction, much stress should be put on precursory changes. In this article, however, coseismic changes will also be reviewed in view of their possible contribution to the prediction research by confirming, at any event, the existence of geomagnetic changes associated with earthquakes. They will be helpful for understanding the mechanism producing a geomagnetic change in association with an earthquake occurrence.

2. 1. 1 Examples in Japan

TAZIMA (1968) examined the results of repeated geomagnetic surveys carried out by the Geographical Survey Institute and pointed out a correlation between anomalous secular changes exceeding 2 nT/yr observed at a number of stations and moderately large earthquakes that occurred nearby the stations. One of the typical examples is an anomalous secular change in the horizontal component observed at Tanabe, central Japan, before the occurrences of the 1960 M 6.1 and the 1961 M 6.4 earthquakes. The epicentral distance for the larger one was 36 km. The anomalous change seems to have appeared at least several years (presumably about 9 yr) before the earthquake and persisted up to the time of the earthquake occurrence. A simple estimation idicates that a stress change of 5–6 bars/yr is required within a sphere having a radius of 10 km to produce the observed anomaly in secular variation, 2–3 nT/yr (TAZIMA, 1968) say. From such examples RIKITAKE (1969) obtained the following relation between the spatial extent of an anomalous area and earthquake magnitude:

$$\log r^3 = 11.4 + 1.1M,$$

where r denotes the mean radius of the anomalous area in centimeters.

The Geographical Survey Institute network also covered the Niigata area where an earthquake of magnitude 7.5, the Niigata earthquake, took place in 1964.   FUJITA (1965) analyzed the data during the period before and after the event and came to a conclusion that the geomagnetic field had changed around the epicentral region during the ten years prior to the Niigata earthquake and seemingly recovered shortly after the shock.   NAGATA (1976) re-examined the data reported by FUJITA (1965) and found out anomalies up to 20 nT in the vertical component appearing during a preseismic period.   Coseismic changes were negatively correlated with the preseismic ones as pointed out by FUJITA (1965).   Taking into account the source mechanism which well accounts for land deformation associated with the earthquake, NAGATA (1976) theoretically estimated changes in the geomagnetic field and showed that the overall distribution of the observed changes was in good agreement with the pattern to be theoretically expected.   It was also shown that magnetization of crustal rock in the area under consideration should amount to 5×10^{-3} emu/cm^3 on the average.   In the calculation, a shear stress of 130 bars derived from the source mechanism was adopted as the average shear stress.

It was suggested from the result of more recent surveys carried out by the Geographical Survey Institute that an anomalous secular change in the vertical component had occurred at a first-order station, Shimoda, before the 1974 Off Izu Peninsula earthquake of magnitude 6.9 (TAZIMA et al., 1976a).   The station is located within the aftershock area. However, the survey repeated immediately after the event revealed no coseismic change there.

One of the most intensive observations of the seismomagnetic effect was undertaken by RIKITAKE and his co-workers (1966a, b, c, d, 1967a, b) in the Matsushiro area where seismic activity was unusually high in 1966.   At Matsushiro, a station where continuous measurement of the total intensity was carried out, an anomalous increase exceeding 5 nT as shown in Fig. 1 was observed in accordance with one of the two peaks in seismic activity (RIKITAKE et al., 1967a).   A decrease of a few nanoteslas was also found at another station located 6 km northeast of Matsushiro.   Three years later, the Matsushiro station was re-estab-

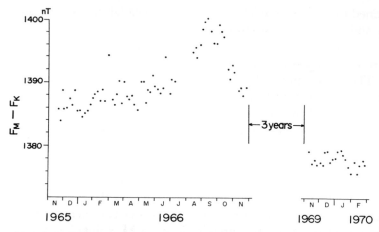

FIG. 1.   Changes in five-day means of the total intensity difference between the Matsushiro (M) and reference (K) stations (after YAMAZAKI and RIKITAKE, 1970).

lished at exactly the same location.   The result revealed that the total intensity had decreased to a level about 10 nT below the previous one as can be seen in Fig. 1 (YAMAZAKI and RIKITAKE, 1970).

The observed changes associated with the Matsushiro earthquake swarm were interpreted as reflecting a change in rock magnetization due to a compressive force acting in the east-west direction (YAMAZAKI and RIKITAKE, 1970).   The existence of such a force was inferred from the seismic source mechanism.   A similar interpretation based on the piezo-magnetic effect was also made by STUART and JOHNSTON (1975), although the origin of the mechanical stress is different.   They proposed a magma intrusion model as a likely cause of geomagnetic changes and ground displacements in addition to high seismicity.   In their model, the stress is caused by the inflation of a shallow magma reservoir.   Futhermore, they ascribed the decrease in the total intensity at Matsushiro during the three-year period to thermal demagnetization due to heat from the magma reservoir.   One of the difficulties in their model is an extremely large stress change, nearly 1 kb, required to explain the observed seismo-magnetic effect.

An interpretation completely different from the previous ones was put forward by MIZUTANI and ISHIDO (1976).   Their idea is based on the electrokinetic effect, which will be described later, in association with the flow of water in the focal region and its vicinity.   The water flow is expected from the dilatancy-diffusion hypothesis (NUR, 1972; SCHOLZ et al., 1973). Moreover, NUR (1974) and KISSLINGER (1975) claimed that the dilatancy-diffusion model also accounts for anomalous crustal uplift and gravity changes observed at Matsushiro besides the seismic events. One of the evidences to support such an interpretation is a good correlation between the geomagnetically anomalous period and the maximum stage of water outflow.

With a view to predicting a future earthquake from a seismic gap in eastern Hokkaido, northern Japan, measurements of the total intensity, declination, and inclination had been repeatedly carried out since 1971 in a region northwest of the expected epicentral area (MORI et al., 1974; YAMASHITA and YOKOYAMA, 1975). As expected in the sense of long-term prediction, an earthquake of magnitude 7.4, the Nemuro-Oki earthquake, occurred about 50 km off-shore in 1974. The survey repeated immediately after the event disclosed a change, presumably co-seismic, of 0.8′ in declination at one of the stations (MORI et al., 1974). However, no precursory changes were found in any components of the geomagnetic field.

Observations of the total intensity before and after the 1970 Southeastern Akita earthquake of magnitude 6.2 suggest further that the seismomagnetic effect will not always be observed. In this occasion, no significant coseismic change was detected at a nearby station with an epicentral distance of 10 km; the difference in the total intensity before and after the event was less than 1 nT (HONKURA et al., 1976). The differences at other sites were also within the accuracy of the observation system. Nothing about precursory changes can be known from the above limited observations.

A survey of the total intensity has repeatedly been conducted in the eastern part of Izu Peninsula, central Japan, where crustal uplift and high micro-seismic acitity have been observed (SASAI and ISHIKAWA, 1977). A semi-permanent station for continuous measurements was also established near the center of the uplift. Anomalous changes up to about 5 nT were observed at many survey sites and their distribution appeared to be correlated with local seismicity and also with time-dependent spatial extent of uplift. One of the survey sites was located over the focal region of the 1976 Kawazu earthquake of magnitude 5.4. A coseismic change of 5 nT was found there. Another station for continuous measurements was set up by the Kakioka Magnetic Observatory group near the western coast of the peninsula. About two months prior to the 1978 Izu-Oshima Kinkai earthquake of magnitude 7.0, the difference in the total intensity between the two stations appears to have changed and then recovered gradually (M. Kawamura, personal communication, 1978). The change amounting to about 5 nT thus observed might be a precursory one.

2.1.2 Examples in the U.S.A.

As a result of total intensity measurements in Nevada with two rubidium vapor magnetometers, a change of about 0.5 nT was observed at the time of a local earthquake of magnitude smaller than 2 (BREINER, 1964). In this case, the change was detected in the difference between two measuring sites, the one being 14 km distant from the other which was regarded as the remote reference station. However, JOHNSTON et al. (1975) claim that the change is unlikely to be of tectonic origin because of its appearance during high geomagnetic activity. Highly sensitive rubidium vapor magnetometers were also used in the San Andreas fault region to investigate a change in the geomagnetic field related to a creep event (BREINER and KOVACH, 1967, 1968). Some anomalous changes in the total intensity were observed some tens of hours before creep events. In one case, a remarkable change was observed simultaneously at four stations established over a span of 25 km. One station exhibited an anomaly amounting to 2.5 nT which persisted for 90 minutes, while anomalies lasted for 36 minutes at other stations. After 16 hours from the appearance of the anomalies a creep displacement of 4 mm occurred along the fault and then an earthquake of magnitude 3.6 took place two days later. That the magnetic event was observed over a 25 km distance and the creep event was not simultaneous with the magnetic event suggests a deep origin for magnetic anomalies: presumably in the plastic-flow region, since it may take 10–20 hours for the effect of plastic flow to appear near the surface. A change of 0.8 nT is explained by a combination of a rock susceptibility of 10^{-3} emu/cm^3 and a stress change of 20 bars.

More intensive field observations have been undertaken in California and western Nevada. A station-pair differential method has been used to eliminate the effect of geomagnetic variations of external origin. Using proton precession magnetometers, the total intensity was measured repeatedly at 50 sites with a separation of 10–15 km (JOHNSTON et al., 1973) and later at 70 sites (JOHNSTON, 1974). The standard deviation in differences between a pair of stations is about 1.8 nT for the distribution of stations. The results from 70 sites revealed anomalous changes slightly above the error level (JOHNSTON, 1974). A correlation was suggested between the observed anomalies and local earthquakes.

The observations have been more intensified. In addition to 120 measuring sites seven stations for continuously measuring the total intensity were established in the seismically active section of the San Andreas fault (SMITH *et al.*, 1974). The continuous data at these seven stations are transmitted through a telemeter system. An earthquake of magnitude 4.3 occurred in 1974 in the Garlock fault survey area and the results of repeated surveys there disclosed an anomalous increase of 2 nT, probably a coseismic one, at a station near the epicenter (JOHNSTON *et al.*, 1975). The change is significant because it well exceeds the error level in the survey area. During the post-earthquake period, the range of anomalous area extended, while anomalous changes were reduced. These anomalies were interpreted as indicating a piezomagnetic effect due to an abrupt stress change and its spatial dispersion. A stress change larger than 10 bars is required to account for the observed changes in the total intensity, if they are of piezomagnetic origin.

An array of 7 stations for continuous measurements has detected no change in association with earthquakes of magnitude up to 3.6 which occurred within 15 km from any stations. In one case, however, a significant change was observed at one of the stations before an earthquake of magnitude 5.2 which occurred on November 28, 1974, at a distance of 11 km from the station (SMITH and JOHNSTON, 1976). Figure 2 shows the averages of measurement for 5 successive days of the total

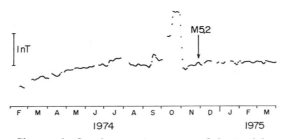

FIG. 2. Changes in five-day running means of the total intensity difference between a pair of stations. An arrow and a bar indicate an earthquake of magnitude 5.2 and a scale length for 1 nT (gamma), respectively (after SMITH and JOHNSTON, 1976).

intensity difference between the station under question and an adjacent one. The total intensity gradually increased from mid-February to late July with the total change amounting to 0.9 nT. It further increased by 1.5 nT in October and this change persisted for about two weeks. Then it abruptly decreased by 1.8 nT on November 1. The major portion of the decrease occurred within several minutes. After that, it remained almost constant even at the time of the earthquake occurrence. It is unlikely that a source causing the anomaly should extend beyond about 10 km from the station, since the anomaly was observed at only one station. This precursory anomaly was interpreted as being due to a piezomagnetic effect. However, the earthquake occurred at the Busch fault, while the station is located on the San Andreas fault. SMITH and JOHNSTON (1976) speculated that the interaction between the faults might have occurred in such a manner that a shear stress change along the San Andreas fault precedes the seismic slip at the Busch fault.

Summarizing the results of observations carried out so far, JOHNSTON et al. (1976) obtained the following conclusions. The spatial extent of a geomagnetic anomaly is one to two orders of magnitude greater than the source dimension of an earthquake. In the San Andreas fault region, anomalies are possibly detectable only for earthquakes of magnitude greater than about 3. The detection level at most of the present sites is 0.5 nT or slightly better for intermediate- to long-term tectonomagnetic effects.

WYSS (1975) analyzed the magnetic data at the Sitka Magnetic Observatory with special reference to the 1972 Sitka earthquake of magnitude 7.2. Two observatories, Victoria in Canada and Colledge in Alaska, relatively close to Sitka were selected as reference stations, although distances between the Sitka and the other stations are about 1,000 km. The small secular change in the difference between the horizontal components observed at the two reference stations suggests little contribution of a spatially nonuniform field originating in the core or outside the earth to the observed secular change in the area under question. Therefore, an anomalous change, if any, in the field difference could be ascribed to one of tectonic origin. It was found, however, that the horizontal field at Sitka first began to decrease about 7.5 years before the 1972 Sitka earthquake and then gradually recovered to the previous level. The maximum change amounted to as much as 20 nT. A similar change was also found in association with the 1958 Yakutat earthquake of magnitude 7.9. WYSS (1975) claims that the change of 20 nT may well be expected as a precursory change if the effect of dilatancy and creep on rock magnetization (MARTIN and WYSS, 1975) is taken into account.

2. 1. 3   Examples in China

According to the report by COE (1971), the difference of the vertical component of the geomagnetic field between Peking and Hongshan changed frequently a few days before the occurrence of earthquakes of magnitude 3 or larger.   More than 150 examples of such anomalies were reported during a period of two years.   Most of the anomalous changes are characterized by a typical pattern; an anomaly appears four or five days before an earthquake, reaches a maximum amounting 1–2 nT, and recovers about two days before the earthquake occurrence.

One of the bases for the successful prediction of an earthquake of magnitude 4.9, which occurred on June 6, 1974, at Ningjiin, was a change in the geomagnetic field (OIKE *et al.*, 1975; PRESS *et al.*, 1975).   An earthquake of magnitude larger than 4 had been anticipated as a medium-term prediction from some factors including a change of long duration in the geomagnetic field.   An anomaly in the geomagnetic field was also found which formed a basis of short-term prediction with a predicted magnitude 4.5.   The U.S. seismological delegation to China also reported on other examples of anomalies in the geomagnetic field (PRESS *et al.*, 1975).   An anomaly in the vertical component appeared 5 months prior to an earthquake of magnitude 5.2 which occurred near Hsingtai in 1972.   Anomalies were also recognized in the vertical field difference between Peking and Hongshan before the occurrence of two earthquakes of magnitude 4.4 and 5.2, respectively, although it is not clear whether these anomalies followed the pattern previously mentioned.

Measurements of the geomagnetic field played an important role on the successful prediction of the 1975 Haicheng earthquake of magnitude 7.3 (SEISMOLOGICAL DELEGATION OF CHINA, 1976).   The geomagnetic field, particularly its vertical component increased 9 months before the earthquake.   The difference in the vertical components between Dairen and Peking amounted to 21.5 nT as shown in Fig. 3.   The anomaly was somewhat reduced four months later but still amounted to 13 nT.   The rapid increase was considered as reflecting a movement of the crust or an abrupt increase of the crustal stress field.   During the period from September, 1974 to February and March, 1975, a 7 nT decrease, perhaps coseismic, was found.   It is interesting to note that a 8 nT decrease was also recognized in the Tangshan area which was destroyed by an earthquake of magnitude 7.8 later in July, 1976 (RALEIGH *et al.*, 1977).

In the case of the successful prediction of the 1976 Lungling earth-

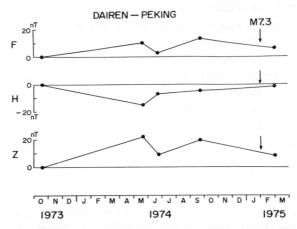

FIG. 3.   Changes in the geomagnetic field difference between Dairen and Peking. *F*, *H*, and *Z* denote the total intensity, the horizontal and vertical components, respectively.   An arrow indicates the occurrence of the Haicheng earthquake of magnitude 7.3.   The Dairen and Peking stations are 460 and 215 km distant from its epicenter (after SEISMOLOGICAL DELEGATION OF CHINA, 1976 and RELEIGH *et al.*, 1977).

quake of magnitude 7.6, much stress seems to have been put on the geomagnetic declination (CHI-YANG, 1978).   A change in the declination was noticed at Yiliang at the medium-term prediction stage, although the station was situated more than 300 km distant from the epicenter. The anomaly still lasted at the short-term stage.   The declination at Tengchung, which is close to the epicenter, also changed at this stage. It is not clear why the geomagnetic field should change in association with an earthquake occurring so far from the measuring site.

NORITOMI (1978) reported on some other Chinese examples of anomalous geomagnetic changes associated with earthquakes.   An abrupt decrease in the geomagnetic field in the Peking area was noticed in April, 1976.   This may be a short-term precursor for the destructive Tangshan earthquake of magnitude 7.8 that took place on July 28, 1976, although no imminent prediction was issued.   In May and July the change reached a maximum value of about 7 nT.   The earthquake occurred during its recovery stage.   Another example was obtained at Ninglang in association with two earthquakes of magnitude 6.9 and 6.8, respectively.   A step-like change in the declination was observed immediately before each earthquake (about 4 and 9 hours before the *M* 6.9 and *M* 6.8 earthquakes, respectively).

2. 1. 4   Examples in other regions

The results of field observations carried out in the Garm area, the U.S.S.R., during a period of a few years disclosed no changes exceeding 0.3 nT for 300 earthquakes of magnitude up to about 4 that occurred within a distance of 20 km (ABDULLABEKOV *et al.*, 1972). Nor were any notable anomalies of long duration in the geomagnetic field observed from the magnetic surveys in the Garm area (SKOVORODKIN *et al.*, 1973). Although rapid variations sometimes appeared several hours before some earthquakes of magnitude 3–3.5, such anomalies were not always observed.

In the Tashkent region, the U.S.S.R., an earthquake of magnitude of about 4.8 occurred in an area where a rapid change was observed in the pattern of the geomagnetic field distribution along a survey line. The survey line lies along an active fault where the 1966 Tashkent earthquake took place. Therefore, the change in the geomagnetic field may be associated with a stress change in the fault area.

In the northwestern part of Turkey a project has been in operation since 1970 aiming at searching for changes in the geomagnetic field associated with earthquakes. The results obtained by the end of July 1971 revealed some local magnetic changes: decrease for a few earthquakes which occurred north or south of the station and increase for an earthquake which occurred to the west (İSPIR and UYAR, 1971). Further instances of geomagnetic changes were obtained during a period from December 1972 to March 1975 (İSPIR *et al.*, 1976). The largest one amounted to as much as 50 nT. A linear relation was found between the ratio of the magnitude of an earthquake to its epicentral distance and the change associated with it. Such a tectonomagnetic effect tends to be enhanced near active faults.

WYSS and MARTIN (1978) analyzed the data at the San Fernando Magnetic Observatory, Spain, in order to examine whether any changes could be found in association with the 1964 and 1969 earthquakes of magnitude 6.2 and 8.0 that occurred below the Altantic Ocean west of Spain. Their epicentral distances are 140 km and 360 km for M 6.2 and M 8.0 earthquakes, respectively. A remarkable decrease in the vertical component amounting to about 100 nT was detected at San Fernando in association with these events, while no similar change was found at other observatories near San Fernando. Although the observed change is too large compared to a change to be expected from the linear elastomagnetic theory, WYSS and MARTIN (1978) claim that the change might be related to a dilatancy process during which the non-linear nature of magnetization change plays an important role as was shown in the experiments for rock samples by MARTIN *et al.* (1978).

2. 2 Problems in detecting small changes in the geomagnetic field

The geomagnetic field is subject to various kinds of variations with their origins in the ionosphere and magnetosphere as well as within the earth. However, these variations can be approximately treated as being spatially uniform over an area of local extent. On the basis of this approximation, it has been considered to be possible to detect changes associated with stress changes or other processes in the earth by taking differences between the fields simultaneously observed at two or more stations. However, a problem arises: whether such an approximation is appropriate or not when a very small change, a few to several nanoteslas, has to be detected.

One of the variations that are by no means spatially uniform even in local nature is the daily solar variation (Sq). The spatial dependence of Sq was investigated by MORI and YOSHINO (1970) on the basis of the total intensity data obtained from eight magnetic observatories in Japan. They brought to light a remarkably non-uniform nature of Sq and pointed out that daytime data should be avoided for the purpose of detecting the tectonomagnetic effect. RIKITAKE (1966b) found that daily fluctuations even in the night-time values are not the same at two stations. A weighted difference method was proposed as a suitable way of removing the effect of such a phenomenon. It was demonstrated that the standard deviation could be reduced by taking weighted differences rather than simple differences. A factor for the weighted difference between a certain pair of stations should be empirically determined from a large amount of data sets.

Short-period geomagnetic variations such as a geomagnetic bay, a storm, and the like are highly affected by local conductivity anomalies in the crust and upper mantle (e.g. RIKITAKE, 1966a; HONKURA, 1977a). Although the weighted difference method also seems to be effective in this case (SASAI and ISHIKAWA, 1976), due to their complicated nature such as polarization dependence, a simple treatment seems unlikely. A practical way then would be a careful selection of data to be used for analysis.

The weighted difference method combined with the use of geomagnetically quiet night-time data seems appropriate for the purpose of avoiding apparent variations in the field difference between stations. However, FUJITA (1973) pointed out that the method should be carefully employed since it is not applicable to a secular variation of long duration. One such variation is the variation in the night-time level which is caused by fluctuations in the equatorial ring current flowing in the magnetosphere. FUJITA (1973) showed from his analysis of data obtained at the Kakioka Magnetic Observatory that some portions of long-term variations in the total intensity disappeared as a result of the equatorial *Dst* correction.

Besides variations originating in the ionosphere and magnetosphere as have been discussed above, those of internal origin also cause a serious problem in detecting changes associated with crustal activity. It has been considered that a secular variation originating in the earth's core is spatially uniform on a local scale and, therefore, it does not affect the field difference. However, detailed analyses of the secular variation of the total intensity revealed a time-dependence of the non-uniform nature of its spatial distribution; the differences in the total intensity between pairs of stations in Japan have exhibited spatial and time dependence (SUMITOMO, 1977). Such dependence was also recognized from the repeated surveys conducted in the Tokai District, Japan. An apparent change due to differences in the secular variation between the survey area and the reference station located about 150 km to the north of the former amounted to about 1 nT/yr during a period from 1971 to 1976 (HONKURA *et al.*, 1977). By analyzing the data obtained at various stations in Japan, HONKURA and KOYAMA (1976) pointed out that the distance between a survey site and a reference station should be less than about 100 km to avoid an apparent change exceeding 1 nT/yr.

2.3   Piezomagnetism

The dependence of various parameters concerning rock magnetism on mechanical stress has been investigated by many workers and some of the results obtained before 1970's were reviewed by NAGATA (1970a, b). The susceptibility changes reversibly, while NRM (natural remanent magnetization) is subject to a change consisting of reversible and irreversible parts. The hard remanent magnetizations such as TRM (thermo-remanent magnetization) and CRM (chemical remanent magnetization) exhibit a reversible change. It is theoretically known that such a reversible change occurs as a result of the rotation of spontaneous magnetization within magnetic domains. An irreversible change can occur in the case of the soft magnetization, IRM (isothermal remanent magnetization), while another irreversible change of magnetization, called PRM (piezo-remanent magnetization), has also been brought to light. These irreversible changes are related to the irreversible displacement of 90° domain walls.

TABLE 1.   Expressions for reversible and irreversible changes of rock magnetization (J) under uniaxial compression (σ) and tension (T).

	Reversible changes		Irreversible changes	
	Susceptibility	TRM, CRM	PRM	Stress demagnetization
$\sigma \| J$	$K_0(1-\beta'\sigma)$	$J_{HR}(1-\beta''\sigma)$	$+CH\sigma^*$	$J_{SR}(1-\alpha'\sigma^*/H)$
$\sigma \| J$	$K_0(1+1/2\beta'\sigma)$	$J_{HR}(1+1/2\beta''\sigma)$	$+(3/4)CH\sigma^*$	$J_{SR}(1-(3/4)\alpha'\sigma^*/H)$
$T \| J$	$K_0(1+\beta'T)$	$J_{HR}(1+\beta''T)$	$+CHT^*$	$J_{SR}(1-\alpha'T^*/H)$
$T \| J$	$K_0(1-1/2\beta'T)$	$J_{HR}(1-1/2\beta''T)$	$+(3/4)CHT^*$	$J_{SR}(1-(3/4)\alpha'T^*/H)$

$\beta'=(0.5\sim5.0)\times10^{-4}\,\mathrm{cm^2/kg}$, $\beta''=(0.3\sim1.2)\times10^{-4}\,\mathrm{cm^2/kg}$, $C=(0.3\sim13.0)\times10^{-6}\,\mathrm{emu\cdot cm^2/kg/Oe}$, and $\alpha'=0.02\sim0.12\,\mathrm{Oe\cdot cm^2/kg}$ for igneous rocks.
A star symbol denotes an irreversible process (after NAGATA, 1969).

Among the various kinds of magnetization, little contribution from IRM is expected in the earth, as long as the earth's magnetic field is maintained. The effects of stress on susceptibility, TRM and CRM, and PRM are known to follow the expressions shown in Table 1. An earthquake occurrence is associated with a sudden stress release which takes place when the strain exceeds the strength of the earth's crust. Since earthquakes have repeatedly occurred, the process of stress accumulation and release should have often been repeated in the crust, so that little contribution from PRM to the seismomagnetic effect is to be expected. Then changes in magnetization due to stress can be expressed by

$$\Delta J_{\|} = -(K_0\beta'H + J_{HR}\beta'')\sigma$$

$$\Delta J_{\perp} = \frac{1}{2}(K_0\beta'H + J_{HR}\beta'')\sigma$$

for magnetization parallel to and perpendicular to the direction of the applied stress, respectively.

STACEY and JOHNSTON (1972) also treated, theoretically, changes in magnetization due to stress. They calculated stress sensitivities of both susceptibility and remanent magnetization for titanomagnetite-bearing rocks in terms of magnetostriction constants and magneto-crystalline anisotropy constants.   It was shown that both sensitivities increase with an increasing composition of titanomagnetite.   Thus, a value of 2×10^{-4} bar^{-1} was suggested for all igneous rocks rather than 1×10^{-4} bar^{-1} which is a reasonable value for pure magnetite.

KEAN et al. (1976) conducted experiments of changes in susceptibility and TRM for various grain sizes.   Results show a large influence of grain size, depending on whether the dominant grain size corresponds to multidomain, pseudo single-domain, or single domain.   For multidomain, the experimental result is quite different from the previously obtained ones.   The result for pseudo single-domain is similar to the previous ones, and the observed changes for single domain are very small. It was pointed out from these results that the theory as put forward by NAGATA (1970a, b) and STACEY and JOHNSTON (1972) on the basis of the rotation of magnetization alone could not account for the complicated nature of magnetization changes of material containing various kinds of domains.   The dependence on the morphology of grains and composition of titanomagnetite was also disclosed as an important factor of stress response to magnetization.

POZZI (1977) investigated the temperature dependence of the piezomagnetic effect and found that the stress sensitivity of induced magnetization is remarkably reduced as temperature rises.   Although the pressure dependence is not taken into consideration, the result appears to be in disagreement with the suggestion made by CARMICHAEL (1977) that the stress sensitivity increases with increasing depth.

In view of the important hypothesis of crustal dilatancy preceding the occurrence of an earthquake, MARTIN and WYSS (1975) investigated the relation between dilatancy and changes in IRM and NRM.   First the change was linear with the elastic deformation and then non-linear behavior appeared after the onset of dilatancy; the rate of change decreased at a dilatancy stage.   Futhermore, a time-dependent change in NRM was well correlated with a time-dependent increase of inelastic volumetric strain at a constant stress.

Such a relation between dilatancy and magnetization changes was confirmed by MARTIN et al. (1978).   However, the behavior is different from one sample to another, and no individual theories existing so far can explain such an experimental result.   Perhaps the effect of grain size might be reflected in such a difference.   The change is mainly characterized by the rotation of the magnetic vector, up to $10°$, although the intensity also changed slightly.   WYSS and MARTIN (1978) also examined the behavior of remanent magnetization during stable sliding for rock samples. No remarkable effect was recognized and the change during stable sliding was similar to that for rock samples without a faulted plane. The non-linear nature of remanent magnetization changes at the stage of dilatancy was tentatively interpreted by WYSS and MARTIN (1978) as reflecting strong local stresses at the tips of newly created cracks. If the elastic behavior during the dilatancy stage is taken into account, the non-linear nature can be ascribed to demagnetization during dilatancy. Provided that the distribution of local stress field is random in the dilatant area, the rotation of remanent magnetization vector will also be random, resulting in net demagnetization. Then the demagnetization due to dilatancy should depend largely on the number of cracks but little on their size and aspect ratios.

2.4   Geomagnetic changes associated with a man-made lake and explosion

If changes in the geomagnetic field due to stress artificially caused in the crust can be observed like the ones observed experimentally for rock samples, applicability of the geomagnetic method to earthquake prediction will increase. As sources of artificial stress, impounding of water into a man-made lake and gun-powder or nuclear explosion have been utilized for the study of tectonomagnetism.

At the time of filling the reservoir caused by the man-made Talbingo dam, Australia, DAVIS and STACEY (1972) made measurements of the total intensity at 15 sites around the reservoir. The maximum depth of the reservoir is about 120 m. Considerable negative changes were observed: up to -5.7 nT at three-quarters full and up to -8.0 nT when completely filled. DAVIS (1974) calculated the piezomagnetic effect from a stress distribution estimated for the reservoir filled with water. The stress sensitivity of rock magnetization was assumed to be 2×10^{-4} bar^{-1}. The average of the observed changes, -5.3 nT, would be well accounted for by the calculation if rock magnetization should amount to 5×10^{-3} emu/cm^3. Magnetization of rock samples taken from the dam site was lower than 1×10^{-3} emu/cm^3, but it is possible that magnetization is much higher at greater depth. The maximum stress change estimated as 14 bars or so is comparable to the stress drop for moderate earthquakes.

Remarkable changes in the total intensity were reported in association with the CANNIKIN nuclear explosion (HASBROUCK and ALLEN,

1972).   A step-like change of 9 nT was observed within 30 sec after the explosion at a measuring site 3 km distant from the epicenter.   Moreover, a 7.0 nT change, relative to another site at an epicentral distance of 9 km, persisted for more than eight days.   Changes were also observed in the preexisting fault area 1.6 km distant from the explosion site. On the side of the explosion site, a 13 nT anomaly was found, while a −11 nT one was found on the opposite side.   In addition, a localized low of 15 nT was found within 15 meters from the fault.   These anomalies in the total intensity were interpreted as reflecting the residual stress or alteration in the remanent magnetization of rocks.

Step-like changes in the vertical component of the geomagnetic field were observed in association with explosions in the Malo-Taskino magnetic deposit area, the Central Urals (UNDZENDOV and SHAPIRO, 1967).   The observed changes ranging 0.2 to 3.8 nT are in good agreement with changes of 1.5–3 nT to be expected from the calculation for a compressive force of 1–2 kg/cm² due to the explosion.

Similar changes in the total intensity were observed in association with artificial explosions in various regions in the U.S.S.R. (KOZLOV et al., 1974).   More than 40 examples were obtained at sites, 50–2,000 meters distant from the explosion sites.   The observed changes range from some tenths of nanoteslas to hundreds, depending on the intensity of rock magnetization, epicentral distances, and explosion power.   These changes can be classified into three types: rapidly reversible, irreversible, and relaxation-type changes.   The irreversible changes could be caused by the plastic flow or destruction of unstable components of remanent magnetization.   Another possibility is that such changes might occur as a result of alteration in the local stress field after the explosions.   The relaxation-type changes seem to reflect redistribution of residual stress within magnetized rocks.

TAZIMA et al. (1976b) reported a step-like change amounting to 0.6–0.7 nT in association with an artificial explosion on Oshima Island, Japan.   It is not clear, however, whether the observed change is a reversible one or an irreversible one, because only one site was set up for the measurement.   The change observed on Oshima Island can be regarded as being larger by an order of magnitude than that observed in association with the CANNIKIN nuclear explosion, if the difference in explosion power is taken into account.   Such a large change could reasonably be interpreted in terms of high magnetization of basaltic rocks constituting Oshima Island.   The same kind of experiment was undertaken one year later at almost the same place, about 10 meters distant from the previous site. A similar step-like change was again observed, although its amplitude was reduced to 0.2 nT (SETO, 1977). The observed changes were interpreted as reflecting fracture of rocks due to the explosions or stress release within rocks caused by fracture of some rocks. This interpretation seems to be in good agreement with the amplitude reduction in the case of the second explosion.

2.5   Model calculations of the seismomagnetic effect

Some earthquakes are associated with the horizontal displacement of preexisting faults. STACEY (1963, 1964) estimated possible changes in the geomagnetic field in association with such a strike-slip type earthquake. Assuming a stress distribution characterized by maximum stress occurring at the center and no stress at each edge, calculations were made for a shear stress drop of 100 bars in various directions of the fault relative to the orientation of rock magnetization. On the average, the difference between positive and negative changes amounts to about 4 nT for a magnetization of 10^{-3} emu/cm^3. Thus the possibility of detecting a seismomagnetic effect was strongly suggested.

More realistic dislocation models were considered by SHAMSI and STACEY (1969) to investigate the seismomagnetic effect associated with the 1906 San Francisco earthquake and the 1964 Alaska earthquake. The results of calculations were rather discouraging for earthquake prediction based on tectonomagnetism. Coseismic changes in the total intensity to be expected for the San Francisco and Alaska earthquakes were about 2 nT and 1 nT, respectively, although larger changes are possible if rock magnetization in the focal regions is higher than 10^{-3} emu/cm^3 assumed in the calculations. A large-scale stress field is important in the above models, but in actual earthquakes the stress field will be more or less inhomogeneous. It was pointed out, therefore, that the inhomogeneity would play an important role on the detectability of seismomagnetic effect.

YUKUTAKE and TACHINAKA (1967) considered a model consisting of an infinitely long cylinder embedded in an elastic half-space. A hydrostatic pressure of 100 bars inside the cylinder was taken as a source of stress within the elastic medium having a magnetization of 3×10^{-3} emu/cm^3. The range of change in total intensity thus obtained for a magnetic inclination of 50° was 2.5–5.3 nT for the cylinder having a radius of 5 km and lying at 10.5–14.5 km depth. However, NAGATA (1969) claims that changes several times larger than expected from the above model would be rather more realistic in view of the small strain obtained at the earth's surface in the above model; a strain of 10^{-4} can be expected in association with the occurrences of earthquakes, while 3.4×10^{-5} was the maximum strain in the above model.

A three-dimensional approach to stress distribution due to internal pressure was undertaken by DAVIS (1976).   The purpose of his calculations was not to estimate the seismomagnetic effect but to compare his results with the volcanomagnetic effect observed over the Kilawea volcano area, Hawaii.   In view of the current hypothesis of dilatancy, such an estimate is important because the high-pressure magma chamber can then be replaced with a dilatant region in the crust.   However, the result obtained for a pressure of 3 kilobars and a magnetization of 5×10^{-3} emu/cm³ suggests that a change to be expected in association with crustal dilatancy would not be large, although a simple dependence of magnetization on stress used in the calculations may not be applicable in the case of dilatancy as was claimed by MARTIN and WYSS (1975) and MARTIN et al. (1978).

A finite dislocation model for a locked fault was considered by TALWANI and KOVACH (1972) with special reference to a magnetic event associated with a creep along the San Andreas fault.   Creep displacement is usually small, about 1–2 mm, compared to a fault movement associated with a great earthquake.   However, stress inhomogeneity, which may be important for a seismomagnetic effect as pointed out by SHAMSI and STACEY (1969), would be high since stress is likely to build up at the edges of creep displacement.   The calculation disclosed a change of 1 nT or so in the total intensity for a magnetization of 10^{-3} emu/cm³ and a shear stress of 10 bars.

Taking into account the pressure and temperature dependence of stress sensitivity of rock magnetization, CARMICHAEL (1977) estimated the piezomagnetic effect within the lithosphere.   The sensitivity itself becomes higher as the source of the piezomagnetic effect lies deeper.   However, rock magnetization becomes weaker as temperature rises, resulting in cancellation of the increasing sensitivity with depth.   Furthermore, geomagnetic changes to be observed on the earth's surface should be small for a deep-seated source.   As a conclusion, it was pointed out from a combination of various factors that the upper portion, down to 15 km depth, plays a dominant role in the seismomagnetic effect due to stress distribution.

3. Electric Phenomena Related to Earthquakes

The electric field derived from the electric potential difference between electrodes buried in the ground or the electric current has been measured for a long time as one of the means of searching for anomalous phenomena associated with earthquakes. Some remarkable changes have been reported from time to time in the literature. YAMAZAKI (1977) reviewed some old reports on changes in the electric field associated with earthquakes and came to the conclusion that no physically meaningful relation between electric field changes and earthquakes has as yet been established.

In Japan, there are many examples of changes in the electric field seemingly associated with earthquakes, but some of them were observed at an epicentral distance of more than a few hundred kilometers. Electrical environments in Japan are by no means simple because of complicated geological structures and noises of natural and artificial origins. Then a question naturally arises; is a change observable in association with a remote earthquake, or isn't the observed change an apparent one due to noises arising from rain falls or other influences? This is one of the reasons why Japanese workers have not been much encouraged from electric work in spite of the numerous examples of anomalous changes in the electric field.

In the meantime, intensive measurements of the electric field or the electric current have been conducted in China and the U.S.S.R. Electric field measurements in China seem to have played an important role in successful prediction for some large earthquakes (SEISMOLOGICAL DELEGATION OF CHINA, 1976; NORITOMI, 1978). Furthermore, an attractive basis for premonitory changes in the electric field was proposed by MIZUTANI et al. (1976) in relation to a dilatancy diffusion hypothesis. It may be high time to positively review the recent work in China, the U.S.S.R., Japan, and the U.S.A., although a physical interpretation has yet to be challenged.

3. 1   Changes in the electric field associated with earthquakes
3. 1. 1   Examples in China

The report of the American Seismological Delegation to China (PRESS et al., 1975) includes some changes in the telluric current associated with earthquakes. A few days before an earthquake of magnitude 4.8, which occurred on July 17, 1972 in Yunnan Province, the telluric current decreased anomalously at a station 20 km distant from the epicenter.   This anomaly was one of the bases for successful prediction in the estimates of magnitude and date.   Another notable anomaly, a 70 μA drop, appeared at an epicentral distance of 90 km a few days prior to the Yunshan-Takuan earthquake of magnitude 7.1 which took place on May 11, 1974.   During about one month from the main shock no striking change was recognized and then an aftershock of magnitude 6.0 occurred.   After this aftershock, the electric current increased by 50 μA and then remained constant for 30 days.   The noise level before the 70 μA drop was a few μA.   Two other examples of anomalous changes in the telluric current were reported, although their statistical significance is not clear.

Before the 1975 M 7.3 Haicheng earthquake, which is very famous because of an almost perfect success in its prediction, the telluric field exhibited anomalous variations (SEISMOLOGICAL DELEGATION of CHINA, 1976).   The anomaly was considered to be one of the forerunners of an impending earthquake and contributed much to its successful prediction. According to the report by the U.S. Seismological Delegation to China (RALEIGH et al., 1977), the NS and EW components of the electric potential measured for electrode distances of 70 and 50 meters, respectively, at an epicentral distance of 25 km underwent remarkable changes before the Haicheng earthquake as shown in Fig. 4.   In particular the potential difference between NS electrodes abruptly dropped by about 100 mV a few days immediately prior to the earthquake occurrence. At another station, at an epicentral distance of 145 km, the NS component showed a 40 μA decrease in the telluric current about 10 hr before the earthquake.   The anomaly lasted for seven hours and then the telluric current recovered to the previous level three hours before the shock. This anomaly appeared after the principal foreshock activity.   Some anomalies were also reported at other stations.   One of them which was found at an epicentral distance of 150 km is similar to the NS component shown in Fig. 4.   The NS and EW components there showed steplike changes about two months before the earthquake; the NS component decreasing rapidly three days before the event.   The changes were incredibly large with their maximum amounting to 500 μA or so.

Telluric field measurements have recently played an important role in medium- and short-term prediction and changes in the telluric field have been considered to be forerunners of an impending earthquake,

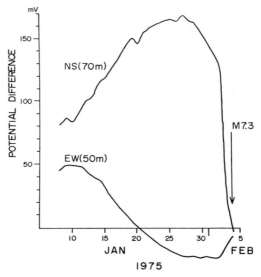

Fɪɢ. 4. Self-potential differences between the NS and EW electrode pairs, respectively. The electrode distance is 70 m for NS and 50 m for EW. An arrow indicates the occurrence of the Haicheng earthquake of magnitude 7.3. The station is located about 25 km from its epicenter (after Seismological Delegation of China, 1976 and Releigh *et al.*, 1977).

contributing to successful prediction in some cases (Noritomi, 1978). Three big earthquakes occurred succesively between Sungpan and Pingwu in Szechwan Province on 16, 22, and 23 August 1976, their magnitudes being 7.2, 6.7, and 7.2, respectively. At the medium-term prediction stage (actually 15–18 months before the earthquake) anomalous behavior in the telluric current was found at several stations. Anomalies tended to be late in appearing at stations far from the epicenter, though even at a station as far as 300 km an anomaly was recognized. Another interesting feature was the systematic distribution of the polarity of anomalous change. Anomalous behavior at the short-term prediction stage (actually 7 days–1 month before the earthquake) is different from that at the medium-term stage; anomalies appeared earlier at remote stations. Contrary to one's expectation, the amplitude of anomalous change was larger at remote stations. Anomalous changes also appeared some hours prior to the earthquakes (about 3 hr and 6 hr before the *M* 7.2 and 6.7 earthquakes).

On May 29, 1976, two big earthquakes took place successively near Lungling in Yunnan Province. Their magnitudes were 7.5 and 7.6, respectively, and the focal depth was 22 km for both. This pair of earthquakes, the Lungling earthquake, was successfully predicted from various kinds of precursory anomalies. Although the telluric current measurements contributed little to the successful prediction, some anomalies were recognized from the examination of data after the occurrence of the Lungling earthquake (CHI-YANG, 1978). These anomalies appeared at the short-term prediction stage (about 2 months before the earthquake) and became more remarkable a few hours before the earthquake occurrence.

However, the Tangshan earthquake of magnitude 7.8 that destroyed the city of Tangshan on July 28, 1976 could not be successfully predicted. The prediction at long-, medium-, and short-term stages was made from various kinds of anomalous phenomena, but the imminent forecast was not issued in time for the occurrence of the earthquake, mainly because of the shortage of clear forerunners at the imminent stage (OIKE, 1978). As for the results of the telluric current measurements, the examination made after the earthquake revealed that anomalies had appeared in 18 records among 51 (35%), although criteria for recognizing anomalies are not clear.

3. 1. 2   Examples in the U.S.S.R.

Electric field measurements have actively been undertaken at the eastern coast in Kamchatka where earthquakes of magnitude greater than 5 have often occurred. According to the results of observations at several stations by December 1968, there are quite a few cases in which anomalous changes in the telluric field were observed at one or two stations near the epicenter several hours before or after the earthquake occurrence (FEDOTOV et al., 1970; SOBOLEV et al., 1972). Anomalies are characterized by an amplitude of 50 mV/km or so and their duration ranges from a few to tens of minutes. The polarization of these anomalies is not systematic, though the electric field induced by geomagnetic variations does tend to be polarized in a direction perpendicular to the coast line. The statistical treatment for the observed anomalies disclosed that most of the anomalies appeared about 8 hr before or after the earthquake occurrences.

An earthquake of magnitude 6 occurred on December 19, 1968 at a depth of 40–50 km. Six days before this event anomalous variations in the telluric field appeared at one of the telluric stations about 30 km distant from the epicenter. The anomaly persisted until the beginning of January and tended to attenuate gradually after the occur-

rence of another earthquake of magnitude 5 on January 2.   Another supplementary measuring system operating at a distance of about 600 m
from the station showed variations similar to those observed by the main
system.   The anomalous change was so large, about 300 mV/km, that
it exceeded the amplitudes of about 30 mV/km for the Sq-field and 60–
70 mV/km due to ocean tides, respectively.

On the assumption that these anomalies were caused by an electric
dipole source appearing in the crust, its location was estimated from the
observed anomalies.   Since the location thus obtained was close to
the epicentral area, it was speculated that such a source would be of
piezoelectric origin, reflecting a change in stress state in the focal area
prior to the earthquake occurrence.   In fact, on the earth's suface near
the telluric station there exists an intrusion of granitic rocks containing
quartz which possesses piezoelectric properties.   If the electric current
flows in the crust, its effect will also appear as a change in the magnetic
field.   The magnetic field calculated for a crustal conductivity of 10^{-3}
S/m amounted to 8 nT which is in good harmony with the observed
result.   Besides the above result, a number of anomalous variations
were observed in association with strong earthquakes occurring within
150 km from the stations.   Most of these anomalies appeared several
days to a few weeks before the earthquake occurrences; a table of these
anomalies is given in the paper by SOBOLEV *et al.* (1972).

Another kind of precursory change in the electric field was reported
in Kamchatka (SOBOLEV and MOROZOV, 1972).   Compared to anomalous
variations as given above, the durations of anomalies belonging to this
kind are very short, being less than an hour.   These anomalies can be
classified into three types: an abrupt change with a duration of 10–40 min,
a sudden burst type with a shorter duration of 10–20 min, and a pulse-
like change.   It was disclosed from the histograms for the first and second
types, respectively, that an anomaly begins to appear 6–8 hr before an
earthquake occurrence and its appearance tends to become more frequent as the earthquake impends.   The last type shows a similar tendency but it also appears quite often even after an earthquake occurrence.
These types of anomalies tend to appear in association with earthquakes
having a focal depth of 10–20 km.   However, precursory changes in the
telluric field were not always observed as is the case also for other factors
in earthquake prediction.

Summarizing the results obtained in the Kamchatka region during
the period from 1967 to 1970, SOBOLEV (1974) pointed out some charac-

teristics of the observed anomalies in the electric field as described in the following. A bay-shaped anomaly appears one to two weeks before an earthquake of magnitude greater than 5. An earthquake takes place during the stage of attenuation of the anomaly or at its maximum stage. The anomaly recovers several days after the earthquake occurrence. A large anomaly can be seen for a stronger earthquake. An anomaly can be observed at an epicentral distance of up to 150 km and for a focal depth shallower than 60 km, and the amplitude of an anomalous change is approximately proportional to the reciprocal of the epicentral distance. If stations are more than 100 km apart from each other, appearance of anomalies is not always simultaneous and a time lag of several days is not unusual. The vector of an anomalous field tends to be polarized in the east-west direction. A source producing an anomalous electric field seems to lie in a focal region.

In December, 1971, a large earthquake of magnitude 7.7 occurred at a depth of 30 km below the sea east of Kamchatka Peninsula. An anomaly in the electric field appeared before the occurrence of this earthquake (one and a half months to about 20 days before the earthquake) and then the field slightly recovered immediately before the event (So-BOLEV, 1974). Many aftershocks having magnitudes greater than 5 occurred and the anomaly lasted for a month or so after the main shock. Anomalies associated with this earthquake before its occurrence were observed in a broad area with a spatial extent of several hundred kilometers: much larger than the aftershock area.

Anomalies in the telluric field have been observed so often that the telluric observation became one of the bases for short-term prediction in Kamchatka (SOBOLEV, 1975a; FEDOTOV et al., 1977). An anomaly generally appears 5–15 days before an earthquake occurrence, its amplitude amounts to about 50 mV/200 m and its time dependence can be characterized by a rapid change and a slow recovery. There are five stations in total and 4–10 independent pairs of electrodes are used at each station. The criterion for judging whether an observed change is anomalous or not is that pairs of measuring system having the same electrode direction show changes similar in time and shape to each other. The data at a station located at the center of the seismic region are mainly used for predicting the time of earthquake occurrence. In estimating the place, the data at other stations are also taken into account.

The data have been analyzed by making use of a computer. The procedure is as follows. High-frequency components and seasonal trends are removed from the raw data. The mean values are determined by averaging the data obtained at a station from several measuring systems having the same electrode direction. The mean value is then smoothed by a weight function designed for effectively detecting anomalies having a duration of 5–20 days. As the final output, the square of the electric field intensity is obtained.

When the output value increases, it is considered that an anomalous stage has begun. Among 12 earthquakes of magnitude greater than 5.5, eight were successfully predicted on the basis of the electric field measurements (FEDOTOV et al., 1977). Therefore, the probability of detection is about 0.65 which is the highest among the three prediction factors; the telluric field, the ratio of seismic wave velocities (V_P/V_S), and the earthquake statistical method. However, the efficiency of prediction is not high partly because of a long warning period and inaccuracy of predicted location.

3. 1. 3   Examples in Japan

FUKUTOMI (1934) reported on an anomalous change in the electric field which appeared 15–16 hr before a shallow earthquake of magnitude 5.5 that occurred in the Izu Peninsula on March 21, 1934. The measuring site was located about 20 km northwest of its epicenter. The electric potential difference between electrodes with 100 m separation rapidly decreased by about 3 mV and gradually recoved. Then a rapid change, though smaller in amplitude, again appeared 2 hr prior to the main shock. More remarkable changes were also observed during a period of high aftershock activity.

NAGATA (1944) carried out observations of the electric field in the aftershock area of the 1943 Tottori earthquake of magnitude 7.4 and found that earthquakes occurred several hours after the appearance of anomalies of short duration in the telluric field. However, there were also many other earthquakes that occurred without anomalies in the telluric field. It was claimed as a conclusion that only some earthquakes would be preceded by a precursory change in the telluric field. Nagata also pointed out that the relation in the electric state between a pair of measuring directions, one across the fault and the other along it, might be disturbed before an earthquake occurrence. Then it was suggested that an active fault might have an important bearing on anomalous behavior in the telluric field.

The result that there are earthquakes that are not preceded by a telluric anomaly was confirmed by the observations of the telluric field carried out immediately after the 1946 Nankai earthquake of magnitude 8.1 (RIKITAKE and YAMADA, 1947).   In the case of the Matsushiro earthquake swarm, no precursory change was observed in association with earthquake occurrences in spite of an enormous number of shocks (RIKITAKE et al., 1966a).

On the other hand, a precursory change in the tellulic field appeared about 65 days prior to the 1978 Izu-Oshima Kinkai earthquake of magnitude 7.0 (KOYAMA and HONKURA, 1978).   The main shock occurred about 30 km southeast of the measuring site.   The electrodes used for measurements were graphitic ones which were buried at a depth of one meter or so.   The distance between the pair of electrodes was 120 m for the NS component and 114 m for EW.   An anomaly appeared in the EW component, although nothing is clear for the NS component because of trouble in the cables connecting the electrodes with the recording system.   It was known beforehand that the electric potential difference between EW electrodes is greatly affected by rain falls, but changes of precipitation origin follow a rather systematic pattern; they are characterized by a rapid increase and a slow recovery within a few days as seen in Fig. 5.   Contrary to such a systematic

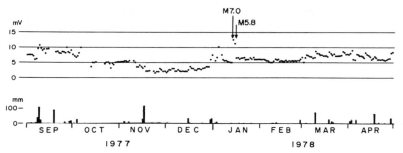

FIG. 5.   Self-potential difference between the EW electrode pair at the NKZ station (see Fig. 15).   Rainfall data are also shown in millimeters.   Arrows denote the main shock (M 7.0) and the largest aftershock (M 5.8) (after KO-YAMA and HONKURA, 1978).

behavior, the potential dfference decreased by about 3 mV around November 10, 1977 and then it tended to recover gradually.   Such a variation of long duration can not be correlated with precipitation and is likely to be a precursory change.   In view of anomalous changes in other factors such as the resistivity, the geomagnetic total intensity, the radon content, and so on, HONKURA (1978) claimed that the anomaly in the telluric field would reflect a streaming potential (see Subsection 3. 3) due to water flow in and around the focal region.

3. 1. 4   Examples in the U.S.A.

The natural electric potential has also been measured in the San Andreas fault area (CORWIN and MORRISON, 1977). Electrodes used for measurements are those buried for the purpose of apparent resistivity measurements based on artificial electric currents. Two notable anomalies were found in association with earthquakes. One of them, which is shown in Fig. 6, has an amplitude of about 90 mV for two pairs of

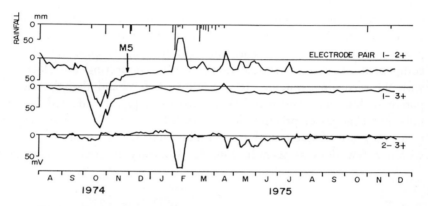

FIG. 6. Self-potential differences between electrode pairs, 1–2, 1–3, 2–3. Rainfall data are shown in millimeters. An arrow denotes a M 5 earthquake (after CORWIN and MORRISON, 1977).

electrodes with separations of 630 m and 640 m, respectively, and appeared 55 days before an earthquake of magnitude 5. These electrode pairs were about 37 km distant from the epicenter. As can be seen in Fig. 6, the potential difference is influenced by precipitation, but it is unlikely that the anomalous change appearing in electrode pairs of 1–2 and 1–3 is due to rain falls. The other having an amplitude of 4 mV/300 m began to appear 110 hr before the occurrence of a smaller earthquake of magnitude 2.4. In this case, however, the epicenter was only 2.5 km distant from the measuring site.

These anomalies appeared rather rapidly and gradually recovered to the previous level, suggesting a diffusion-like process occurring in and around the focal region. If dilatancy takes place and water flows into a dilatant zone, a streaming potential can be expected to appear. Assuming that the hydraulic diffusivity amounts to 5.8×10^{-4} cm²/sec, a change in the electric field was calculated from the streaming potential. The agreement between the observed and calculated values suggests that the observed anomaly in the electric field is a precursory phenomenon probably resulting from water flow into an impending earthquake region.

3. 2 Earthquake lightning

The phenomenon of earthquake lightning has been reported for centuries as shown in review papers by YASUI (1973) and DEER (1973). However, it is still controversial whether an earthquake lightning is really related to an earthquake occurrence, because most of the reports are based on eye-witness of local people at the time of an unusual event. YASUI (1973) speculated that the increase in the electrical conductivity of the atmosphere would manifest itself as an earthquake lightning under the existence of a strong electric field of some origin. Such an increase in electrical conductivity is likely to be caused by radon driven out by an earthquake from an acidic rock such as quartz-diorite containing much radon.

KONDO (1968) made observations of the atmospheric electric field during the Matsushiro earthquake swarm to examine if there exists any relation between an earthquake lightning and an anomalous change in atmospheric electricity. In many occasions, a notable decrease, up to 10–20%, in the electric field was observed in association with a locally felt earthquake, although no luminous phenomenon was observed during his observation period. Such a decrease was ascribed to the increase in the atmospheric electrical conductivity given rise to by radon released in the atmosphere during an earthquake. More detailed examination was made by PIERCE (1976), who pointed out that the electrical conductivity is more sensitive to a change in radon content in the atmosphere than the electric field. Technically, however, the latter is more easily measured and the discussion was focused on the detectability of a 20–40% change in the electric field. However, noises due to various factors such as clouds, precipitation, natural and artificial agencies of many kinds are as large as or even much larger than a change to be expected in connection with an earthquake process. The atmospheric electric-field has also been measured in Kamchatka, the U.S.S.R. No significant result, however, was obtained in spite of the report that atmospheric glow was sometimes observed in association with earthquakes (SADOVSKY and NERSESOV, 1974).

With regard to the electric potential which has been believed to be the basic cause of earthquake lightning, two probable mechanisms of its generation were put forward; the streaming potential of subterranean capillary flow of water (TERADA, 1931) and piezoelectricity (FINKELSTEIN and POWELL, 1970; FINKELSTEIN et al., 1973). In the light of the dilatancy-diffusion model (see Subsection 5. 4), the former mechanism was reconsidered in more detail by MIZUTANI et al. (1976) as the origin of an anomalous electric current associated with an earthquake occurrence.

Since a luminous phenomenon, if it really exists, is likely to be related to an anomalous change in the electric field, in the ground and/or the atmosphere, the practical approach to earthquake prediction would be the measurements of the electric field rather than the luminous phenomenon itself. However, this phenomenon is by no means fully understood yet and the mechanism of earthquake lightning still remains of great interest.

3. 3   Piezoelectric and electrokinetic effects

As a likely cause of an anomalous change in the electric field associated with an earthquake occurrence, two different kinds of mechanisms have been put forward: piezoelectricity related to a stress change and an electrokinetic effect due to water flow through cracks and pores.

3. 3. 1   The piezoelectric effect

The piezoelectric effect of quartz-bearing rocks has been reported mostly in the U.S.S.R. literature as referred to by TUCK et al., (1977). In an attempt to verify the reports obtained so far, TUCK et al. (1977) carried out an experiment with a highly aligned quartzite. However, the observed effect was small and is within the effect to be statistically expected for the finite number of discrete quartz grains. This result suggests that an anomalous electric field of piezoelectric origin would be very small even in an area where quartz-bearing rocks exist in abundance.

On the other hand, recent Russian literature, as was reviewed by DMOWSKA (1977), disclosed some piezoelectric effect which is observable as a telluric signal. The coefficient of piezoelectric activity ($K = \Delta V/\Delta\sigma$) was experimentally shown to amount to 10^{-8} V·m^2·N^{-1}, where ΔV and $\Delta\sigma$ represent the voltage and the applied stress. Therefore, an electric field of 10^4 mV/km can be expected for a stress change of a few bars. However, piezoelectric effect depends on the rate of stress change. The above result was obtained for a stress rate of 10^3 N·m^{-2}·s^{-1}. Now, it is necessary that the piezoelectric effect should produce a field stronger than 1 mV/km for a change in the electric field to be significantly detected. Thus it was pointed out that the stress rate must be higher than 10^{-1} N·m^{-2}·s^{-1} in order for the piezoelectric effect to be useful from the viewpoint of earthquake prediction.

The rate of stress change is much higher at the time of seismic-wave propagation than at a stress build-up stage before an earthquake occurrence. Therefore, it may not be unlikely that the electric field due to piezoelectric effect should appear at the time of the earthquake occurrence, resulting in its manifestation as earthquake lightning as presumed by FINKELSTEIN and POWELL (1970). Such presumption seems to be in good agreement with the report that earthquake lightning tends to appear above the top of mountains containing quartz-bearing rocks such as quartz-diorite (YASUI, 1973).

3.3.2 The electrokinetic effect

The important role of water has been pointed out regarding a process within a region of an impending earthquake as was claimed in the dilatancy-diffusion hypothesis (NUR, 1972; SCHOLZ et al., 1973). If underground water flows through cracks and pores, a streaming potential would be created as was first pointed out by TERADA (1931). According to the recent paper by MIZUTANI et al. (1976), the electric current density I and the flux of fluid flow J can be obtained from the following basic equations:

$$-I = \phi\sigma \operatorname{grad} E - \frac{\phi\varepsilon\zeta}{\eta} \operatorname{grad} P$$

$$-J = -\frac{\phi\varepsilon\zeta}{\eta} \operatorname{grad} E + \frac{k}{\eta} \operatorname{grad} P$$

where ϕ and k denote the porosity and the bulk permeability of the medium, and σ, ε, and η indicate the electrical conductivity, the dielectric constant, and the viscosity of the fluid, respectively. P and E indicate the pressure of the fluid and the electric potential. ζ is the zeta potential and usually amounts to tens of millivolts. Assuming that the pore pressure gradient ranges from 1 to 100 bars/km, MIZUTANI et al. (1976) estimated the electric current density and obtained a value of $0.7 \times 10^{-7} \sim 0.7 \times 10^{-4}$ A/m^2. Since the electrical conductivity is assumed to be 0.1 S/m, the electric field due to the streaming potential ranges from 1 to 1,000 mV/km. The amplitude of the observed anomalies in the electric field associated with earthquakes is of the order of

100 mV/km on the average. Therefore, it does not seem to be un-reasonable to claim that some of the anomalies reported so far might have been of electrokinetic origin.

A similar work on the electrokinetic effect was undertaken by a Russian group headed by Sobolev, according to the review paper by DMOWSKA (1977). They estimated the maximum electric field to be expected from the streaming potential and concluded that their estima-tion, 1–10 mV/km, is two orders of magnitude lower than that by MI-ZUTANI et al. (1976). However, they also claim that the estimated elec-tric field should be detectable.

It should be noticed that the intensity of the electric field due to the electrokinetic effect strongly depends on values of various parameters assumed for estimation (DMOWSKA, 1977). Another problem concern-ing the streaming potential is whether the process suggested in the di-latancy-diffusion model actually takes place before earthquake occurrences. Even if dilatancy actually occurs and underground water flows into a dilatant region, the process would be highly time-dependent and, there-fore, the time-dependent behavior of an anomaly should be examined as done by CORWIN and MORRISON (1977).

SOBOLEV (1975b) carried out an experiment on rock fracture under biaxial compression, and found that the electric potential changed dras-tically prior to the fracture. Mizutani and his co-workers (personal communication, 1978) obtained similar results and interpreted them in terms of the streaming potential. According to their interpretation, a pore pressure gradient is created as a result of the opening of new cracks and water contained in a rock sample would flow toward the new cracks, resulting in the creation of a streaming potential. The asser-tion that a change in the electric field due to a streaming potential can be observed also seems to be supported by the reports that the electric potential difference between electrodes changed when underground water was pumped out (H. Mizutani, J. Miyakoshi, personal communi-cation, 1978).

4. Ground Resistivity Changes and Strain

4. 1   The Yamazaki effect

The electrical resistivity of rocks depends on various factors and changes according to a change in them. Pressure is one of them as experimentally shown by BRACE and ORANGE (1966, 1968b). The dependence of a resistivity change on pressure is highly influenced by the degree of water saturation.   In a dry condition, virtually no change is observed, while the resistivity increases monotonously with pressure in a completely saturated condition.   A partially saturated rock has a peculiar pressure dependence; the resistivity first decreases and then gradually increases.   These phenomena have been reasonably interpreted in terms of the amount of conducting water within a rock sample and its degree of interconnection.   Some water would be squeezed out from a saturated rock when the pressure is applied, which results in a resistivity increase.   For a partially saturated rock, the initial decrease is due to an enhancement in the degree of interconnection as a result of porosity decrease.   A saturation condition will eventually be attained as porosity reduction continues, and thereafter the situation should be the same as in the case of a saturated rock.

However, a strange resistivity change of the order of 0.1% was reported in association with ground extension and contraction of the order of 10^{-6} due to ocean-tide loading (I. Yokoyama, unpublished, 1952).   An experimental study of the relation between the resistivity and linear strain was then undertaken by YAMAZAKI (1965, 1966, 1973) for rock samples of pumice tuff, lapilli tuff, and tuffaceous sandstone.

These samples have unusually large porosity: 0.408, 0.514, and 0.567 for lapilli tuff, tuffaceous sandstone, and pumice tuff samples, respectively.   The relation between the resistivity change ($\Delta\rho/\rho$) and the linear strain ($\Delta L/L$) was characterized by the ratio $\dfrac{\Delta\rho}{\rho}\bigg/\dfrac{\Delta L}{L}$ which Yamazaki called the ratio of rock strain amplification.   The ratio was 30 for a tuffaceous sandstone sample at a water content of 9.5% and 18 for a pumice tuff sample at 5.6%, respectively.   It was also found that the ratio for these samples depends little on water content and linear strain ($\Delta L/L$).

On the other hand, the behavior of a lapilli tuff sample is characterized by its peculiar nature.   The ratio of rock strain amplification depends remarkably on water content, and at a low water content the ratio becomes one order of magnitude higher than that for tuffaceous sandstone and pumice tuff samples.   The ratio also depends in a rather complicated manner on the linear strain.   The ratio increases rapidly at a very small strain depending on the water content as shown in Fig. 7 (YAMAZAKI, 1973).   For example, at a 13.9% water content the ratio exceeds 300 for a strain of an order of 10^{-5}.   An amplification ratio of

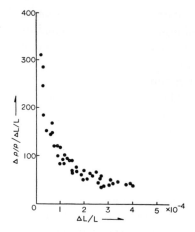

FIG. 7.   The ratio of a resistivity change ($\Delta\rho/\rho$) to strain ($\Delta L/L$) plotted against strain for a lapilli tuff sample having a water content of 18% (after YAMAZAKI, 1966).

a 10^3 order for a strain of 10^{-6} associated with ocean-tide loading at Aburatsubo seems quite reasonable because the lapilli tuff sample used for the experiment was taken from the Aburatsubo station.   Furthermore, the ratio estimated from coseismic resistivity steps amounts to an order of 10^4 for a very small strain of 10^{-8}–10^{-9} as will be described later.

Such an unusual nature of rock resistivity was also noticed by MORROW et al. (1977) for porous rocks and they called it the " Yamazaki effect ".   The effect was found for a number of tuffs, several sandstones, and even for a granite sample.   The amplification ratio for these samples amounted to 10^4 or so at a strain of 10^{-4}.   The conditions necessary for the appearance of the Yamazaki effect seem to be a porosity higher than 10%, small strain, and a special nature of pores which remains unknown.   The water content, or the degree of saturation, also seems to be one of the conditions, although details are unknown as yet.

YAMAZAKI (1966) pointed out that permeability might characterize the special mechanism of electric conduction within a lapilli tuff sample, since its permeability was much higher than that for the tuffaceous sandstone and pumice tuff samples.   It has been speculated that water arrangement within the rock structure characterized by a high permeability would be highly affected by a slight deformation of the structure.

Recently, field measurements of ground resistivity were carried out in China by artificially applying compressional force to the tuffaceous surface-layer (Yü-LIN and Fu-YE, 1978). The porosity and water content of the layer were about 0.024 and 20%, respectively. According to the result, the ratio of strain amplification amounted on the average to 1.5×10^3 against a linear strain of about 3×10^{-5}: almost the same as that observed at Aburatsubo in association with a strain of 10^{-6} due to ocean tide loading. Based on this result, Yü-LIN and Fu-YE (1978) interpreted the precursory decreases in apparent resistivity observed at some stations before the 1976 Tangshan earthquake of magnitude 7.8 (see Section 5), about 5% in the epicentral area and about 2% around it, as being due to a strain of 3×10^{-5} or so before this event. Such a conclusion appears to be supported by observations of strain based on the base line surveys. However, it remains questionable to the author why resistivity steps were not clearly recognized at the time of the earthquake occurrence if precursory changes are considered to be of strain origin.

4. 2   Ground resistivity changes as recorded by the resistivity variometer

A highly sensitive instrument is required to detect changes in resistivity of 0.1% associated with ocean tides and perhaps smaller than that in association with earthquakes. YAMAZAKI (1967) constructed a resistivity variometer for this purpose. The method of measuring the ground resistivity is based on the four pole arrangement called the Wenner method. Four graphite electrodes with an equal separation of 1.6 m were buried along a N81°W direction in an observation vault at Aburatsubo. A 67 Hz AC current is driven into the ground through the outer pair of electrodes and the potential difference between the inner pair is picked up and sent to the recording system.

After some improvements were made to his original variometer (YAMAZAKI, 1968), the present variometer having a very high sensitivity was constructed. Figure 8 shows the block diagram of the present Yamazaki variometer. The potential difference between P_1 and P_2 due to the current driven into the ground through I_1 and I_2 is amplified, band-pass filtered and again amplified. The amplified signal is sent to the servo-motor. The rotation of the motor is transferred through gear-work to a change in resistance at the balancer and also at the output stage. The variometer is always kept in balance and the voltage corresponding to a certain balance stage is recorded. Thus the ground

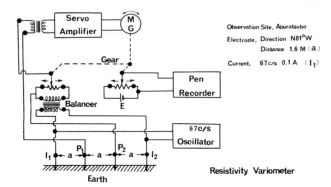

FIG. 8.   The block diagram of the Yamazaki resistivity variometer (after YA-
MAZAKI, 1977).

resistivity is continuously monitored.

The resistivity variometer clearly recorded resistivity changes due to cyclic ocean-tide loading and so the Yokoyama's preliminary result was confirmed; a resistivity change of the order of 0.1% was observed against a strain of 10^{-6}.   A seiche having a period of about 13 min sometimes occurs in the Aburatsubo Bay and causes a strain of the order of 10^{-7} at the Aburatsubo station.   The resistivity variometer also recorded a resistivity change due to such a seiche.

As is shown in Fig. 9, the resistivity variometer recorded a step-like change amounting to about 2.2×10^{-4} (in $\Delta\rho/\rho$) in association with the occurrence of the 1968 Tokachi-Oki earthquake of magnitude 7.9 which occurred at a distance of about 680 km from the Aburatsubo station (RIKITAKE and YAMAZAKI, 1969, 1970).   However, the change was not very rapid but rather gradual, about 5 min having been required for its completion.   In addition to the coseismic change, a precursory change was also recognized.   About two hours prior to the shock the ground resistivity started to decrease: in the opposite sense to the co-seismic change.

Since then, similar resistivity steps have been observed in association with some earthquakes of magnitude larger than about 5.   Among them is included the one which occurred more than 1,000 km distant from Aburatsubo.   Comparing these resistivity steps with strain steps to be expected at Aburatsubo from the dislocation theory, YAMAZAKI (1974) pointed out that the amplification ratio amounts to 3×10^4 or thereabouts for a small strain of 10^{-8}–10^{-9}: two orders of magnitude higher than that for a strain of the order of 10^{-6} due to tidal load-

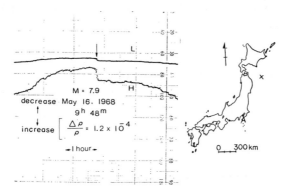

FIG. 9.   The record showing a resistivity change observed in association with the 1968 Tokachi-Oki earthquake of magnitude 7.9. *H* and *L* on the record indicate channels for high and low gains, respectively. Symbols A and × on the map denote the Aburatsubo station and the earthquake epicenter (after YAMAZAKI, 1977).

ing.   It was also found that coseismic resistivity changes tend to become larger for greater earthquakes, if the observed changes are transferred for comparison to hypothetical ones at an epicentral distance of 100 km.

YAMAZAKI (1975a) further examined from 20 examples of coseismic resistivity steps the relation between the rise time of a resistivity step and earthquake magnitude and found that the rise time tends to be longer for larger earthquakes.   YAMAZAKI (1975a, b) also examined the detectability of resistivity steps associated with earthquakes.   Earthquakes causing resistivity steps could clearly be separated from others according to their magnitudes combined with their epicentral distances.   A resistivity step is generally observed if the earthquake magnitude is larger than M given by the following expression:

$$M = -12.5 + 2.5 \log_{10} \varDelta$$

where $\varDelta$ denotes the epicentral distance in centimeters.   Precursory changes appeared in 9 among 20 examples with their precursor time ranging from 1 to 7 hr.   The time seems to be independent of the earthquake magnitude.

RIKITAKE and YAMAZAKI (1977a, b) made statistical analysis on the basis of more examples, 21 in total, of precursory resistivity changes with their precursor time ranging from half an hour to 7 hours.   The result revealed that their nature is very similar to that of other short-term precursors as found in land deformation, tilt, strain, and underground water (RIKITAKE, 1975b). These short-term precursors can be characterized by a Weibull distribution. The precursor time scatters around some hours regardless of earthquake magnitudes. Rikitake and Yamazaki claim that the ground resistivity precursors as well as other short-term anomalies might have something to do with preliminary failure prior to the main rupture in the focal region.

A similar study was undertaken in the San Andreas fault region. In association with local earthquakes, coseismic resistivity steps were observed at the Stone Canyon Geophysical Observatory (BUFE et al., 1973). Such resistivity steps were also some orders of magnitude larger than the strain steps. A resistivity change associated with the 1972 Melendy Ranch earthquake of magnitude 5, which took place at a distance of 6 km from the observatory, was about four orders of magnitude larger than the strain step observed at the same observatory. As at Aburatsubo, therefore, the amplification ratio amounts to 10^4. However, no precursory change in ground resistivity has been observed at this observatory.

An attempt to detect resistivity changes associated with creep events was made by FITTERMAN and MADDEN (1977) along the San Andreas fault at Melendy Ranch. The method used for this purpose was a quasi-Schlumberger array with a 200-meter transmitting dipole and a 60-meter receiver dipole. However, no change exceeding 0.005% was found in association with the observed creep events. The result suggests that the electrical behavior of clay, main rock at the observation site, might be insensitive to strain and the creep zone could be narrow compared to the measuring geometry.

5. Electrical Resistivity Changes in and around Focal Regions

The electrical resistivity, one of the physical parameters of a crustal material, may be subject to a change associated with an earthquake. In view of the current hypothesis of dilatancy preceding an impending earthquake, much attention has been paid to resistivity changes. Crack opening and closure alone would not appreciably change the resistivity of a medium. However, if underground water is involved in a dilatancy process, the situation is quite different. Underground water usually contains various kinds of ions and, therefore, its resistivity is some orders of magnitude lower than crustal rocks themselves. Thus the bulk resistivity of the medium is controlled by the amount of water and its state within cracks and pores, as will be described in Subsection 5.4.

As for the means of detecting resistivity changes, three different kinds of measurements have been carried out. An apparent resistivity can be directly determined by driving artificial electric currents into the ground at one site and measuring the electric potential at another site. A change in resistivity somewhere in the crust will manifest itself as a change in apparent resistivity determined from such a resistivity sounding. The other two methods are based on the electric current induced by external geomagnetic variations and a change in resistivity can be indirectly detected through the perturbation of the current due to such a resistivity anomaly. The effect of such a perturbation should appear in the electric field observed on the earth's surface. Thus the magnetotelluric method can be used for the present purpose. Such a perturbation will also be reflected in the secondary magnetic field due to the induced currents. This effect could possibly be observed by examining the parameters for anomalies of short-period geomagnetic variation.

5. 1   The resistivity sounding method based on artificial electric currents
5. 1. 1   Examples in the U.S.S.R.

Apparent resistivity measurements based on the so-called dipole-dipole method have been intensively carried out in the Garm area, the U.S.S.R., by Barsukov and his co-workers. At the early stage of field operation an electric current of 17–25 amperes was driven into the ground through electrodes 1,000 m apart from each other. The electric potential difference was measured at two sites 7–8 km west and east-northeast of the current dipole, respectively. The distance between electrodes for potential measurements was 400–600 m. According to a preliminary trial, a diurnal variation with an amplitude of several percent was observed at both stations (AL'TGANZEN and BARSUKOV, 1972). Since the observed diurnal change in apparent resistivity was similar to the crustal deformation due to a tidal force, it was speculated that the change might be related to a change in rock porosity due to tidal loading. Another possibility was also considered; that is, the change may be due to fluctuations in ground water table.

In the meantime, a close correlation was found between changes in apparent resistivity and in the water level of the river running near the current dipole (BARSUKOV, 1970, 1973). When melted fresh water flows into the river during the daytime and the water level rises, the resistivity around the current dipole increases because of the decrease in salinity.   The current density is then reduced in the current dipole area, resulting in an enhancement of current density around the receiver site. Thus the reason for diurnal variations in the apparent resistivity was reasonably understood.   A drift-like trend was also found to be due to fluctuations of the water level in the river.

In order to avoid such an effect of the river the survey site was removed to a new place which is far removed from the river and 2,400 m above sea level (BARSUKOV, 1970, 1973).   The distance between electrodes for the current dipole was taken as 1,000 m and two receiver sites were established 6 km southeast and 3.5 km northwest of the source dipole, respectively.   No diurnal variation exceeding the accuracy of the system was then observed at this new measuring site.   The result of measurements carried out during a period from June 1967 to December 1968 revealed a remarkable change in apparent resistivity (BARSUKOV, 1970). It was found that earthquakes of magnitude 4 or so occurred at resistivity minimum stages.   The decrease in apparent resistivity amounted to 7–11% depending on the distance between the epicenter and the measuring site.

Some more examples of anomalous variations in apparent resistivity were obtained in the same area (BARSUKOV, 1972a).   The result obtained up to the beginning of 1971 suggests that earthquakes of magnitude larger than 3.5 are preceded by an anomalous change in apparent resistivity.   Anomalies are characterized by a duration of 1.5–2.0 months and a decrease of 15–18%.   A resistivity change in the focal area should exceed the observed change in apparent resistivity value, probably reaching 30–40% (BARSUKOV, 1972a).   The important role of pore pressure in such a precursory phenomenon was emphasized, because increase in pore pressure leads to reduction in strength of the medium and decrease in its resistivity (BARSUKOV, 1972a, b).

In case the observed changes are not actually related to earthquakes, two other possibilities were considered as a likely cause of resistivity changes: variations in the resistivity of the surface layer down to 2 m or so caused by precipitation, temperature, and so on and the effect of underground water at depth down to several hundred meters (BARSUKOV and SOROKIN, 1973).   The effect of the surface layer is negligibly small, because the distance between the receiver site and the current dipole (6 km) is much longer than the depth of the layer (2 m).   In order to examine the other possibility AC currents of different frequencies, 128 and 450 Hz, were used besides ordinary DC currents (0.1 Hz square wave). The penetration depths for such frequencies are about 1,400 m and 800 m, respectively, in the survey area (BARSUKOV and SOROKIN, 1973). The result showed no appreciable change for A Ccurrents of either frequency, while a decrease of several percents was observed for DC currents. Therefore, it can be reasonably concluded that the resistivity does change deep in the crust.

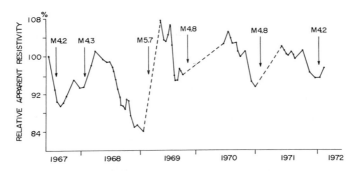

FIG. 10. Changes in apparent resistivity in percents observed at Garm. Arrows indicate the occurrences of earthquakes with respective magnitudes (after BARSUKOV, 1974 and RIKITAKE, 1976a).

It was disclosed from the results obtained up to 1972 that resistivity variations had been observed in association with earthquakes occurring at a depth shallower than 10–15 km and at a distance within 10 km from the center of the measuring system (BARSUKOV, 1974). The anomalous changes amounted to up to 18% for large and near earthquakes and up to 10% for small and remote ones. The precursor time for these anomalies ranges from about 2 to 7 months for earthquakes of magnitude 4.2–5.7 and it tends to be longer for larger earthquakes (see Fig. 10). One of the most typical examples is the anomaly observed in early 1972 as shown in Fig. 10. No variation exceeding 2% was observed from the spring of 1971 to January, 1972. Then the apparent resistivity began to decrease and an earthquake occurred at a distance of 3–4 km from the receiver dipole. The resistivity of the surface layer was also examined with special reference to meteorological factors. It was found that the resistivity changed substantially at a depth of 30–40 cm and less remarkably at a depth of 1.5 m. However, the resistivity did not change at a depth of 1 km or so as already mentioned.

An attempt to estimate the spatial extent of an anomalous area was made from anomalies of apparent resistivity observed at two receiver sites 5 km and 10 km distant, respectively, from the current dipole (BARSUKOV et al., 1974). The apparent resistivity decreased by 7% at the near site and by 4% at the remote site in association with an earthquake of magnitude 3.8 which occurred at a depth of 5–10 km. It was shown that the anomalous region in this case had a horizontal stretch of at least 15 km and its top surface lay at a depth of 2–3 km. The observed change in apparent resistivity could be accounted for by a resistivity decrease of 20% or thereabouts in the anomalous region.

5. 1. 2   Examples in the U.S.A.

Measurements of apparent resistivity similar to those in the Garm region have been carried out in the San Andreas fault area (MAZZELLA and MORRISON, 1974). A 1.5 km long current dipole was established on the northeastern side of the fault. Two receiver sites are located across the fault at distances of 10 km and 15 km, respectively, from the current dipole. A square-wave current with a peak to peak amplitude of 200 amperes has been used as a source current. As shown in Fig. 11, two anomalous variations in apparent resistivity were observed in as-

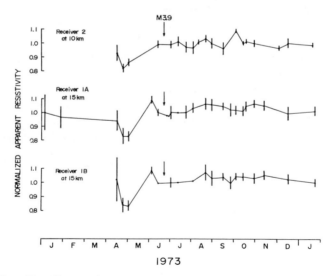

FIG. 11.   Changes in apparent resistivity observed in the San Andreas fault region. An arrow indicate an earthquake of magnitude 3.9 (after MAZZELLA and MORRISON, 1974).

sociation with earthquakes.

In one of the cases, the apparent resistivity decreased by 10%, then increased, reaching 10–15% above the initial level, and finally recovered to its original value. After this anomaly with a duration period of 60 days, an earthquake of magnitude 3.9 occurred near the receiver sites. Its focal depth was about 9. 5 km. The possibility that this anomaly was caused by a near-surface phenomenon can be ruled out, since no variation was observed at a receiver site established for monitoring near-surface effects. The other example, though less remarkable, was obtained 30 days before an earthquake swarm equal in energy to an earthquake of magnitude 3.5.

These changes can be accounted for by an 80 percent change in the resistivity of the portion with 4 km in width and 2–6 km in depth in the fault region. MAZZELLA and MORRISON (1974) claim that such an anomalous area was created as a result of new cracks and water inflow in them as required by the dilatancy-diffusion model. It was also pointed out that a duration of 60 days for an earthquake magnitude of 3.9 is in good agreement with the relation between precursor time and magnitude as put forward by SCHOLZ et al. (1973). However, an alternative interpretation is also suggested to avoid the difficulty of the unusually large uplift to be expected from a resistivity change of 80% due to porosity increase; namely; the electrically anisotropic nature of the vertical slab was considered (H. F. Morrison, personal communication, 1977).

5. 1. 3 Examples in China

A prediction was successfully made for an earthquake of magnitude 4.9 which occurred at Ningjiin near Hsintai on June 6, 1974. The bases for this prediction consist of various anomalies (OIKE et al., 1975; PRESS et al., 1975). A resistivity change detected by the resistivity sounding method was used as one of the bases for medium- and short-term predictions (OIKE et al., 1975). However, no significant attention was reportedly paid to the anomalous decrease in apparent resistivity which appeared 2 to 3 months prior to the earthquake (PRESS et al., 1975).

Another example of resistivity change was obtained at an epicentral distance of 50 km prior to an earthquake of magnitude 7.8 (PRESS et al., 1975). The apparent resistivity started to increase about 13 months before the shock and took on a maximum value, 9% higher than the previous level, about 7 months after the anomaly first appeared. Then the apparent resistivity recovered to its previous level just before the shock. Besides this anomaly, there are many anomalies presumably related to moderate-size earthquakes, although the magnitude of the observed changes is about 10%, which is comparable to an error level.

According to the report by the 1977 Delegation of the Seismological Society of Japan to China, resistivity measurements based on the resistivity sounding method have been carried out extensively (NORI-TOMI, 1978). Resistivity anomalies were observed at some of the stations before a few major earthquakes, as is described in the following. In the Peking, Tientsin, and Tangshan area measurements of resistivity have been carried out at 13 stations. The measurement system consists of the Wenner method with a separation of 1–3 km for a pair of electrodes through which an artificial current of 1–5 amperes is driven into the ground. Some of the stations exhibit a seasonal change probably reflecting fluctuations in the underground water table.

In association with the destructive Tangshan earthquake of magnitude 7.8 which occurred on July 28, 1976, many anomalies in resistivity were observed at medium- and short-term prediction stages (No-RITOMI, 1978; Yü-LIN and FU-YE, 1978). At the Tangshan station in the epicentral region a decrease in apparent resistivity superposed on a seasonal trend was found to have started 27 months before the earthquake occurrence. The change in this long-duration anomaly amounted to 2–3%. A gradual decrease was noticed at another station located about 70 km northwest of the epicenter. The anomaly lasted for 28 months and began to disappear after the shock. These examples are shown in Fig. 12. Anomalies of shorter duration also appeared at two other stations. The apparent resistivity started to decrease 3–4 months before the shock, reaching the minimum value at the time of the shock and gradually recovering during the six months after the shock. These four examples contributed to the establishment of a criterion for recognizing a change as a precursor at the medium-term prediction stage; that is, the precursory anomaly must exceed 3% in relative change. Medium-term resistivity precursors observed prior to the Tangshan earthquake are characterized by the following. Precursory changes in resistivity are more remarkable at stations near the epicenter and drastic changes like the frequently occurring step-like ones tend to appear there, while changes at remote stations are rather smooth.

Some anomalies are reported to have appeared immediately before the Tangshan earthquake. A rapid decrease of 3% during 6 days before the earthquake was observed at the Changli station having an epi-

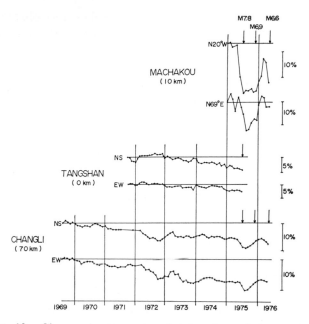

Fig. 12. Changes in apparent resistivity observed in association with the Tangshan earthquake of magnitude 7.8. The stations, Machakou, Tangshan, and Changli are 10, 0, and 70 km distant from its epicenter, respectively (after Yü-lin and Fu-ye, 1978).

central distance of 70 km where anomalies at the medium- to short-term prediction stages were also found. Another rapid decrease in apparent resistivity appeared about 3 days before the shock at the Machakou station located 10 km distant from the epicentral region. There was another type of resistivity anomaly, namely, a rapid 17% increase in apparent resistivity 13 months after the earthquake occurrence (Noritomi, 1978). Some other anomalies, though less remarkable, were also reported in the recent paper by Yü-lin and Fu-ye (1978).

Anomalous variations in apparent resistivity were also observed in association with the 1976 Sungpan-Pingwu earthquake of magnitude 7.2 (Noritomi, 1978). In this case, measurements are based on the quasi-Schlumberger method with distances of 1 km and 200 m for current and receiver electrodes, respectively. Anomalies at the medium-term prediction stage were reportedly observed even at a station located as far as 300 km from the epicenter, although at remote stations changes tend to be small and times of their first appearance tend to be late. At some of the stations where medium-term anomalies were observed, about 1.5–3.5 yr before the earthquake, short-term ones with precursor times of about 4–45 days also appeared.

5. 2 The method of short-period geomagnetic variations
5. 2. 1 Changes in the transfer function

YANAGIHARA (1972) reported that transfer functions characterizing the empirical relation between the three components of short-period geomagnetic variations underwent remarkable secular changes at the Kakioka Magnetic Observatory. The data used for analysis are ssc's (sudden storm commencement), si's (sudden impulse), and similar variations. These variations observed at Kakioka are known to follow the following empirical relation reasonably well:

$$\varDelta Z = A \varDelta H + B \varDelta D$$

where $\varDelta Z$ and $\varDelta H$ denote variations in the vertical and horizontal fields, and $\varDelta D$ is a variation in declination. A and B have been called transfer functions. Generally they are complex functions of frequency, or period. As is shown in Fig. 13, the A-value decreased until around 1920, increased to a maximum around 1940, and tended to decrease gradually thereafter. Large changes can be seen in the B-value but these would be less significant since geomagnetic variations used for analysis have

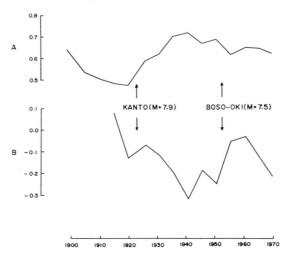

FIG. 13. Changes in transfer functions A and B at the Kakioka Magnetic Observatory. Arrows indicate the occurrences of the M 7.9 Kanto and M 7.5 Boso-Oki earthquakes, respectively (after YANAGIHARA, 1972).

in most cases a north-south polarization and therefore little contribution to B-value estimates.

No evidence is known to support an external origin for such changes in transfer functions. Since Kakioka is located in the region of the Central Japan Conductivity Anomaly, YANAGIHARA (1972) suggested an internal origin: alteration in an underground conductivity anomaly giving rise to anomalous variations in the Z component in the central part of Japan. YANAGIHARA (1972) further pointed out that the changes in A-value at Kakioka appeared to be correlated with the occurrence of the 1923 Kanto earthquake of magnitude 7.9 and speculated an association of an underground conductivity anomaly with the Kanto earthquake.

In Yanagihara's analysis, phase difference between the components and period dependence of transfer functions were not taken into consideration. However, SHIRAKI and YANAGIHARA (1977) calculated complex transfer functions from magnetic storm records by making use of the spectral analysis method. One storm record of 24-hr duration was chosen each year from 1917 to 1974 except for 1918 and 1919. Their result shows that a general trend in Au-values (in-phase part of A) agrees well with Yanagihara's previous result for A-values. Since different types of geomagnetic variations were used and different kinds of analyses were made, changes in transfer functions at Kakioka seem to have actually taken place.

If a change in electrical conductivity amounting to one order of magnitude actually takes place in the earth as suggested by the experiment of BRACE and ORANGE (1968a), it is not unreasonable to expect changes in transfer functions. However, the Kakioka Magnetic Observatory is about 100 km distant from the focal region of the Kanto earthquake and a change in A-value amounting to as much as 0.2 can hardly be accounted for by merely a conductivity change in the focal region.

In view of the high-seismicity zone existing to the south of the Kakioka Observatory and the unusually high seismic activity there before the Kanto earthquake, HONKURA (1977b) showed that there could exist a mechanism which would give rise to a change in A-value at Kakioka in association with the Kanto earthquake. It was found from model calculations based on a two-dimensional approximation that an A-value change of 0.1 at Kakioka (about a half of the observed change) would have been possible if the conductivity had changed by one order of magnitude in the high-seismicity zone before the occurrence of the Kanto earthquake. This earthquake occurred at the boundary between the Asia and Philippine Sea plates as a result of strain accumulation due to the motion of the Philippine Sea plate relative to the Asia plate. It is not unlikely that strain accumulation within the Asia plate would eventually result in activity enhancement in the high-seismicity zone. Then, a conductivity change, if it actually occurs, should have something to do with the strain accumulation. The electrical conductivity of lapilli tuff taken from Aburatsubo, Japan, exhibits a remarkable change, under certain conditions, for a very small strain (see Section 4). It is not clear, however, that rocks constituting the crust and upper mantle have similar nature.

YANAGIHARA and NAGANO (1976) found a good correlation between changes in transfer functions at Kakioka and earthquakes of magnitude greater than 5 which occurred in the above-mentioned high-seismicity zone. Decrease in A-value and also in B-value seems to precede earthquake occurrences. This is in good agreement with Yanagihara's result in A-value, supporting the speculation that the decrease in A-value at Kakioka before the Kanto earthquake might be associated with the high activity in the seismicity zone prior to this event. The long duration of the change reported by YANAGIHARA (1972) could be the result of frequent occurrences of short-duration decrease in A-value associated with a number of seismic events in the high seismicity zone.

The study of changes in transfer functions has been very active at the Kakioka Magnetic Observatory since the pioneer work of YOSHI-MATSU (1963, 1964, 1965). He first pointed out the existence of a good correlation between changes in $\Delta Z/\Delta H$-value or $\Delta Z/\Delta D$-value and earthquake occurrences. However, earthquakes with epicentral distances of 300 km or more are also included in his investigation. Moreover, Yoshimatsu's conclusion was not confirmed from further analyses made by KUBOKI and YOSHIMATSU (1973). They suggested from detailed analyses of data between 1924 and 1938, however, that a remarkable change in B-value at Kakioka appearing around 1932–1933 might be associated with the 1933 Sanriku earthquake of magnitude 8.3. In recent work on the time dependence of transfer functions at Kakioka, more stress is put on the examination of various factors affecting transfer function estimates (SHIRAKI, 1977; M. Shiraki, Y. Sano, personal communication, 1978). These studies may contribute to the reliability of time changes of transfer functions at Kakioka as percursors associated with earthquakes.

MIYAKOSHI (1975) analyzed magnetograms at Tashkent and Askhabad in Central Asia with special reference to the 1966 Tashkent earthquake of magnitude 5.5 and the 1970 Askhabad earthquake of magnitude 6.6, respectively. No notable change in transfer function was found in association with the Askhabad earthquake, while at Tashkent a significant increase in yearly mean values of $(A^2+B^2)^{1/2}$ could be seen about a year before the Tashkent earthquake. Its focal region was about 30 km from the magnetic observatory. The duration of the anomaly is in agreement with a precursory change in radon content at a well station in Tashkent (SADOVSKY et al., 1972). MIYAKOSHI (1975) presumed that the electrical conductivity might have increased as a result of fluid flow into a focal region in such a manner as described by SCHOLZ et al. (1973).

RIKITAKE (1979) analyzed short-period geomagnetic variations at the Sitka Magnetic Observatory in order to examine if any changes in transfer functions could be seen in association with the 1972 Sitka earthquake of magnitude 7.2. The focal region is located only about 40 km west of the observatory. The analysis was made by scaling individually selected events on the magnetograms as in the method used by YANAGIHARA (1972) and MIYAKOSHI (1975). Transfer function A increased by about 0.1 a year before the earthquake occurrence. Since the Sitka Observatory is located at the geomagnetic latitude of 60°, geomagnetic variations there tend to be influenced more or less by overhead electric currents in the auroral zone. However, the data used for analysis were carefully selected to avoid serious influence from such currents and it was found that the change in transfer function A might reflect a precursory one rather than an apparent one due to fluctuations of the electric current system in the auroral zone.

WEN-YAO et al. (1978) reported changes in the transfer function A at one of the stations in the conductivity anomaly region in the Kansu Province, China. The station is located north of an underground conducting belt extending almost in the east-west direction. They suggested that a high level of A-value during a period of a few to several months prior to earthquakes of magnitude 5.1, 6.5, and 7.2, respectively, might reflect precursory changes in the electrical conductivity of their focal regions. Immediately after the earthquakes of magnitude 6.5 and 7.2, the A-value rapidly decreased by 0.3–0.4.

In 1978 two large earthquakes occurred in Japan: the Izu-Oshima

Kinkai earthquake of magnitude 7.0 and the Miyagi-Ken Oki earthquake of magnitude 7.4.   Three components of short-period geomagnetic variation have been measured at Nakaizu near the epicentral region
of the Izu-Oshima Kinkai earthquake and also at Onagawa located west
of the epicenter of the Miyagi-Ken Oki earthquake.   Preliminary results
of data analyses indicate that transfer functions at both stations seem
to have changed before the earthquakes (Y. Honkura, in preparation;
M. Seto, personal communication, 1978).

5. 2. 2   Changes in the amplitude of the horizontal field

It is well known that the amplitude of the horizontal component
of short-period geomagnetic variation is enhanced over a conducting
body as a result of concentration of induced electric currents.   It is useful, therefore, to monitor the amplitude of the horizontal component as
a means of detecting an underground conductivity change.   An amplitude change can be monitored by calculating the ratio of the amplitude
at a station of interest to that at another remote station as the reference
station.   If the separation between the two stations is much less than
the scale length of the external source field, the amplitude of the external
field can be regarded as the same at both stations, and changes in the
amplitude ratio, if any, can be attributed to those of internal origin.

When an earthquake of magnitude 6.2 named the Southeastern
Akita earthquake took place in the northeastern part of Japan, some
temporary stations happened to be in operation near its epicentral area.
Honkura has examined amplitude ratios in H and D variations of short
period, respectively (NIBLETT and HONKURA, 1979).   At a station located
about 10 km west of the focal region, the amplitude of the H component
seems to have been enhanced at the time of earthquake occurrence, while
the amplitude of D was slightly depressed, although these changes are
not very significant at the 95% confidence limit.   Depression in amplitude was recognized for both component at another station about 15 km
southeast of the focal area.   The observed changes at both stations of
interest can be explained qualitatively if the electrical conductivity had
increased in the focal region.   This result suggests that the observed
changes might have some association with the earthquake, although the
duration of the anomalies (about one and a half months) is very short
compared to that for a premonitory change in the ratio of compressional
to shear wave velocities (V_P/V_S) which appeared two years prior to the
same earthquake (HASEGAWA et al., 1975).

HONKURA and KOYAMA (1978) conducted a similar research in an

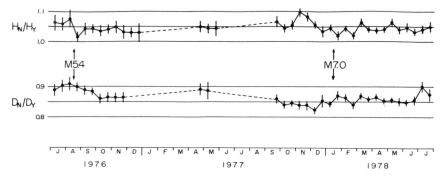

FIG. 14.   Changes in the ratio of the amplitude of short-period geomagnetic variation at NKZ (see Fig. 15) to that at the reference station for the H and D components, respectively.   Arrows indicate M 5.4 and M 7.0 earthquakes, respectively (after HONKURA and KOYAMA, 1978).

area where anomalous crustal uplift and anomalously high micro-earthquake activity have been observed.   Three components of short-period geomagnetic variations have been measured with a flux-gate magnetometer at NKZ (Fig. 15) since July, 1976.   In August 1976, about one and a half months later, an earthquake of magnitude 5.4 occurred approximately 15 km south of NKZ.   The Yatsugatake Magnetic Observatory, which is about 140 km distant from NKZ, was chosen as the reference station.   The amplitude ratio was determined every half a month for the H and D components, respectively.   The results suggest that slight changes may have occurred in association with the earthquake as shown in Fig. 14, although they are not very significant at the 95% confidence limit.   The large error bars (two standard errors) are due partly to the simple method of the analysis; a maximum amplitude was read for individual disturbances having a duration of 10–40 min. In January 1978, another large earthquake of magnitude 7.0 occurred below the sea about 30 km southeast of NKZ.   In Fig. 15 are shown epicenters of the main shock (M 7.0), a large foreshock (M 4.9) which occurred near the main shock, and large aftershocks (M 4.9, 5.1, 5.8). About two months before the main shock, the amplitude of H at NKZ increased by about 5% and gradually returned to the previous level as shown in Fig. 14.   On the other hand, the amplitude of D began to decrease gradually about three months before the main shock.   After the occurrence of the main shock, the amplitudes of both H and D appear to have undergone oscillatory changes.

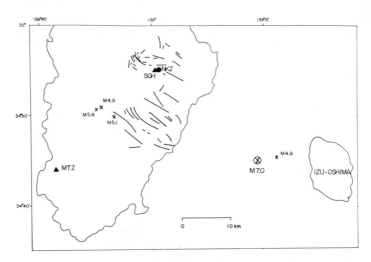

FIG. 15.   Localities of the main shock (M 7.0) of the 1978 Izu-Oshima Kin-kai earthquake and its major aftershocks.   SGH and MTZ denote stations for total intensity measurements and NKZ for measurements of geomagnetic variation and self-potential.   Solid lines indicate active faults as found by MATSUDA (1977) (after HONKURA and KOYAMA, 1978).

It is not clear whether or not these changes reflect an underground conductivity change.   However, there are some indications which seem to support the assertion that the amplitude enhancement in the H component around the middle of November, 1977, reflects a precursory change in the conductivity of the crust near the station.   First, the natural electric field at NKZ began to decrease around November 10; the change amounted to 30 mV/km or so.   Secondly, the total magnetic intensity at SGH (very close to NKZ) relative to MTZ (about 30 km southwest of SGH) decreased by about 5 nT (M. Kawamura, personal communication, 1978).   These two anomalies were very similar to each other not only in time but also in shape.   Thirdly, the radon content also changed in early November at a well station near NKZ (H. Wakita, personal communication, 1978).   The coincidence in the time of appearance of the various anomalies in virtually the same area suggests that changes are likely to be precursory phenomena.   Most of the changes are probably related to fluid flow along the active faults having a nearly NWW-SEE trend as shown by solid lines in Fig. 15 (HONKURA, 1978).

5. 2. 3   *Theoretical work*

RIKITAKE (1976b) examined through two-dimensional model calculations whether any detectable change in short-period geomagentic variations could be expected in association with the appearance of a conductivity change of one order of magnitude in the crust. A fairly large change, up to 100%, in the horizontal field was obtained for a certain combination of the period of geomagnetic variation and the spatial extent of an anomalous area. Expected changes in the vertical component is smaller than those in the horizontal one for his model. From a relation between the magnitude of an earthquake and the spatial extent of an anomalous area, the period of geomagnetic variation for which changes are most effectively detected can be estimated for various magnitudes. In the case of magnitude 7, for instance, variations having a period of 30 sec or so should give the most significant result. For magnitude 8, a period of a few minutes is the most suitable period for detecting changes. If the period is too long or too short for a certain size of an anomaly, large changes are not expected from his model.

Some of the precursory changes reported so far are fairly large even at periods of a few minutes to several tens of minutes. Such long periods are not likely to be very effective as suggested by the above result of model calculations. However, inductive or conductive response of the crust and upper mantle highly depends on the local conductivity structure. The important point derived from the model calculations is, therefore, that it is not impossible to detect changes in short-period geomagnetic variations if the conductivity changes appreciably in association with earthquakes.

5. 3   The magnetotelluric method

Changes in magnetotelluric parameters were examined by HONKURA (1976b) from electric and magnetic variations of short period observed at stations near the epicentral region of the 1970 Southeastern Akita earthquake of magnitude 6.2. Although no clear conclusion was obtained, slight changes were noticed in the amplitude of the induced electric field and also in its direction. That the changes are in good agreement with those to be expected for a resistivity decrease in the focal region suggests an association of the observed changes with the earthquake.

Magnetotelluric measurements have been conducted in Southern California for the purpose of detecting changes, if any, in apparent resistivity (REDDY et al., 1976; PHILLIPS et al., 1977). Some significant changes were detected and it appears that crustal resistivity changes might have occurred, although no clear relation between apparent resistivity changes and tectonic activity has been reported. Meanwhile, WHITCOMB (1977) claimed that changes up to 15% in apparent resistivity estimates were observed prior to earthquake occurrences.

Studies of the time dependence of magnetotelluric parameters have been undertaken in a seismically active region in eastern Canada. Remarkable changes at periods of 5 and 10 min were significantly detected (HONKURA et al., 1976a; KURTZ and NIBLETT, 1978). A remarkable long-term increase of 14% per year in apparent resistivity has been observed for three and a half years. Superposed on it, some variations with shorter duration have also been observed. Nonetheless no major earthquakes have occurred and seismic activity has been rather low in the anomalous region during the observation period. KURTZ and NIBLETT (1978) speculated that continuous low-level seismicity suggests an on-going crustal instability, with which the observed changes in apparent resistivity may be associated.

In the Garm region, the U.S.S.R., BARSUKOV (1970, 1973) also attempted the use of the magnetotelluric method for the purpose of earthquake prediction. However, it was found that the accuracy of magnetotelluric parameter estimates is poorer than that of apparent resistivity determined from the artificial current method. No interesting result regarding the magnetotelluric field has been reported, while the latter method turned out to be successful as described in Section 5.1.

Theoretical estimation on changes in magnetotelluric parameters was made by HONKURA (1976a) on the assumption that the induced electric field is approximately uniform in the crust and the effect of secondary induction is negligibly small. This assumption is valid only if the period of variation field is sufficiently long and their penetration depth is much deeper than the crustal depth. Calculations were made for three cases corresponding to anomalies associated with very shallow, shallow and deep earthquakes, respectively. Except for the deep earthquake, changes can be expected for magnetotelluric parameters such as the amplitude and direction of the induced electric field, although they are small unless the measuring site is located within an anomalous area. For short periods the above result is not applicable. If the penetration depth of the varying fields is comparable to the depth of an anomalous area, much larger changes will be expected as shown by REDDY et al. (1976) who made one- and two-dimensional calculations for a short period variation.

5. 4   *Experimental study of electrical resistivity changes*

It is obvious that the pressure dependence of electrical resistivity is the most important factor concerning a resistivity change associated with an earthquake.   BRACE *et al.* (1965) investigated the pressure effect of the resistivity of crystalline rocks under a condition of water-saturation.   According to the results of their experiments, the resistivity increased very rapidly as the confining pressure rose to 2 kb and then followed an almost linear trend of gradual increase at pressures up to about 10 kb.   The resistivity at this stage was 3 orders of magnitude higher than the initial one.   This result could be interpreted as reflecting a reduction in porosity with increasing pressure, because the resistivity of crystalline rocks at low temperatures would be controlled mostly by the content of conducting water in cracks and pores.   The rapid change at low pressure corresponds to closure of cracks and the gradual change occurring thereafter reflects reduction in the cross section of pores. The dependence of resistivity on porosity (η) has been known to follow approximately the empirical relation:

$$\rho = \rho_{\mathrm{w}} \eta^{m}$$

where ρ_{w} denotes the resistivity of water and m is a constant which usually ranges from 1 to 2, depending on the state of the conducting paths within the medium.

BRACE *et al.* (1966) experimentally showed that many rocks become dilatant at one-third or two-thirds of the fracture stress as a result of creation of new cracks.   Decrease in resistivity is then expected under a condition of saturation.   If a rock saturated with water is stressed in compression, the resistivity will first increase because of crack closure and then rapidly decrease as dilatancy proceeds at one-third or two-thirds of the fracture stress.   Such a behavior in resistivity was clearly demonstrated by the experiments of BRACE and ORANGE (1966); a change in resistivity amounted to as much as one order of magnitude.   The micro-crack dilatancy is one of the characteristics of intact rocks and repeatedly occurs after many cycles of stress to within 60 to 90% of the fracture stress (BRACE, 1977).

The resistivity of a rock partially saturated with water behaves in a way quite different from that of a saturated rock.   The initial resistivity under partial saturation is quite low.   According to the experimental results obtained by BRACE and ORANGE (1968b), the resistivity decreased drastically during pressure increase at a low pressure range, then remained almost constant, and tended to follow the same pattern as a saturated rock as the pressure exceeded 6 kb.   Such a resistivity change is interpreted as follows.   At low pressure most cracks are still open and some of them contain little water and therefore contribute little to the formation of conducting paths.   As pressure increases, cracks tend to close and some water is expelled into unsaturated cracks. The decrease of resistivity is a result of increase in the number of effective conduction paths thus created.   Further increase in pressure will eventually cause a saturation condition and the situation will become the same as that of an initially saturated rock.

If a rock contains conducting minerals, its resistivity depends little on the degree of saturation because mineral conduction becomes more important.   The resistivity of such a rock decreases rapidly as pressure increases and remains almost unchanged thereafter (BRACE and ORANGE, 1968b).   This is because of closer contact between mineral grains contributing to electric conduction.   It is obviously expected that the resistivity of a dry rock without conducting minerals depends very little on pressure.   Therefore, little resistivity variation will be expected for dilatancy models without water diffusion such as the ones proposed by MOGI (1974), STUART (1974), and BRADY (1974).

Detailed examination of a resistivity decrease due to dilatancy disclosed that m in the above relation between porosity and resistivity is rather close to 1, while m usually amounts to about 2 for saturated rocks (BRACE and ORANGE, 1968a).   Porosity increase at a dilatancy stage results from creation of new cracks.   Therefore, the above result seems quite reasonable in view of the nature of the cracks characterized by their long and narrow cavity (BRACE et al., 1972).   Such cracks will contribute to effective creation of new conduction paths not only through a high degree of interconnection of new cracks themselves but also through their role in connecting pre-existing isolated-pores with other cracks and pores.   It is theoretically known that m is close to 1 if the degree of interconnection is high (WAFF, 1974; HONKURA, 1975).

Another important result of the experiment by BRACE and ORANGE (1968a) is the difference in resistivity change between virgin and faulted rock samples.   Once a fault is formed, it remains as a permanent conduction path and a resistivity decrease due to further dilatancy tends to be masked.   This is not a discouraging fact from the viewpoint of earthquake prediction, since the possibility that the resistivity still changes remarkably in the vicinity of a fault in the earth is not precluded. The discouraging fact derived from the experiment is rather the observed result that sliding occurs along a pre-existing fault before the applied stress reaches a level high enough to cause a significant resistivity decrease.

Recently a mechanism of stick slip during frictional sliding has been investigated with special reference to shallow earthquakes occurring along a pre-existing fault (BRACE, 1972, 1977). Aiming at examining whether any resistivity change occurs during frictional sliding or not, WANG *et al.* (1975) made laboratory experiments for a Westerly granite sample. The observed resistivity change is characterized by an exponential function of time in response to a sudden change of normal stress, which suggests that a diffusion process occurs probably when some cracks, containing water, close with increasing normal stress. Another important result is that the resistivity decreases as the shear stress increases and it suddenly returns to the previous level at the time of a stick-slip, that is, a sudden release of shear stress. The minimum level of resistivity coincides with the occurrence of the stick-slip. Similar behavior could be observed repeatedly for each cycle. The amount of resistivity change tends to increase with increasing normal stress.

WANG *et al.* (1975) interpreted these changes in terms of opening and closing of cracks in the vicinity of the sliding surface. When the shear stress increases, cracks will open and water will flow into them under the applied pressure. Thus the resistivity decreases with increasing shear stress. If the shear stress is suddenly released, cracks will close and water within them will be expelled into the sliding surface. Then the resistivity should return to the previous level. However, porosity increase due to dilatancy associated with frictional sliding is less than 5×10^{-5} at low confining pressure (HADLEY, 1973) which corresponds to a resistivity change of only a few to several percents as supported by the experiment of WANG *et al.* (1975). This fact appears to be rather discouraging for an earthquake prediction study based on resistivity measurements in spite of a precursory decrease in resistivity before a stick-slip. However, a decrease of 10–20% has been observed in field measurements as a precursory resistivity change. The discrepancy may be due to the difference between conditions in laboratory experiments and the real earth, or there may be another mechanism which gives rise to a large resistivity change prior to an earthquake.

In the above discussions a change in the state of water within cracks and pores has not been taken into consideration.   On the other hand, ANDERSON and WHITCOMB (1973) proposed a dilatancy model in which pore water is subject to a liquid-to-vapor transition.   Since the resistivity of water in its liquid state is much lower than that of its vapor (QUIST and MARSHALL, 1969; QUIST *et al.*, 1970), the resistivity may suddenly change if some vapor becomes mixed with pore water during a dilatancy process.   However, the result of the experiment made by BRACE (1975) shows that resistivity changes little even if some water is converted to vapor at the liquid-to-vapor transition.   BRACE (1975) presumed that electric conduction along the surface of cracks would be important and, therefore, a relatively small amount of water should be sufficient for such a surface conduction.

6.   Statistics of Electric and Magnetic Earthquake Precursors

Precursor time, which is defined as the time from the first appearance of an anomalous change to an earthquake occurrence, for the observed precursors ranges from tens of minutes to several years or more as shown in the preceding chapters.   From various kinds of precursors including resistivity changes, SCHOLZ *et al.* (1973) examined the dependence of precursor time on an earthquake magnitude and obtained a linear relation between logarithmic precursor time ($\log_{10} T$) and earthquake magnitude (M).   The obtained relation shows that the precursor time is long for larger earthquakes and SCHOLZ *et al.* (1973) interpreted this in terms of a dilatancy diffusion model.   RIKITAKE (1975a) estimated the spatial extent of dilatant region for various magnitudes on the basis of empirical relations between M and the spatial extent of land deformation and between M and aftershock area.   Then it was assumed that precursor time corresponds to the time required for completion of a diffusion process in a dilatant region.   From calculations of the process within a region corresponding to M, a relation between M and precursor time was derived.   The result is in good agreement with the relation obtained by SCHOLZ *et al.* (1973) from various field data.

RIKITAKE (1975b) collected further examples of earthquake precursors and also examined some possible relations between M and precursor time.   As a result it was found that precursors could be separated into three types.   In one of these, called the A_2 type, the precursor time depends on M in such a manner as can be expressed by a definite relation between precursor time and M.   The precursor time of the

other two types, A_1 and B, is independent of M.   The B type consists
of precursors in tilt, strain, and foreshock, while the A_1 type is characterized by a short precursor time of several hours for other elements in
earthquake prediction.

It will be worthwhile making a similar statistical analysis for electric and magnetic data.   For this purpose anomalous changes observed
before earthquakes were chosen from the data presented in this article,
under the condition that the precursor time is known for an earthquake
of known magnitude.   Thus 92 samples in total were obtained as shown
in Table 2: 13 for the geomagnetic field (G), 32 for the electric field
(E), 21 for the ground resistivity as observed by the resistivity variometer $(R1)$, and 26 for the resistivity in and around a focal region $(R2)$.
Not all of these samples may really be associated with earthquakes, but
for the time being there is no criterion for judging which one is a real
precursor and which one is not.   In this chapter all of the selected
data will be regarded as earthquake precursors.

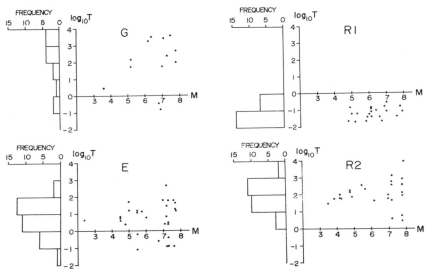

FIG. 16.   Histograms of logarithmic precursor time ($\log_{10} T$) and plots of
$\log_{10} T$ against the magnitude for the respective group.

Figure 16 shows the histogram of $\log_{10} T$ for each one of the four
groups, where T denotes the precursor time in days.   The plot of $\log_{10} T$
against M is also given for each group.   It is interesting to notice that
the precursor time most frequently observed is quite different between

TABLE 2.   Precursor data.

Earthquake			Precursor		Epicentral distance (km)	Reference
Place	Year	M	Amount	Precursor time (day)		
G-type precursors						
off Tanabe	1961	6.4	2 nT/year	3,285	36	Tazima (1968)
Niigata	1964	7.5	20 nT	3,685	40	Fujita (1965)
Izu-Oshima Kinkai	1978	7.0	5 nT	60	30	Honkura (1978)
San Andreas fault area	1967	3.6	2.5 nT	3	—	Breiner and Ko- vach (1967)
Busch fault	1975	5.2	1.5 nT	56	11	Smith and Johnston (1976)
Sitka	1972	7.2	20 nT	2,740	40	Wyss (1975)
Hsingtai	1972	5.2	—	150	—	Press *et al.* (1975)
Haicheng	1975	7.3	21.5 nT	270	215	Seismolo- gical Delega- tion of China (1976)
Tangshan	1976	7.8	8 nT	480	0	Raleigh *et al.* (1977)
Tangshan	1976	7.8	7 nT	105	160	Noritomi (1978)
Ninglang	1976	6.9	—	0.17	—	Noritomi (1978)
Ninglang	1976	6.8	—	0.38	—	Noritomi (1978)
off San Fernando	1964	6.2	100 nT	1,825	140	Wyss and Martin (1978)
E-type precursors						
Yunnan Province	1972	4.8	—	2.5	20	Press *et al.* (1975)
Yunshan-Takuan	1974	7.1	70 A	2.5	90	Press *et al.* (1975)
Haicheng	1975	7.3	100 mV/70 m	2.5	25	Raleigh *et al.*

TABLE 2. (continued)

Earthquake			Precursor		Epicentral distance (km)	Reference
Place	Year	M	Amount	Precursor time (day)		
						(1977)
Haicheng	1975	7.3	40 μA	0.42	145	RALEIGH et al. (1977)
Haicheng	1975	7.3	40 μA	0.13	145	RALEIGH et al. (1977)
Haicheng	1975	7.3	500 μA	60	150	RALEIGH et al. (1977)
Haicheng	1975	7.3	500 μA	3	150	RALEIGH et al. (1977)
Sungpan-Pingwu	1976	7.2	—	450	<300	NORITOMI (1978)
Sungpan-Pingwu	1976	7.2	—	540	<300	NORITOMI (1978)
Sungpan-Pingwu	1976	7.2	—	7	<300	NORITOMI (1978)
Sungpan-Pingwu	1976	7.2	—	30	<300	NORITOMI (1978)
Sungpan-Pingwu	1976	7.2	—	0.13	—	NORITOMI (1978)
Sungpan-Pingwu	1976	6.7	—	0.25	—	NORITOMI (1978)
Lungling	1976	7.6	—	60	—	CHI-YANG (1978)
Lungling	1976	7.6	—	0.13	—	CHI-YANG (1978)
Kamchatka	1968	6	300 mV/km	6	30	SOBOLEV et al. (1972)
Kamchatka	1959	7 3/4	~50 mV/200 m	15	<150	SOBOLEV et al. (1972)
Kamchatka	1965	5 3/4	~50 mV/200 m	14	<150	SOBOLEV et al. (1972)
Kamchatka	1968	5	~50 mV/200 m	17	<150	SOBOLEV et al. (1972)
Kamchatka	1969	5 1/2	~50 mV/200 m	16	<150	SOBOLEV et al. (1972)
Kamchatka	1969	5 1/2	~50 mV/200 m	13	<150	SOBOLEV et al. (1972)
Kamchatka	1969	4 1/2	~50 mV/200 m	7	<150	SOBOLEV et al. (1972)
Kamchatka	1969	4 1/2	~50 mV/200 m	4	<150	SOBOLEV et al. (1972)
Kamchatka	1969	4 1/2	~50 mV/200 m	5	<150	SOBOLEV et al. (1972)

TABLE 2.   (continued)

Earthquake			Precursor		Epicentral distance (km)	Reference
Place	Year	M	Amount	Precursor time (day)		
Kamchatka	1969	4 1/2	$\sim$50 mV/200 m	6	<150	SOBOLEV *et al.* (1972)
Kamchatka	1971	7.7	$\sim$50 mV/200 m	45	<150	SOBOLEV (1974)
Kamchatka	1971	7.7	$\sim$50 mV/200 m	20	<150	SOBOLEV (1974)
Izu Peninsula	1934	5.5	3 mV/100 m	0.65	20	FUKUTOMI (1934)
Izu Peninsula	1934	5.5	3 mV/100 m	0.08	20	FUKUTOMI (1934)
Izu-Oshima Kinkai	1978	7.0	30 mV/km	65	30	KOYAMA and HONKURA (1978)
San Andreas fault area	1974	5	140 mV/km	55	37	CORWIN and MORRISON (1977)
San Andreas fault area	1975	2.4	4 mV/300 m	4.6	2.5	CORWIN and MORRISON (1977)
R1-type precursors Tokachi-Oki	1968	7.9	0.72 ($\times 10^{-4}$) in $\Delta\rho/\rho$	0.09	713	RIKITAKE and YAMAZAKI (1977a, b)
E off Northern Honshu	1968	6.7	0.2	0.02	582	RIKITAKE and YAMAZAKI (1977a, b)
Middle of Saitama Pref.	1968	6.1	1.10	0.13	96	RIKITAKE and YAMAZAKI (1977a, b)
E off Hokkaido	1969	7.8	0.57	0.17	1,094	RIKITAKE and YAMAZAKI (1977a, b)
Middle of Gifu Pref.	1969	6.6	0.80	0.05	206	RIKITAKE and YAMAZAKI (1977a, b)
Eastern Yamanashi Pref.	1971	5.3	0.47	0.06	76	RIKITAKE and

TABLE 2. (continued)

Earthquake			Precursor		Epicentral	Reference
Place	Year	M	Amount	Precursor time (day)	distance (km)	
off Erimomisaki	1971	7.0	0.10	0.28	780	YAMAZAKI (1977a, b) RIKITAKE and YAMAZAKI (1977a, b)
Northwestern Chiba Pref.	1971	4.8	0.17	0.03	66	RIKITAKE and YAMAZAKI (1977a, b)
Northern Chiba Pref.	1971	5.2	0.13	0.02	122	RIKITAKE and YAMAZAKI (1977a, b)
Fukui-Gifu Border	1972	6.0	0.17	0.08	267	RIKITAKE and YAMAZAKI (1977a, b)
E off Hachijo-jima	1972	7.2	0.43	0.08	252	RIKITAKE and YAMAZAKI (1977a, b)
Tokyo Bay	1973	4.9	0.52	0.14	50	RIKITAKE and YAMAZAKI (1977a, b)
Near Choshi	1973	5.9	0.28	0.04	116	RIKITAKE and YAMAZAKI (1977a, b)
Southern Chiba Pref.	1973	5.0	0.10	0.07	59	RIKITAKE and YAMAZAKI (1977a, b)
E off Chiba Pref.	1974	6.1	0.29	0.15	119	RIKITAKE and YAMAZAKI (1977a, b)
Near South coast of Izu Peninsula	1974	6.9	0.48	0.17	100	RIKITAKE and YAMAZAKI (1977a, b)
Near Hachijo-jima	1974	6.1	0.27	0.05	150	RIKITAKE and YAMAZAKI (1977a, b)

TABLE 2. (continued)

Earthquake			Precursor		Epicentral distance (km)	Reference
Place	Year	M	Amount	Precursor time (day)		
SW Ibaraki Pref.	1974	5.8	0.20	0.02	99	RIKITAKE and YAMAZAKI (1977a, b)
E off Hachijo-jima	1974	6.4	0.26	0.10	242	RIKITAKE and YAMAZAKI (1977a, b)
off Choshi	1974	6.1	0.13	0.09	163	RIKITAKE and YAMAZAKI (1977a, b)
South of Honshu	1974	7.6	0.15	0.04	505	RIKITAKE and YAMAZAKI (1977a, b)
R2-type precursors Garm area	1967	4.2	12%	66	<10	BARSUKOV (1974) and RIKITAKE (1975b)
Garm area	1968	4.3	3%	57	<10	BARSUKOV (1974) and RIKITAKE (1975b)
Garm area	1969	5.7	18%	225	<10	BARSUKOV (1974) and RIKITAKE (1975b)
Garm area	1969	4.8	14%	141	<10	BARSUKOV (1974) and RIKITAKE (1975b)
Garm area	1970	4.8	12%	180	<10	BARSUKOV (1974) and RIKITAKE (1975b)
Garm area	1972	4.2	8%	102	<10	BARSUKOV (1974) and RIKITAKE (1975b)
San Andreas fault area	1973	3.9	24%	60	5	MAZZELLA and MORRISON (1974)

TABLE 2. (continued)

Earthquake			Precursor		Epicentral distance (km)	Reference
Place	Year	M	Amount	Precursor time (day)		
San Andreas fault area	1973	3.5	6%	30	2	MAZZELLA and MORRISON (1974)
Red Mountain, China	1974	4.9	—	75	—	PRESS et al. (1975)
Kansu Province	1973	7.8	9%	390	50	PRESS et al. (1975)
Tangshan	1976	7.8	2–3%	810	0	YÜ-LIN and FU-YE (1978)
Tangshan	1976	7.8	—	840	70	YÜ-LIN and FU-YE (1978)
Tangshan	1976	7.8	—	120	10	YÜ-LIN and FU-YE (1978)
Tangshan	1976	7.8	—	90	70	YÜ-LIN and FU-YE (1978)
Tangshan	1976	7.8	3%	6	70	YÜ-LIN and FU-YE (1978)
Tangshan	1976	7.8	—	3	10	YÜ-LIN and FU-YE (1978)
Sungpan-Pingwu	1976	7.2	—	540	—	NORITOMI (1978)
Sungpan-Pingwu	1976	7.2	—	1,260	—	NORITOMI (1978)
Sungpan-Pingwu	1976	7.2	—	4	—	NORITOMI (1978)
Sungpan-Pingwu	1976	7.2	—	45	—	NORITOMI (1978)
Kanto	1923	7.9	0.2 in A-value	9,125	100	YANAGIHARA (1972)
Tashkent	1966	5.5	0.2 in $(A^2+B^2)^{1/2}$-value	365	30	MIYAKOSHI (1975)
Sitka	1972	7.2	0.1 in A-value	365	40	RIKITAKE (1979)
Southeastern Akita	1970	6.2	5%	45	15	NIBLETT and HONKURA (1979)
Izu-Oshima Kinkai	1978	7.0	5%	60	30	HONKURA and KOYAMA (1978)
Izu-Oshima Kinkai	1978	7.0	4%	90	30	HONKURA and KOYAMA (1978)

G and E, and also between $R1$ and $R2$; some hundreds to thousands of days for G, some days to some tens of days for E, some hours for $R1$, and some tens to hundreds of days for $R2$. Except for some data, G and $R2$ seem to be characterized by the relation between $\log_{10} T$ and M as obtained by SCHOLZ et al. (1973) and by RIKITAKE (1975b) for the A_2 type.

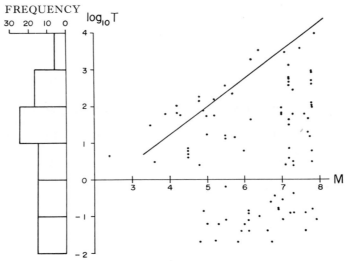

FIG. 17.   Histogram of logarithmic precursor time and plots of $\log_{10} T$ against the magnitude for all the data.   A straight line shows the relation between $\log_{10} T$ and M determined by RIKITAKE (1975b).

In Fig. 17 are shown the histogram and the plot of $\log_{10} T$ against M for all the data.   One of the interesting characteristics is that there is no remarkable peak in the histogram, which seems to imply that the distribution of precursor time is rather random as long as time duration of some hours to a few years is concerned.   No clear relation between $\log_{10} T$ and M is recognizable from the plot of $\log_{10} T$ against M.   One of the reasons for these results might be that most of the observed precursors are not real ones but rather apparent changes due to various kinds of noises which are likely to be of random nature.

Another interpretation is possible and one can still claim that earthquake precursors are essentially classified into the A_1 and A_2 types. The precursor time for the A_2 type is controlled by a dilatancy-diffusion process in the crust.   However, the assumption that the process is uniform in time and space does not necessarily hold good. It may be possible that a precursor appears much later at one station than at another. In such a case, it is the longest precursor time that should be controlled by the dilatancy-diffusion process. Such a presumption seems to be in agreement with the result that the longest precursor time for a certain magnitude is well described by the relation between $\log_{10} T$ and M obtained, for example, by RIKITAKE (1975b) as shown by the straight line in Fig. 17.

Another possibility is the dependence of the time required for a diffusion process under different geological environments; that is, the difference in hydraulic diffusivity. For instance, an area where a network of active faults exists (for example the Izu Peninsula as shown in Fig. 15) is likely to be characterized by a high diffusivity because water easily moves through fracture zones along faults. This could be the reason why the precursor time, about 2 months, for the M 7.0 Izu-Oshima Kinkai earthquake was shorter than expected for its magnitude.

Why does each group have a peak in the histogram different from others, if the above presumption is true? It seems to the author that each peak reflects the difference in the most probable chance of detecting anomalous changes by the observation system of each group. For example, it will not be easy to detect anomalies of short duration in the geomagnetic field because of large fluctuations of external origin. The telluric field, on the other hand, will not be stably observable as an anomaly of long duration because of instability in the electric condition between an electrode and the ground. The $R1$-type measurement is highly affected by tidal loading, rainfall, temperature, and so on, hence an anomaly of long duration is not easily observed. In the $R2$-type measurement a number of data are required to obtain a high accuracy, resulting in its inability as a detector of anomalies of short duration.

7. Problems and Perspective

Many precursory changes have been reported in the geomagnetic field (G), the electric field (E), and the resistivity $(R1, R2)$ which have all been analyzed from a statistical viewpoint in the preceding section. It seems to be claimed then that electric and magnetic methods should be very promising means of predicting an earthquake. However, there are many problems to be solved. Since only a slight change in the geomagnetic field is expected prior to an earthquake, careful attention should be paid to an effective method of its detection. In this connection, continuous measurements of the total intensity and preferably three components at an array of measuring sites should be most effective. A station separation of less than 50 km or so is recommended to avoid the effect of spatially non-uniform secular variation of the geomagnetic field.

The most serious problem is the magnetic noise arising from artificial electric currents, particularly currents leaking from DC-operated electric railways. Therefore, the geomagnetic method is of no use in a highly industrialized region. Another problem, though less serious, is that some of the destructive earthquakes occur in trench or trough regions below oceans. It is now possible, however, to make observations of the geomagnetic field on the sea-floor; and sea-floor magnetometers combined with an observation system on land may provide important information on an impending earthquake of large magnitude in subduction zones.

The natural electric-field method is also seriously affected by artificial electric currents in an industrialized region. Another problem in this method is the noise due to precipitation, temperature, and the like. However, this difficulty can be overcome by a carefully arranged measuring system. For example, an apparent vairation due to alteration in contact condition between a certain electrode and the ground should be distinguished from other signals if some pairs of electrodes, of course in the same direction, are simultaneously in operation. As in the Kamchatka region, simultaneous observations at several stations operating as mentioned above will greatly increase the reliability of an observed change.

As for the resistivity measurement of $R2$ type, perturbation of the electric current driven into the earth through a pair of current electrodes should be utilized besides apparent resistivity estimates. If a resistivity anomaly appears in the crust, it will perturb the current and the direction in which the current flows at a certain receiver site will also change as a result of the perturbation. Thus, monitoring of the current direction at some receiver sites will provide additional information on a resistivity change. The method based on the principle of electromagnetic induction should be further studied particularly for short-period (shorter than a few minutes) variations, even though the use of such a short period is limited in an industrialized area because of artificial noises. If an array of stations is established, noise reduction in analysis will greatly improve and slight changes may become detectable as in the artificial current method.

One of the characteristics of electric and magnetic methods for earthquake prediction is a wide range of precursor time: tens of minutes to several years as was shown in Section 6. This is very important since its implication is that electric and magnetic methods are useful at medium-, short-, and imminent-stages of prediction as was the case for the successful predictions of some large earthquakes in China. A practical prediction of moderate to large earthquakes may not be impossible, if the restuls of electric and magnetic measurements are combined with other factors such as seismic velocity ratio (V_P/V_S), land deformation including tilt and strain, seismic activity, gravity, underground-water level and its chemical composition, and so on. In this sense, electric and magnetic measurements with highly sensitive instruments at a number of stations should be encouraged over an area where a seismic event is expected in the near future. In addition, basic studies in seismically active regions should be made continuously to obtain useful information on changes in electric and magnetic parameters.

Acknowledgements

I express my sincere thanks to Prof. T. Rikitake for reading the manuscript and for his encouragement throughout the work. I am also thankful to Dr. Y. Yamazaki for his figures used in this article and to Mrs. M. Takamori for typing the manuscript.

REFERENCES

ABDULLABEKOV, K. N., L. S. BEZUGLAYA, V. P. GOLOVKOV, and YU. P. SKO-
VORODKIN, On the possibility of using magnetic methods to study tectonic
processes, *Tectonophysics*, **14**, 257–262, 1972.

AL'TGAUZEN, N. M. and O. M. BARSUKOV, On temporal variations of electrical
conductivity, in *Physical Bases of Seeking Methods of Predicting Earthquakes*,
edited by M. A. Sadovsky, Acad. Sci. U.S.S.R. (English edition, U.S.
Geological Survey, 1972).

ANDERSON, D. L. and J. H. WHITCOMB, The dilatancy-diffusion model of
earthquake prediction, in *Proceedings of the Conference on Tectonic Problems
on the San Andreas Fault System*, edited by R. L. Kovach and A. Nur,
Stanford Univ. Publ., *Geol. Sci.*, **13**, 417–426, 1973.

BARSUKOV, O. M., Relationship between the electrical resistivity of rocks and tectonic causes, *Izv. Acad. Sci. U.S.S.R. (Phys. Solid Earth)*, **5**, 55–59, 1970 (English edition).

BARSUKOV, O. M., Variations of electric resistivity of mountain rocks connected with tectonic causes, *Tectonophysics*, **14**, 273–277, 1972a.

BARSUKOV, O. M., The hypothesis of a relationship between electrical resistivity and the strength of sedimentary rocks, *Izv. Acad. Sci. U.S.S.R. (Phys. Solid Earth)*, **7**, 578–585, 1972b (English edition).

BARSUKOV, O. M., The search for electrical criteria for predicting earthquakes, in *Experimental Seismology*, edited by M. A. Sadovsky, Science Press, Moscow (English edition, U.S. Geological Survey, 1973).

BARSUKOV, O. M., Variations in the electrical resistivity of rocks and earthquakes, in *Earthquake Precursors*, edited by M. A. Sadovsky, I. L. Nersesov, and L. A. Latynina, Acad. Sci. U.S.S.R. (English edition, U.S. Geological Survey, 1974).

BARSUKOV, O. M. and O. N. SOROKIN, Variations in apparent resistivity of rocks in the seismically active Garm region, *Izv. Acad. Sci. U.S.S.R. (Phys. Solid Earth)*, **8**, 685–687, 1973 (English edition).

BARSUKOV, O. M., P. D. KRASNIUK, N. A. LISTOV, and O. M. SOROKIN, Approximate estimation of the size of the zone of earthquake buildup from measurement of the electrical resistivity of a mountain massif, in *Earthquake Precursors*, edited by M. A. Sadovsky, I. L. Nersesov, and L. A. Latynina, Acad. Sci. U.S.S.R. (English edition, U.S. Geological Survey, 1974).

BRACE, W. F., Laboratory studies of stick-slip and their application to earthquakes, *Tectonophysics*, **14**, 189–200, 1972.

BRACE, W. F., Dilatancy-related electrical resistivity changes in rocks, *Pure Appl. Geophys.*, **113**, 207–217, 1975.

BRACE, W. F., Recent laboratory studies of earthquake mechanics and prediction, *J. Phys. Earth.* **25**, Suppl., S185–S202, 1977.

BRACE, W. F. and A. S. ORANGE, Electrical resistivity changes in saturated rock under stress, *Science*, **153**, 1525–1526, 1966.

BRACE, W. F. and A. S. ORANGE, Electrical resistivity changes in saturated rocks during fracture and frictional sliding, *J. Geophys. Res.*, **73**, 1433–1445, 1968a.

BRACE, W. F. and A. S. ORANGE, Further studies of the effects of pressure on electrical resistivity of rocks, *J. Geophys. Res.*, **73**, 5407–5420, 1968b.

BRACE, W. F., A. S. ORANGE, and T. R. MADDEN, The effect of pressure on the electrical resistivity of water-saturated crystalline rocks, *J. Geophys. Res.*, **70**, 5669–5678, 1965.

BRACE, W. F., B. W. PAULDING, Jr., and C. SCHOLZ, Dilatancy in the fracture of crystalline rocks, *J. Geophys. Res.*, **71**, 3939–3953, 1966.

BRACE, W. F., E. SILVER, K. HADLEY, and C. GOETZE, Cracks and pores: a closer look, *Science*, **178**, 162–164, 1972.

BRADY, B. T., Theory of earthquakes, 1. A scale independent theory of rock failure, *Pure Appl. Geophys.*, **112**, 701–725, 1974.

BREINER, S., Piezomagnetic effect at the time of local earthquakes, *Nature*, **202**, 790–791, 1964.

BREINER, S. and R. KOVACH, Local geomagnetic events associated with displacements on the San Andreas fault, *Science*, **158**, 116–118, 1967.

BREINER, S. and R. KOVACH, Local magnetic events associated with displacement along the San Andreas fault (California), *Tectonophysics*, **6**, 69–73, 1968.

BUFE, C. G., W. H. BAKUN, and D. TOCHER, Geophysical studies in the San Andreas fault zone at the Stone Canyon Observatory, California, in *Proceedings of the Conference on Tectonic Problems on the San Andreas Fault System*, edited by R. L. Kovach and A. Nur, Stanford Univ. Publ., *Geol. Sci.*, **13**, 86–93, 1973.

CARMICHAEL, R. S., Depth calculation of piezomagnetic effect for earthquake prediction, *Earth Planet. Sci. Lett.*, **36**, 309–316, 1977.

CHI-YANG, T., Bases for the prediction of the Lungling earthquake and the temporal and spatial characteristics of precursors, in *Proceedings on the Chinese Earthquake Prediction by the 1977 Delegation of the Seismological Society of Japan*, pp. 13–32, Seismol. Soc. Japan, 1978 (in Japanese).

COE, R. S., Earthquake prediction program in the People's Republic of China, *Trans. Am. Geophys. Union*, **52**, 940–943, 1971.

CORWIN, R. F. and H. F. MORRISON, Self-potential variations preceding earthquakes in central California, *Geophys. Res. Lett.*, **4**, 171–174, 1977.

DAVIS, P. M., The piezomagnetic computation of magnetic anomalies due to ground loading by a man-made lake, *Pure Appl. Geophys.*, **112**, 811–819, 1974.

DAVIS, P. M., The computed piezomagnetic anomaly field for Kilawea volcano, Hawaii, *J. Geomag. Geoelectr.*, **28**, 113–122, 1976.

DAVIS, P. M. and F. D. STACEY, Geomagnetic anomalies caused by a man-made lake, *Nature*, **240**, 348–349, 1972.

DERR, J. S., Earthquake lights: a review of observations and present theories, *Bull. Seismol. Soc. Am.*, **63**, 2177–2187, 1973.

DMOWSKA, R., Electromechanical phenomena associated with earthquakes, *Geophys. Surv.*, **3**, 157–174, 1977.

FEDOTOV, S. A., N. A. DOLBILKINA, V. N. MOROZOV, V. I. MYACHKIN, V. B. PREOBRAZENSKY, and G. A. SOBOLEV, Investigation on earthquake prediction in Kamchatka, *Tectonophysics*, **9**, 249–258, 1970.

FEDOTOV, S. A., G. A. SOBOLEV, S. A. BOLDYREV, A. A. GUSEV, A. M. KONDRATENKO, O. V. POTAPOVA, L. B. SLAVINA, V. D. THEOPHYLAKTOV, A. A.

KHRAMOV, and V. A. SHIROKOV, Long- and short-term earthquake prediction in Kamchatka, *Tectonophysics*, **37**, 305–321, 1977.

FINKELSTEIN, D. and J. POWELL, Earthquake lightning, *Nature*, **228**, 759–760, 1970.

FINKELSTEIN, D., R. D., HILL, and J. R. POWELL, The piezoelectric theory of earthquake lightning, *J. Geophys. Res.*, **78**, 992–993, 1973.

FITTERMAN, D. V. and T. R. MADDEN, Resistivity observations during creep events at Melendy Ranch, California, *J. Geophys. Res.*, **82**, 5401–5408, 1977.

FUJITA, N., The magnetic disturbances accompanying the Niigata earthquake, *J. Geod. Soc. Japan*, **11**, 8–25, 1965 (in Japanese with English abstract).

FUJITA, N., Secular change of the geomagnetic total force in Japan for 1970. 0, *J. Geomag. Geoelectr.*, **25**, 181–194, 1973.

FUKUTOMI, T., Report of the strong Izu earthquake of March 21, 1934, *Bull. Earthq. Res. Inst., Univ. Tokyo*, **12**, 527–538, 1934 (in Japanese with English abstract).

HADLEY, K., Laboratory investigation of dilatancy and motion on fault surface at low confining pressure, in *Proceedings of the Conference on Tectonic Problems on the San Andreas Fault System*, edited by R. L. Kovach and A. Nur, Stanford Univ. Publ., *Geol. Sci.*, **13**, 427–435, 1973.

HASBROUCK, W. P. and J. H. ALLEN, Quasi-static magnetic field changes associated with the CANNIKIN nuclear explosion, *Bull. Seismol. Soc. Am.*, **62**, 1479–1487, 1972.

HASEGAWA, A., T. HASEGAWA, and S. HORI, Premonitory variation in seismic velocity related to the Southwestern Akita earthquake of 1970, *J. Phys. Earth*, **23**, 189–203, 1975.

HONKURA, Y., Partial melting and electrical conductivity anomalies beneath the Japan and Philippine Seas, *Phys. Earth Planet. Inter.*, **10**, 128–134, 1975.

HONKURA, Y., Perturbation of the electric current by a resistivity anomaly and its application to earthquake prediction, *J. Geomag. Geoelectr.*, **28**, 47–57, 1976a.

HONKURA, Y., Perturbation of the induced electric current by a resistivity anomaly in the crust and its application to earthquake prediction, Proc. Conductivity Anomaly Symp. 1976, Geol. Survey Japan, 175–181, 1976b (in Japanese).

HONKURA, Y., Electrical conductivity anomalies in the earth, *Geophys. Surv.*, **3**, 225–253, 1977a.

HONKURA, Y., An interpretation of the anomalous change in $\Delta Z/\Delta H$-value observed at Kakioka in association with the Kanto earthquake, Proc. Conductivity Anomaly Symp. 1977, 177–187, 1977b (in Japanese).

HONKURA, Y., On a relation between anomalies in the geomagnetic and telluric fields observed at Nakaizu and the 1978 Izu-Oshima Kinkai earthquake,

Bull. Earthq. Res. Inst., Univ. Tokyo, **53**, 931–937, 1978.

HONKURA, Y. and S. KOYAMA, On a problem in earthquake prediction research based on the survey of the geomagnetic total intensity, Proc. Conductivity Anomaly Symp. 1976, Geol. Survey Japan, 145–150, 1976 (in Japanese).

HONKURA, Y. and S. KOYAMA, Observations of short-period geomagnetic variations at Nakaizu (1), *Bull. Earthq. Res. Inst., Univ. Tokyo*, **53**, 925–930, 1978.

HONKURA, Y., E. R. NIBLETT, and R. D. KURTZ, Changes in magnetic and telluric fields in a seismically active region of eastern Canada: preliminary results of earthquake prediction studies, *Tectonophysics*, **34**, 219–230, 1976a.

HONKURA, Y., T. YOSHINO, and T. YUKUTAKE, Results of the total intensity surveys before and after the 1970 Southeastern Akita earthquake, Proc. Conductivity Anomaly Symp. 1976, Geol. Survey Japan, 139–144, 1976b (in Japanese).

HONKURA, Y., S. KOYAMA, T. YOSHINO, and T. YUKUTAKE, Results of the total intensity surveys in the Tokai District, Proc. Conductivity Anomaly Symp. 1977, Kakioka Mag. Obs., 137–143, 1977 (in Japanese).

İSPIR, Y. and O. UYAR, An attempt in determining the seismo-magnetic effect in NW Turkey, *J. Geomag. Geoelectr.*, **23**, 295–305, 1971.

İSPIR, Y., O. UYAR, Y. GÜNGÖRMÜS, N. ORBAY, and B. ÇAĞLAYAN, Some results from studies on tectonomagnetic effect in NW Turkey, *J. Geomag. Geoelectr.*, **28**, 123–135, 1976.

JOHNSTON, M. J. S., Preliminary results from a search for regional tectonomagnetic effects in California and Western Nevada, *Tectonophysics*, **23**, 267–275, 1974.

JOHNSTON, M. J. S., B. E. SMITH, and R. MULLER, Tectonomagnetic experiments and observations in western U.S.A., *J. Geomag. Geoelectr.*, **28**, 85–97, 1976.

JOHNSTON, M. J. S., G. D. MYREN, N. W. O'HARA, and J. H. RODGERS, A possible seismomagnetic observation on the Garlock fault, California, *Bull. Seismol. Soc. Am.*, **65**, 1129–1132, 1975.

JOHNSTON, M. J. S., B. E. SMITH, J. R. JOHNSTON, and F. J. WILLIAMS, A search for tectonomagnetic effects in California and Western Nevada, in *Proceedings of the Conference on Tectonic Problems on the San Andreas Fault System*, edited by R. L. Kovach and A. Nur, Stanford Univ. Publ., *Geol. Sci.*, **13**, 225–238, 1973.

KEAN, W. F., R. DAY, M. FULLER, and V. A. SCHMIDT, The effect of uniaxial compression on the initial susceptibility of rocks as a function of grain size and composition of their constituent titanomagnetites, *J. Geophys. Res.*, **81**, 861–872, 1976.

KISSLINGER, C., Process during the Matsushiro, Japan, earthquake swarm as revealed by leveling, gravity, and spring-flow observations, *Geology*, **3**, 57–

62, 1975.

KONDO, G., The variation of the atmospheric electric field at the time of earthquake, *Mem. Kakioka Mag. Obs.*, **13**, 11–23, 1968.

KOYAMA, S. and Y. HONKURA, Observations of electric self-potential at Naka-izu (1), *Bull. Earthq. Res. Inst., Univ. Tokyo*, **53**, 939–942, 1978.

KOZLOV, A. N., A. N. PUSHKOV, R. Sh. RAKHMATULLIN, and Yu. P. SKOVORO-KIN, Magnetic effects during explosions in rocks, *Izv. Acad. Sci. U.S.S.R., Earth Phys.*, 66–71, 1974 (English edition).

KUBOKI, T. and T. YOSHIMATSU, Secular change of transfer functions at Ka-kioka, Proc. Conductivity Anomaly Symp. 1973, Geograph. Survey Inst., 99–103, 1973 (in Japanese).

KURTZ, R. D. and E. R. NIBLETT, Time-dependence of magnetotelluric fields in a tectonically active region in eastern Canada, *J. Geomag. Geoelectr.*, **30**, 561–577, 1978.

MARTIN III, R. J. and M. WYSS, Magnetism of rocks and volumetric strain in uniaxial failure tests, *Pure Appl. Geophys.*, **113**, 51–61, 1975.

MARTIN III, R. J., R. E. HABERMANN, and M. WYSS, The effect of stress cycling and inelastic volumetric strain on remanent magnetization, *J. Geophys. Res.*, **83**, 3485–3496, 1978.

MATSUDA, T., Active faults in the Amagisan area, Izu Peninsula, *Bull. Earthq. Res. Inst., Univ. Tokyo*, **52**, 223–234, 1977 (in Japanese with English abstract).

MAZZELLA, A. and H. F. MORRISON, Electrical resistivity variations associated with earthquakes on the San Andreas fault, *Science*, **185**, 855–857, 1974.

MIYAKOSHI, J., Secular variation of Parkinson vectors in a seismically active region of Middle Asia, *J. Fac. Gen. Educ., Tottori Univ.*, **8**, 209–218, 1975.

MIZUTANI, H. and T. ISHIDO, A new interpretation of magnetic field variation associated with the Matsushiro earthquakes, *J. Geomag. Geoelectr.*, **28**, 179–188, 1976.

MIZUTANI, H., T. ISHIDO, Y. YOKOKURA, and S. OHNISHI, Electrokinetic phenomena associated with earthquakes, *Geophys. Res. Lett.*, **3**, 365–368, 1976.

MOGI, K., Rock failure and earthquake prediction, *J. Soc. Mater. Sci. Japan*, **23**, 320–331, 1974 (in Japanese).

MORI, T. and T. YOSHINO, Local difference in variations of the geomagnetic total intensity in Japan, *Bull. Earthq. Res. Inst., Univ. Tokyo*, **48**, 893–922, 1970.

MORI, T., Y. MIZUNO, and K. HASEGAWA, On the results of geomagnetic observations for earthquake prediction research in the eastern part of Hokkaido, *Mem. Kakioka Mag. Obs.*, **16**, 59–68, 1974 (in Japanese with English abstract).

MORROW, C. A., W. F. BRACE, and E. CARTER, Dramatic electrical changes in porous rocks at low stress, *Trans. Am. Geophy. Union*, **58**, 1235, 1977 (abstract).

NAGATA, T., Changes in the electric potential difference in the vicinity of the Shikano Fault, *Bull. Earthq. Res. Inst., Univ. Tokyo*, **22**, 72–82, 1944 (in Japanese).

NAGATA, T., Tectonomagnetism, *Int. Assoc. Geomag. Aeron. Bull.*, **27**, 12–43, 1969.

NAGATA, T., Effects of a uniaxial compression on remanent magnetizations of igneous rocks, *Pure Appl. Geophys.*, **78**, 100–109, 1970a.

NAGATA, T., Basic magnetic properties of rocks under the effects of mechanical stresses, *Tectonophysics*, **9**, 167–195, 1970b.

NAGATA, T., Tectonomagnetism in relation to seismic activities of the earth's crust: seismo-magnetic effect in a possible association with the Niigata earthquake in 1964, *J. Geomag. Geoelectr.*, **28**, 99–111, 1976.

NIBLETT, E. R. and Y. HONKURA, Time-dependence of electromagnetic transfer functions and their association with tectonic activity, *J. Geomag. Geoelectr.*, **31**, 1979 (in press).

NORITOMI, K., Geoelectric and geomagnetic observations and phenomena associated with earthquake in China, Proceedings on the Chinese earthquake prediction by the 1977 delegation of the Seismological Society of Japan, *Seismol. Soc. Japan*, 57–87, 1978 (in Japanese).

NUR, A., Dilatancy, pore fluids, and premonitory variations of t_S/t_P travel times, *Bull. Seismol. Soc. Am.*, **62**, 1217–1222, 1972.

NUR, A., The Matsushiro earthquake swarm: a confirmation of the dilatancy-fluid flow model, *Geology*, **2**, 217–221, 1974.

OIKE, K., Precursors and predictions of large earthquakes in China, Proceedings on the Chinese earthquake prediction by the 1977 delegation of the Seismological Society of Japan, *Seismol. Soc. Japan*, 57–87, 1978 (in Japanese with English abstract).

OIKE, K., R. SHICHI, and T. ASADA, Earthquake prediction in the People's Republic of China, *Zisin 2*, **28**, 75–94, 1975 (in Japanese).

PHILLIPS, R. J., I. K. REDDY, and J. H. WHITCOMB, Investigation into the relationship between resistivity changes and tectonic activity in Southern California, *Trans. Am. Geophys. Union*, **58**, 731, 1977 (abstract).

PIERCE, E. T., Atmospheric electricity and earthquake prediction, *Geophys. Res. Lett.*, **3**, 185–188, 1976.

POZZI, J. P., Effects of stresses on magnetic properties of volcanic rocks, *Phys. Earth Planet. Inter.*, **14**, 77–85, 1977.

PRESS, F., M. BULLOCK, R. M. HAMILTON, W. F. BRACE, C. KISSLINGER, M. G. BONILLA, C. R. ALLEN, L. R. SYKES, C. B. RALEIGH, L. KNOPOFF, R. W. CLOUGH, and R. HOFHEINZ, Jr., Earthquake research in China, *Trans.*

Am. Geophy. Union, **56**, 838–881, 1975.

QUIST, A. S. and W. L. MARSHALL, The electrical conductances of some alkali metal halides in aqueous solutions from 0 to 800 °C and pressure to 4,000 bars, *J. Phys. Chem.*, **73**, 978–985, 1969.

QUIST, A. S., W. L. MARSHALL, E. V. FRANK, and W. VON OSTEN, A reference solution for electrical conductance measurements to 800° and 12,00 bars. Aqueous 0.01 demal potassium chloride, *J. Phys. Chem.*, **74**, 2241–2243, 1970.

RALEIGH, B., G. BENNETT, H. CRAIG, T. HANKS, P. MOLNAR, A. NUR, J. SAVAGE, C. SCHOLZ, R. TURNER, and F. WU, Prediction of the Haichen earthquake, *Trans. Am. Geophys. Union*, **58**, 236–272, 1977.

REDDY, I. K., R. J. PHILLIPS, J. H. WHITCOMB, D. M. COLE, and R. A. TAYLOR, Monitoring of time-dependent resistivity by magnetotellurics, *J. Geomag. Geoelectr.*, **28**, 165–178, 1976.

RIKITAKE, T., *Electromagnetism and the Earth's Interior*, 308 pp., Elsevier, Amsterdam, 1966a.

RIKITAKE, T., Elimination of non-local changes from total intensity values of the geomagnetic field, *Bull. Earthq. Res. Inst., Univ. Tokyo*, **44**, 1041–1070, 1966b.

RIKITAKE, T., Geomagnetism and earthquake prediction, *Tectonophysics*, **6**, 59–68, 1968.

RIKITAKE, T., Prediction of magnitude and occurrence time of earthquakes, *Bull. Earthq. Res. Inst., Univ. Tokyo*, **47**, 107–128, 1969 (in Japanese with English abstract).

RIKITAKE, T., Dilatancy model and empirical formulas for an earthquake area, *Pure Appl. Geophys.*, **113**, 141–147, 1975a.

RIKITAKE, T., Earthquake precursors, *Bull. Seismol. Soc. Am.*, **65**, 1133–1162, 1975b.

RIKITAKE, T., *Earthquake Prediction*, 357 pp., Elsevier, Amsterdam, 1976a.

RIKITAKE, T., Crustal dilatancy and geomagnetic variations of short period, *J. Geomag. Geoelectr.*, **28**, 145–156, 1976b.

RIKITAKE, T., Changes in the direction of magnetic vector of short-period geomagnetic variations before the 1972 Sitka, Alaska, earthquake, *J. Geomag. Geoelectr.*, **31**, 441–448, 1978.

RIKITAKE, T. and J. YAMADA, Earth-current observations at Tomioka, Tokushima Prefecture after the Nankai earthquake, *Spec. Bull. Earthq. Res. Inst., Univ. Tokyo*, **5**, 186–188, 1947 (in Japanese).

RIKITAKE, T. and Y. YAMAZAKI, Electrical conductivity of strained rocks (the fifth paper)—residual strains associated with large earthquakes as observed by a resistivity variometer—, *Bull. Earthq. Res. Inst., Univ. Tokyo*, **47**, 99–105, 1969.

RIKITAKE, T. and Y. YAMAZAKI, Strain steps as observed by a resistivity variometer, *Tectonophysics*, **9**, 197–203, 1970.

RIKITAKE, T. and Y. YAMAZAKI, Resistivity changes as a precursor of earthquake, *J. Geomag. Geoelectr.*, **28**, 497–505, 1977a.

RIKITAKE, T. and Y. YAMAZAKI, Precursory and coseismic changes in ground resistivity, *J. Phys. Earth*, **25**, Suppl., S161–S173, 1977b.

RIKITAKE, T., Y. YAMAZAKI, Y. HAGIWARA, K. KAWADA, M. SAWADA, Y. SASAI, T. WATANABE, and Y. SANZAI, Geomagnetic and geoelectric studies of the Matsushiro earthquake swarm, 1, *Bull. Earthq. Res. Inst., Univ. Tokyo*, **44**, 363–408, 1966a.

RIKITAKE, T., Y. YAMAZAKI, Y. HAGIWARA, K. KAWADA, M. SAWADA, Y. SASAI, and T. YOSHINO, Geomagnetic and geoelectric studies of the Matsushiro earthquake swarm, 2, *Bull. Earthq. Res. Inst., Univ. Tokyo*, **44**, 409–418, 1966b.

RIKITAKE, T., T. YUKUTAKE, Y. YAMAZAKI, M. SAWADA, Y. SASAI, Y. HAGIWARA, K. KAWADA, T. YOSHINO, and T. SHIMOMURA, Geomagnetic and geoelectric studies of the Matsushiro earthquake swarm, 3, *Bull. Earthq. Res. Inst., Univ. Tokyo*, **44**, 1335–1370, 1966c.

RIKITAKE, T., Y. YAMAZAKI, M. SAWADA, Y. SASAI, T. YOSHINO, S. UZAWA, and T. SHIMOMURA, Geomagnetic and geoelectric studies of the Matsushiro earthquake swarm, 4, *Bull. Earthq. Res. Inst., Univ. Tokyo*, **44**, 1735–1758, 1966d.

RIKITAKE, T., Y. YAMAZAKI, M. SAWADA, Y. SASAI, T. YOSHINO, S. UZAWA, and T. SHIMOMURA, Geomagnetic and geoelectric studies of the Matsushiro earthquake swarm, 5, *Bull. Earthq. Res. Inst., Univ. Tokyo*, **45**, 395–416, 1967a.

RIKITAKE, T., T. YUKUTAKE, M. SAWADA, Y. SASAI, T. WATANABE, and H. TACHINAKA, Geomagnetic and geoelectric studies of the Matsushiro earthquake swarm, 6, *Bull. Earthq. Res. Inst., Univ. Tokyo*, **45**, 919–944, 1967b.

SADOVSKY, M. A. and I. L. NERSESOV, Forecasts of earthquakes on the basis of complex geophysical features, *Tectonophysics*, **23**, 247–255, 1974.

SADOVSKY, M. A., I. L. NERSESOV, S. K. NIGMATULLAEV, L. A. LATYNINA, A. A. LUKK, A. N. SEMENOV, I. G. SIMBIREVA, and V. I. ULOMOV, The process preceding strong earthquakes in some regions of Middle Asia, *Tectonophysics*, **14**, 295–307, 1972.

SASAI, Y. and Y. ISHIKAWA, Repeated magnetic survey of the total force intensity in the Boso Peninsula: 1968–1976, *Bull. Earthq. Res. Inst., Univ. Tokyo*, **51**, 83–113, 1976 (in Japanese with English abstract).

SASAI, Y. and Y. ISHIKAWA, Changes in the geomagnetic total force intensity associated with the anomalous crustal activity in the eastern part of the Izu Peninsula (1), *Bull. Earthq. Res. Inst., Univ. Tokyo*, **52**, 173–190, 1977 (in Japanese with English abstract).

SCHOLZ, C. H., L. R. SYKES, and Y. P. AGGARWAL, Earthquake prediction:

a physical basis, *Science*, **181**, 803–810, 1973.

SEISMOLOGICAL DELEGATION OF CHINA, *Proceedings of the Lectures*, 83 pp., Seismol. Soc. Japan, 1976 (in Japanese).

SETO, T., Observation of geomagnetic changes associated with the explosion experiment on Izu-Oshima Island, Proc. Conductivity Anomaly Symp. 1977, Kakioka Mag. Obs., 151–153, 1977 (in Japanese).

SHAMSI, S. and F. D. STACEY, Dislocation models and seismomagnetic calculations for California 1906 and Alaska 1964 earthquakes, *Bull. Seismol. Soc. Am.*, **59**, 1435–1448, 1969.

SHIRAKI, M., Monitoring of CA transfer functions at Kakioka, Proc. Conductivity Anomaly Symp. 1977, Kakioka Mag. Obs., 163–168, 1977 (in Japanese).

SHIRAKI, M. and K. YANAGIHARA, Transfer functions at Kakioka (part II): reevaluation of their secular changes, *Mem. Kakioka Mag. Obs.*, **17**, 19–25, 1977 (in Japanese with English abstract).

SKOVORODKIN, Yu. P., L. S. BEZUGLAYA, and V. N. VADKOVSKIY, Magnetic investigations in an epicentral zone, in *Experimental Seismology*, edited by M. A. Sadovsky, Science Press, Moscow (English edition, U.S. Geological Survey, 1973).

SMITH, B. E. and M. J. S. JOHNSTON, A tectonomagnetic effect observed before a magnitude 5.2 earthquake near Hollister, California, *J. Geophys. Res.*, **81**, 3556–3560, 1976.

SMITH, B. E., M. J. S. JOHNSTON, and G. D. MYREN, Results from a differential magnetometer array along the San Andreas fault in central California, *Trans. Am. Geophys. Union*, **56**, 1113, 1974 (abstract).

SOBOLEV, G. A., Prospects for routine prediction of earthquakes on the basis of electrotelluric observations, in *Earthquake Precursors*, edited by M. A. Sadovsky, I. L. Nersesov, and L. A. Latynina, Acad. Sci. U.S.S.R. (English edition, U.S. Geological Survey, 1974).

SOBOLEV, G. A., Application of electric method to the tentative short-term forecast of Kamchatka earthquakes, *Pure Appl. Geophys.*, **113**, 229–235, 1975a.

SOBOLEV, G. A., The study of precursors of failure under biaxial compression, *Pure Appl. Geophys.*, **113**, 45–49, 1975b.

SOBOLEV, G. A. and V. N. MOROZOV, Local disturbances of the electrical field on Kamchatka and their relation to earthquakes, in *Physical Bases of Seeking Methods of Predicting Earthquakes*, edited by M. A. Sadovsky, Acad. Sci. U.S.S.R. (English edition, U.S. Geological Survey, 1972).

SOBOLEV, G. A., V. N. MOROZOV, and N. I. MIGUNOV, The electrotelluric field and a strong earthquake on Kamchatka, *Izv. Acad. Sci. U.S.S.R. (Phys. Solid Earth)*, **7**, 108–113, 1972 (English edition).

STACEY, F. D., Seismo-magnetic effect and the possibility of forecasting earthquakes, *Nature*, **200**, 1083–1085, 1963.

STACEY, F. D., The seismomagnetic effect, *Pure Appl. Geophys.*, **58**, 5–22, 1964.

STACEY, F. D. and M. J. S. JOHNSTON, Theory of the piezomagnetic effect in titanomagnetite-bearing rocks, *Pure Appl. Geophys.*, **97**, 146–155, 1972.

STUART, W. D., Diffusionless dilatancy model for earthquake precursors, *Geophys. Res. Lett.*, **1**, 261–264, 1974.

STUART, W. D. and M. J. S. JOHNSTON, Intrusive origin of the Matsushiro earthquake swarm, *Geology*, **3**, 63–67, 1975.

SUMITOMO, N., Geomagnetism in relation to tectonic activities of the earth's crust in Japan, *J. Phys. Earth*, **25**, Suppl., S147–S160, 1977.

TALWANI, P. and R. L. KOVACH, Geomagnetic observations and fault creep in California, *Tectonophysics*, **14**, 245–256, 1972.

TAZIMA, M., Accuracy of recent magnetic survey and a locally anomalous behavior of the geomagnetic secular variation in Japan, *Bull. Geograph. Survey Inst.*, **13**, 1–78, 1968.

TAZIMA, M., H. MIZUNO, and M. TANAKA, Geomagnetic secular change in Japan, *J. Geomag. Geoelectr.*, **28**, 69–84, 1976a.

TAZIMA, M., T. SETO, and M. YOSHIDA, Observation of geomagnetic changes associated with the explosion experiment on Izu-Oshima Island, Proc. Conductivity Anomaly Symp. 1976, Geol. Survey Japan, 155–160, 1976b (in Japanese).

TERADA, T., On luminous phenomena accompanying earthquakes, *Bull. Earthq. Res. Inst., Univ. Tokyo*, **9**, 225–255, 1931.

TUCK, G. J., F. D. STACEY, and J. STARKEY, A search for piezoelectric effect in quartz-bearing rocks, *Tectonophysics*, **39**, T7–T11, 1977.

UNDZENDOV, B. A. and V. A. SHAPIRO, Seismomagnetic effect in a magnetite deposit, *Izv. Acad. Sci. U.S.S.R., Earth Phys.*, 121–126, 1967 (English edition).

WAFF, H. S., Theoretical considerations of electrical conductivity in a partially molten mantle and implications for geothermometry, *J. Geophys. Res.*, **79**, 4003–4010, 1974.

WANG, C. Y., R. E. GOODMAN, P. N. SUNDARAM, and H. F. MORRISON, Electrical resistivity of granite in frictional sliding: application to earthquake prediction, *Geophys. Res. Lett.*, **2**, 525–528, 1975.

WEN-YAO, X., Q. KUI, and W. SHI-MING, On the short period geomagnetic variation anomaly of the eastern Kansu Province, *Acta Geophysica Sinica*, **21**, 218–224, 1978 (in Chinese with English abstract).

WHITCOMB, J. H., Earthquake prediction-related research at the seismological laboratory, California Institute of Technology, 1974–1976, *J. Phys. Earth*, **25**, Suppl., S1–S11, 1977.

WYSS, M., A search for precursors to the Sitka, 1972, earthquake: sea level, magnetic field, and P-residuals, *Pure Appl. Geophys.*, **113**, 297–309, 1975.

WYSS, M. and R. J. MARTIN III, Tectonomagnetism and magnetic changes in rocks prior to failure, *Pure Appl. Geophys.*, **116**, 1978 (in press).

YAMASHITA, H. and I. YOKOYAMA, Annual measurements of geomagnetic total intensity in the eastern part of Hokkaido (part 1), *Geophys. Bull. Hokkaido Univ.*, **33**, 31–39, 1975 (in Japanese with English abstract).

YAMAZAKI, Y., Electrical conductivity of strained rocks (the first paper) — laboratory experiments on sedimentary rocks—, *Bull. Earthq. Res. Inst., Univ. Tokyo*, **43**, 783–802, 1965.

YAMAZAKI, Y., Electrical conductivity of strained rocks (the second paper)— further experiments on sedimentary rocks—, *Bull. Earthq. Res. Inst., Univ. Tokyo*, **44**, 1553–1570, 1966.

YAMAZAKI, Y., Electrical conductivity of strained rocks (the third paper)— a resistivity variometer—, *Bull. Earthq. Res. Inst., Univ. Tokyo*, **45**, 849–860, 1967.

YAMAZAKI, Y., Electrical conductivity of strained rocks (the fourth paper)— improvement of the resistivity variometer—, *Bull. Earthq. Res. Inst., Univ. Tokyo*, **46**, 957–964, 1968.

YAMAZAKI, Y., Electrical resistivity of strained rocks (construction of a resistivity variometer), *Zisin 2*, **26**, 55–66, 1973 (in Japanese with English abstract).

YAMAZAKI, Y., Coseismic resistivity steps, *Tectonophysics*, **22**, 159–171, 1974.

YAMAZAKI, Y., Precursory and coseismic resistivity changes, *Pure Appl. Geophys.*, **113**, 219–227, 1975a.

YAMAZAKI, Y., A detectability of the resistivity variometer system at Aburatsubo, *Zisin 2*, **28**, 31–40, 1975b (in Japanese with English abstract).

YAMAZAKI, Y., Tectonoelectricity, *Geophys. Surv.*, **3**, 123–142, 1977.

YAMAZAKI, Y. and T. RIKITAKE, Local anomalous changes in the geomagnetic field at Matsushiro, *Bull. Earthq. Res. Inst., Univ. Tokyo*, **48**, 637–643, 1970.

YANAGIHARA, K., Secular variation of the electrical conductivity anomaly in the central part of Japan, *Mem. Kakioka Mag. Obs.*, **15**, 1–11, 1972.

YANAGIHARA, K. and T. NAGANO, Time change of transfer function in the Central Japan Anomaly, *J. Geomag. Geoelectr.*, **28**, 157–163, 1976.

YASUI, Y., A summary of studies on luminous phenomena accompanied with earthquakes, *Mem. Kakioka Mag. Obs.*, **15**, 127–139, 1973.

YOSHIMATSU, T., Results of geomagnetic routine observations and earthquakes (II), *Mem. Kakioka Mag. Obs.*, **11**, 71–83, 1963 (in Japanese with English abstract).

YOSHIMATSU, T., Results of geomagnetic routine observations and earthquakes (III), *Mem. Kakioka Mag. Obs.*, **11**, 55–68, 1964 (in Japanese with English abstract).

YOSHIMATSU, T., Results of geomagnetic routine observations and earthquakes (IV), *Mem. Kakioka Mag. Obs.*, **12**, 123–130, 1965 (in Japanese with English abstract).

YUKUTAKE, T. and H. TACHINAKA, Geomagnetic variation associated with stress change within a semi-infinite elastic earth caused by a cylindrical force source, *Bull. Earthq. Res. Inst., Univ. Tokyo*, **45**, 785–798, 1967.

YÜ-LIN, Z. and Q. FU-YE, Electrical resistivity anomaly observed in and around the epicentral area prior to the Tangshan earthquake of 1976, *Acta Geophysics Sinica*, **21**, 181–190, 1978 (in Chinese with English abstract).